Oceanography

An Illustrated Guide

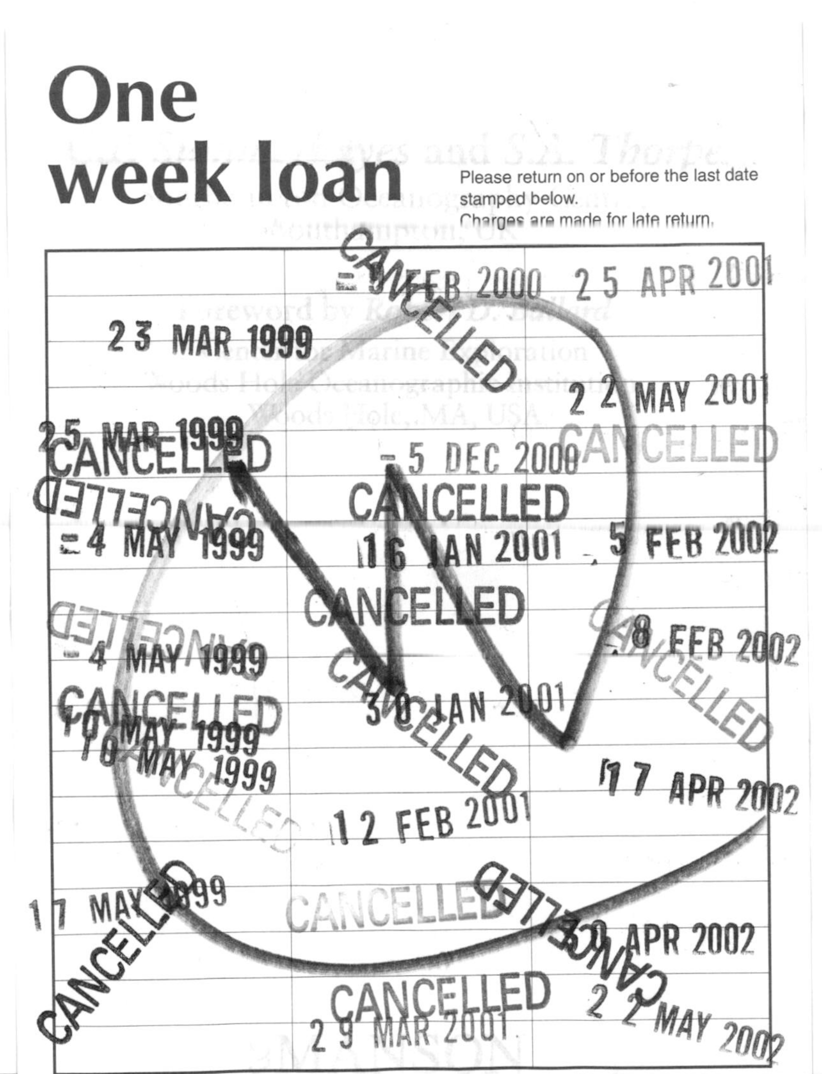

Southampton Oceanography Centre

The Contributors

All contributors are based at the Southampton Oceanography Centre, apart from Dr A. Gebruk (P.P. Shirshov Institute of Oceanography, Moscow, Russia), and Dr C.M. Young (Harbor Branch Oceanographic Institution, Ft Pierce, Florida, USA).

ISBN 1–874545–38–3 (hard cover edition)
ISBN 1–874545–37–5 (soft cover edition)

Second impression 1998

A CIP catalogue record for this book is available from the British Library.

For full details of all Manson Publishing Ltd titles please write to Manson Publishing Ltd, 73 Corringham Road, London NW11 7DL, UK.

Text organisation and supervision: John Ormiston
Design and layout: Patrick Daly
Line artwork: Kate Davis
Proofreading and index: Michael Forder
Colour separations by: Tenon & Polert, Hong Kong
Printed by: Grafos SA, Barcelona, Spain

Preface

The original motivation for producing this book was a comment and question from a young student visiting the Department of Oceanography at the University of Southampton; 'I couldn't find anything in the bookshops to tell me what oceanography is, so what is it?' Our purpose here is to provide some answers, to explain the science, and to express some of the delight and the privilege we feel in being involved in it. Chapters are written in a variety of styles, but all, we hope, are at a level at which a science undergraduate should have no difficulty in understanding; the book contains a wealth of information and guides to further reading, which all should find of interest and informative. This book is not directed at any particular teaching course, but provides much which will supplement courses in environmental science.

So, what is Oceanography? Oceanography begins at the shoreline. It is the science of the oceans, their interaction with the atmosphere above and with the underlying sea-floor sediments and oceanic crust, their chemical and biological components, their physical properties and motion, their geology, their creation, past history, and development, their present state, and their future. Oceanography is founded on the basic scientific disciplines of biology, chemistry, geology, physics and mathematics. Many of the problems addresse by oceanographers are interdisciplinary, so thei solution demands a breadth of knowledge tha crosses conventional scientific boundaries an requires multidisciplinary team collaboration. Th science uses the range of facilities of all these bas sciences, for example advanced computers and t analytical laboratory instruments of biologists a chemists. Oceanography also relies heavily upon range of technologies, for example computing, el tronics, optics, and acoustics which, together w the engineering involved in the design and c struction of winches and wires, enable the scien to make observations and sample the remote oc depths – it is this remoteness which ma Oceanography akin to Space Science. It requ elaborate and robust instruments to survive hostile environment, one which by its very na spans the Globe. Because the collection of mat and measurements from the sea is essential, it i expensive and high-risk science, using ships, craft, satellites, and submersibles. International laboration is often necessary to provide resources and ships for major experiments and grammes of research.

Oceanography is a science with applicati areas of considerable economic importanc example, to fisheries, to hydrocarbon or m

Contents

Foreword

The latter half of the twentieth century has seen the science and technology of Oceanography progress from a primitive state of understanding to a highly sophisticated science, although many believe oceanography is still in its infancy.

It was not until 1960, for example, that oceanographers first recognized the size and significance of the Mid-Ocean Ridge, a great mountain range that stretches through the ocean basins of the world for a distance of 70,000 km and covers close to 23% of the Earth's total surface area. Even more ironic is the fact that astronauts walked on the surface of the Moon before Earth scientists explored the Ridge's rift valley for the first time, in 1973, using manned submersibles.

Following this lowly start, oceanographers have not only learned the significance of this great undersea mountain range to the genesis of oceanic crust, but have also discovered the existence of hydrothermal events along the axis of the Ridge, surrounded by important mineral deposits and exotic life forms that live independently of the Sun's life-supporting energy.

The discovery of volcanism and hydrothermal circulation within the Mid-Ocean Ridge not only had an impact on the biological and geological sciences, it also helped us to better understand th chemistry of the oceans. We now know, for example, that the entire volume of the world's ocean

Prologue

This book is published to mark the occasion o major event in Oceanography in Britain, the form tion of the Southampton Oceanography Cen (SOC). This Centre brings together the Natu Environment Research Council's Institute Oceanographic Sciences and Research Ves Services, with its three ships, *Discovery*, *Cha Darwin*, and *Challenger*. In a fine new build alongside the Empress Dock in Southampton, SOC is one of the largest institutions devoted to study of the Earth and its oceans in Europe an the world.

The book is a collection of contributions la by the staff of the SOC and their colleague reflects the range of their interests, and I hope it conveys something of their excitement and e siasm for their subject.

All of us were saddened by the death of ou nent colleague John Swallow during the pre tion of the book. We are sure that he would approved of this venture, which aims to bri

The Global Oceans

1. Bahamas
2. Baltic Sea
3. Bay of Biscay
4. Bermuda
5. Black Sea
6. Bosporus
7. Canary Islands
8. East Pacific Rise
9. East African Rift Valley
10. Florida
11. Grand Banks
12. Gulf of Mexico
13. Gulf Stream
14. Hatteras Abyssal Plain
15. Hawaiian Islands
16. Hawaiian–Emperor Seamounts
17. Iberian Peninsula
18. Indian Ocean
19. Irish Sea
20. Isthmus of Panama
21. Kuroshio
22. Labrador Sea
23. Madeira Abyssal Plain
24. Mediterranean Sea
25. Mid-Atlantic Ridge
26. Monterey, California
27. Newfoundland
28. Ninetyeast Ridge
29. North Sea
30. North Atlantic Ocean
31. Norwegian Sea
32. Pacific Ocean
33. Philippines
34. Poole Bay, UK
35. Porcupine Seabight
36. Porcupine Abyssal Plain
37. Red Sea
38. Rio Grande Rift
39. River Zaire
40. River Seine
41. Rockall Bank
42. Santa Barbara Basin
43. Sardinia
44. Somali Current
45. Southampton Water
46. Southern Ocean
47. Storegga slide
48. Weddell Sea

Based on the General Bathymetric Chart of the Oceans (GEBCO), published by the Canadian Hydrographic Service, Ottawa, Canada, 1984; reproduced with permission of the International Hydrographic Organization and the Intergovernmental Oceanographic Commission (of UNESCO).

The Geological Time Scale[a,b]

Era	Sub-era, Period, Sub-period	Epoch	Age (Myr)
Cenozoic	Quaternary Sub-era	Holocene	0.01
		Pleistocene	1.64
	Tertiary Sub-era, Neogene Period	Pliocene	5.20
		Miocene	23.3
	Palaeogene	Oligocene	35.4
		Eocene	56.5
		Palaeocene	65.0
Mesozoic	Cretaceous	Senonian	88.5
		Gallic	131.8
		Neocomian	145.6
	Jurassic Period	Malm	157.1
		Dogger	178.0
		Lias	208.0
	Triassic Period	Triassic 3	235.0
		Triassic 2	241.1
		Scythian	245.0
Palaeozoic	Permian Period	Zechstein	256.1
		Rotliegendes	290.0
	Carboniferous Period – Pennsylvanian Sub-period	Gzelian	295.1
		Kasimovian	303.0
		Moscovian	311.3
		Bashkirian	322.8
	Carboniferous Period – Mississippian Sub-period	Serpukhovian	332.9
		Visean	349.5
		Tournaisian	362.5
	Devonian Period	Devonian 3	377.4
		Devonian 2	386.0
		Devonian 1	408.5
	Silurian Period	Pridoli	410.7
		Ludlow	424.0
		Wenlock	430.4
		Llandovery	439.0
	Ordovician Period	Ashgill	443.1
		Caradoc	463.9
		Llandeilo	468.6
		Llanvirn	476.1
		Arenig	493.0
		Tremadoc	510.0
	Cambrian Period	Merioneth	517.2
		St. David's	536.0
		Caerfai	570.0

[a] After Harland, W.B., Armstrong, R.L., Cox, A.V., Craig, L.E., Smith, A.G., and Smith, D.G. (1990), *A Geologic Time Scale*. Cambridge University Press, Cambridge.

[b] These Eras comprise the Phanaerozoic Eon. Preceding it is the PreCambrian, dating back to the origin of the Earth, at around 4600.0 M.

Standard International (SI) Units

Wherever possible the units used are those of the International System of Units known as SI. Oceanographers have traditionally used other units, such as the litre, which often cannot be avoided because of their common usage. Despite the recommendations periodically published by international committees as to what constitutes a standardised scientific terminology, agreement is still rather poor. Conversion between units often requires great care.

SI Unit Prefixes

Name	*Symbol*	*Multiplying factor 10^N, N is given below*
peta	P	15
tera	T	12
giga	G	9
mega	M	6
kilo	k	3
hecto	h	2
deca	da	1
deci	d	–1
centi	c	–2
milli	m	–3
micro	µ	–6
nano	n	–9
pico	p	–12
femto	*f*	–15
atto	a	–18

Commonly Used SI Units

SI units

Name	*Symbol*	*Name*	*Equivalent cgs*
Force	N	Newton	$kg\ m/s^2$
Pressure	Pa	Pascal	$kg/m\ s^2$
Energy/work	J	Joule	$kg\ m^2/s^2$
Power/energy flux	W	Watt	$kg\ m^2/s^3$
Irradiance	E/m^2s	Einstein/m^2s	mol photons/m^2s

The expression of gas concentrations is a particularly problematic area. The SI unit for pressure is the Pascal (1 Pa = 1 N/m^2). Although the bar (1 bar = 10^5 Pa) is also retained for the time being, it does not belong to the SI system. Various texts and scientific papers still refer to gas pressure in units of the torr (symbol: Torr), the bar, the conventional millimetre of mercury (symbol: mmHg), atmospheres (symbol: atm), and pounds per square inch (symbol: psi) – although these units will gradually disappear. Irradiance is also measured in W/m^2. Note; 1 mol photon = 6.02×10^{23} photons.

The SI unit used for the amount of substance is the mole (symbol: mol), and for volume the SI unit is the cubic metre (symbol: m^3). It is technically correct, therefore, to refer to concentration in units of mol/m^3. However, because of the volumetric change that sea water experiences with depth, marine chemists prefer to express sea water concentrations in molal units, mol/kg.

CHAPTER 1:

How the Science of Oceanography Developed

M.B. Deacon

Early Ideas about the Sea

Oceanography is a young science with a long history. Scientists in the seventeenth and eighteenth centuries tried to study the sea, but were often frustrated by its sheer size and complexity and the practical difficulties involved. During the nineteenth century technological advances made systematic exploration of the deep sea possible for the first time, and oceanography became an independent scientific discipline. However, the greatest strides toward understanding the sea and its importance, both as a feature which governs the Earth as we know it and one that influences human activities in many ways, have been made during the twentieth century. Oceanography today is very different from what it was 50 years ago, let alone 200 years or more, so is there any good reason why anyone, apart from historians, should be interested in its past? Science is a continuum; as an activity it grows out of its past, even if that sometimes means rejecting outmoded ideas or unreliable data. A look at how oceanography has developed can be a valuable way of helping to understand the modern science, both in terms of a set of ideas and as an institution.

Primitive societies developed complex mythologies to explain the workings of the universe, and invoked deities to account for natural phenomena. By Greek and Roman times, however, philosophers were beginning to look for natural causes for things about them, from the movement of the heavens[8] to the waves of the sea[5]. Of these, Aristotle (in the fourth century BC) most influenced later European science. He wrote widely on natural science, as well as politics and philosophy, and his works contain much to interest oceanographers, including the first known observations on marine biology. We find him considering such diverse topics as how winds cause waves, water movements in straits, and the water balance of the ocean. Aristotle believed that the presence of water vapour in the atmosphere, the source of rain, was due to evaporation, principally from the sea. Rainfall supplied rivers and these flowed into the sea, so the level of the ocean was maintained. This seems obvious today, but other suggestions were plausible given the state of knowledge at the time. Up to the end of the seventeenth century it was widely held that ground water was absorbed by the land directly from the sea, and that this formed the source of wells and springs. Then measurements made by the French scientist Edmé Mariotte, showing that the rainfall in the Paris area was, in fact, sufficient to account for the flow of water in the Seine, removed the objection that rainfall was insufficient to account for large rivers.

The Age of the Discoveries

To account for the movement of the heavens, Aristotle suggested that the Sun, Moon, and stars revolved around the Earth attached to concentric, crystalline (and therefore invisible) spheres. These, he said, derived their movement from an outermost sphere, which became known to medieval science as the *primum mobile*, or prime mover. When Columbus made his voyages of discovery across the Atlantic in the 1490s, he experienced a westward-flowing current in the tropics, and similar currents were also identified in the Pacific and Indian Oceans. Only in the Indian Ocean, north of the equator, where the proximity of the Asian land mass creates strong seasonal variations in winds and weather (the monsoons), is this flow subject to periodic reversals in direction, a fact already known to Arab geographers of the ninth century AD. The westward flow near the equator (there is actually an eastward-flowing equatorial counter-current dividing it into two streams, but this was not identified until the early nineteenth century) was thought by many sixteenth century writers to be due to motion transmitted to the Earth's fluid envelope by the *primum mobile*[3]. However, after Copernicus suggested that the Sun was at the centre of the Universe and that the Earth rotated around it, this explanation had to be adapted. The new version supposed that the westward movement, thought to exist throughout the oceans, although most marked near the equator, was due to inertia – as the Earth rotated daily on its axis, the sea lagged behind.

During the seventeenth century we find alternative explanations beginning to appear[4]. These were

1.1

Figure 1.1 Chart showing ocean currents from *Mundus Subterraneus* by Athanasius Kircher[14], a Jesuit mathematician who taught at Rome. This chart was probably the first to attempt to show ocean currents in the major oceans, whose geographical limits were by then quite well-understood, except in the polar regions (where far more land is shown than actually exists). Kircher speculated that the ocean was connected with water masses in the interior of the Earth (his subterranean world) through openings in the sea floor. The siting of the abysses he shows on his map was not entirely fanciful, some being located on sites of reputed whirlpools. For example, Kircher supposed that water was sucked in at the Maelstrom – in the Atlantic off Norway – and flowed via a subterranean tunnel into the Baltic. This had no actual basis in fact, but modern oceanographers are finding that sea water under the high pressures that exist at the sea bed penetrates the rocks and sediments of the sea floor and migrates through them. At mid-ocean rift systems, this sea water, charged with chemicals leached from the rocks through which it has passed at high temperatures, is forced back into the ocean in a scenario that would, if suggested, have seemed as exotic as Kircher's abysses until relatively recently (see Chapter 10). (Courtesy of The Royal Society, London, England.)

mainly the work of Roman Catholic philosophers who, after Galileo's condemnation, were forbidden to express Copernican ideas. One of these, Athanasius Kircher[14] (*Figure 1.1*) suggested that as the Sun travels over the sea its heat evaporates the water below. This creates a depression in the sea surface so that currents flow in from either side to restore its level. The evaporated water falls as rain in higher latitudes, so that the circulation is maintained. Isaac Vos, a Dutch scholar who later settled in England, objected that this would actually lead to a current flowing eastward, from the part of the ocean which the Sun had not yet reached[34]. He agreed that it was the Sun's heat which was responsible for currents, but said it operated by expanding the water, so that the level of the sea rose slightly as the Sun moved across it. This was sufficient to cause a flow toward a lower level. Vos' theory was one of the most original and well-supported to appear up to that time (1663). He was less convincing when he tried to introduce effects from the Earth's annual rotation in its orbit around the Sun in order to explain tides as a by-product. The idea that surface currents are actually caused by winds occurs in seventeenth-century literature, but few scientists took it seriously at that time.

The cause of tides had been keenly discussed since Greek philosophers first learned of their existence (because the Mediterranean is so enclosed, its tides are generally small and unnoticed.) The much larger rise and fall of the tides on ocean coasts (twice daily in most places) and the monthly springs and neaps (high and low tidal ranges) seemed to be linked to the Moon and its phases, but how did this come about? Throughout the Middle Ages and the Renaissance numerous explanations were made, usually linked to contemporary thinking on cosmology[8] but without, apparently, any attempt to obtain more accurate knowledge of tides themselves. However, these were clearly well-known to seafarers and during the sixteenth century we find details, usually of the 'establishment' of ports (i.e., time of high water relative to the Moon's passage overhead), appearing in printed works on navigation.

The Scientific Revolution of the Seventeenth Century

The Treasure for Travellers by William Bourne[2], which contains an interesting account of the geography of the sea as seen by a representative of the newly educated professional classes, is a good example of a change in attitude. Bourne takes his own observations as a starting point to suggest how common coastal features, such as cliffs, beaches, and stacks, might have come into existence instead of, as conventional scholars would have done, concentrating on discussion of previous ideas, which could often be traced back to Aristotle. The achievement of the scientific revolution of the seventeenth century was to highlight the need to advance science by experiment and observation, in conjunction with theory.

This was the philosophy which lay behind the foundation of the Royal Society in 1660, but the Fellows' interest in the science of the sea[5] was also influenced by the growing importance of maritime affairs in English national life during the sixteenth and seventeenth centuries. The projects they undertook, therefore, had a dual purpose. The wish to further knowledge of the natural world, in collaboration with scientists in other countries, was combined with the hope of information which would be of practical benefit to seafarers, an intention that was partly humanitarian, but which also had its roots in the desire for national economic and strategic advantage.

Much of the work done by individual Fellows of the Society was directed toward devising apparatus that sailors could use on voyages. They particularly hoped to obtain information about the depth of the sea, but were worried that soundings made in the ordinary way, with lead and line, might be inaccurate because of subsurface water movements bending the line out of the vertical. To overcome this a device suggested in earlier literature was adopted, consisting of a weight with a float attached to it in such a way that when the weight hit the sea bed the float disengaged and rose to the surface through its own buoyancy. The depth of water was calculated from the time this took. Robert Hooke improved the basic design of the apparatus (*Figure 1.2*); during the next 150 years much energy was devoted to improving this method. Although Hooke's device performed well when tested in shallow water, unfortunately both it and its successors suffered from an unsuspected design fault that made it useless in deep water. Pressure of water increases in proportion to its depth, so that in the open sea the float, which was made of wood, became waterlogged and could not rise to the surface. Hooke also designed a sampler to collect sea water from different depths. This was intended to find out if the sea was only salty at the surface, as Aristotle was supposed to have said. If this idea was correct (it was not) then fresh water would be brought up from the depths.

By the provision of instructions and encouragement, the Royal Society[31] had some success in persuading its followers to collect information about the sea, in spite of difficulties experienced with the apparatus. Notably, some of the tidal observations thus obtained were used by Sir Isaac Newton to illustrate his theory of the Universe. He was able to show that tides are due to the gravitational attraction of the Moon and, to a lesser extent, the Sun. This claim proved highly controversial, since the idea of gravity operating through empty space had been rejected by thinkers earlier in the seventeenth century. They were trying to show that nature is governed by physical laws and were therefore reluctant to employ a concept that appeared no more soundly based than discredited ideas about astrological 'influence'. The fact that Newton was able to express the effect of gravity in mathematically demonstrable laws led to the gradual acceptance of his views, but his tidal theory was also crucial in this process. His friend Edmond Halley[11] felt so strongly about it that he wrote an article summarising Newton's arguments in language and terms that could be understood by the layman (the *Principia* had been published in Latin). Information on tides and currents (see *Figure 1.3*) also came from British travellers abroad. Another famous scientist, Robert Boyle, relied on such observations when he wrote three essays on the salinity, depth, and temperature of the sea in the 1670s, still worth reading by anyone interested in the science of the sea[1]. However, Newton percipiently remarked in one of his letters that, rather than sailors sending information to mathematicians at home, it would be more fruitful to send the mathematicians to sea.

Early scientific interest in the sea was not, of course, confined to Britain. Perhaps the most interesting contribution at this time was made by L.F. Marsigli[33]. A native of Bologna, his university studies seem to have given direction and method to a boundless curiosity about the world in general. As a

1.2

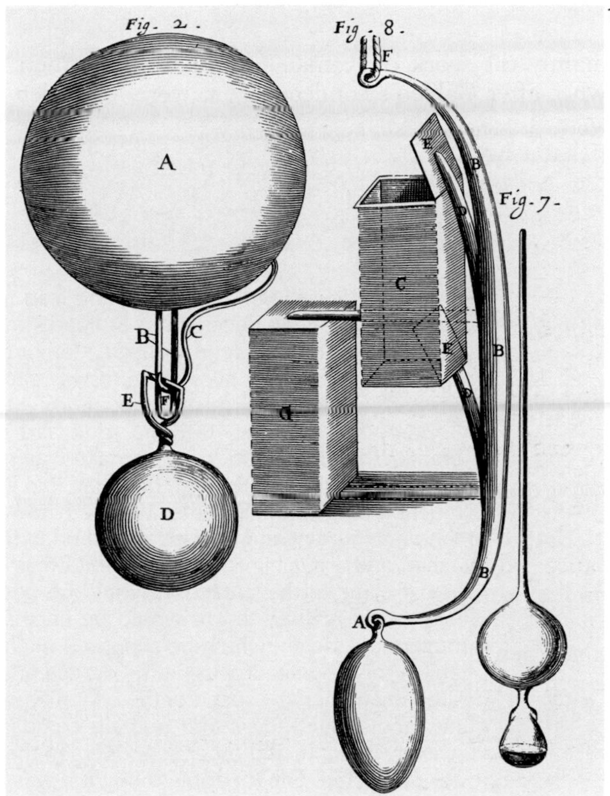

Figure 1.2 Sounding machine (Fig 2), water sampler (Fig. 8), and hydrometer (Fig. 7); this plate was published in 1667 by The Royal Society[31], founded in 1660, which wanted to collect information on scientific topics as widely as possible, so these instructions were drawn up to show overseas travellers the kind of observations that were wanted, and how to make them. Among other things, they hoped to find how the depth of the sea altered from place to place, and whether the sea was salt throughout, or only at the surface. The Society experimented with apparatus for measuring depth and bringing up water from the lower layers of the sea, but had variable success. Robert Hooke, the society's curator, produced designs which were an improvement on earlier models. Anita McConnell[16] has pointed out, however, that these woodcuts, which appeared in the journal, were not faithful copies of his drawings and would not actually work! (Courtesy of The Royal Society, London, England.)

1.3

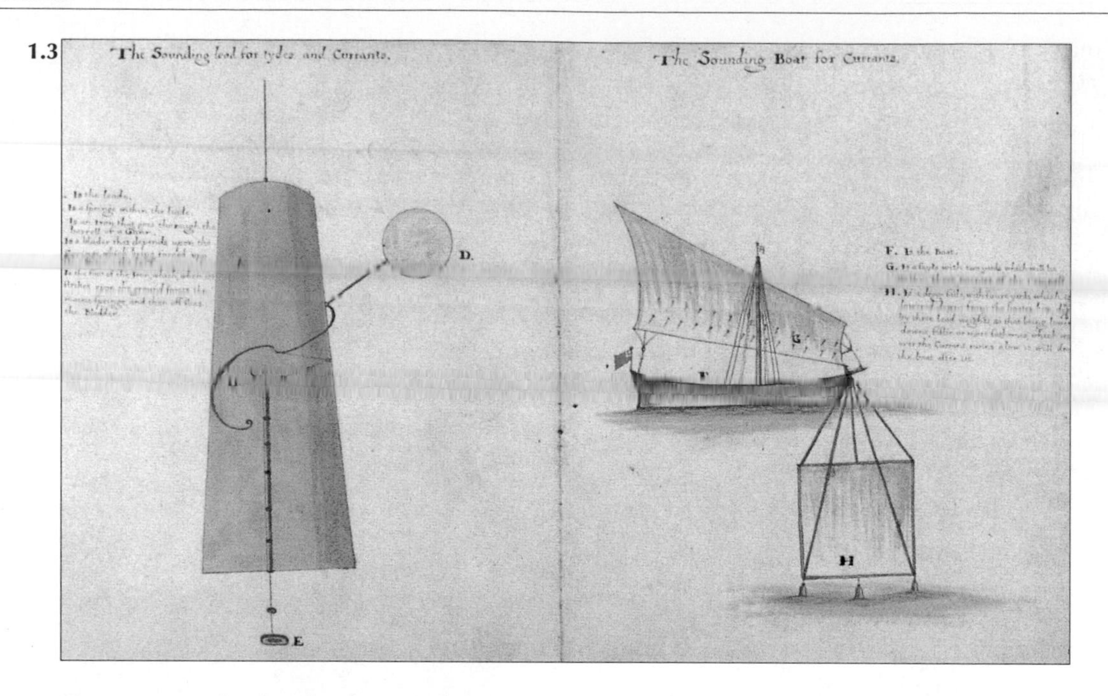

Figure 1.3 Richard Bolland's sounding lead for tides and currents, and sounding boat for currents. These are reproduced from Bolland's manuscript *Mediterranean Journall* of 1675 (in the Pepys Library at Magdalene College, Cambridge). Bolland was stationed at Tangier during its brief period as a British possession (it was part of the dowry of Charles II's queen, Catherine of Braganza). During the British occupation, extensive works were carried out to provide a safe anchorage for shipping. Bolland took part in this work and made a chart of the Strait of Gibraltar, showing its tides and surface-water movements[5]. The existence of a current flowing into the Mediterranean through the Strait had long been known. Sailors believed that there was a compensatory outflow into the Atlantic below. Bolland wanted to obtain proof of the existence of the undercurrent and devised a method of doing this. The sounding lead had a small float (D) attached and a mechanism to release it on striking the bottom. He hoped that by seeing where the float came to the surface, it would be possible to work out the speed of the undercurrent. However, he realised that it would be necessary to allow for the effect of the surface current throughout its depth (which he supposed might be as much as 100 fathoms – about 180 m). To establish this he intended using the drogue attached to the boat, which could be lowered to the desired depth, and the speed of the boat measured relative to that of the surface water. These are quite sophisticated ideas and nothing comparable was achieved until the nineteenth century. However, Bolland, and other supporters of the undercurrent, could offer no explanation of how it was generated. Other people thought it was only a seaman's yarn. A colleague of Bolland's at Tangier, Sir Henry Sheeres, wrote an essay to prove that the inflow from the Atlantic into the Mediterranean was maintained by the climate of the area. Low rainfall and hot sun meant that evaporation exceeded the input of water from rain and rivers (see Chapter 2), and the sea's level would otherwise have fallen below that of the ocean outside. Edmond Halley used this explanation, widely accepted during the next two centuries. However, it did not account for what happened to the salt that is left behind when sea or other salt water evaporates. A German scientist, J.S. von Waitz, who was connected with the salt industry (which relies on this principle), pointed out in the mid-eighteenth century that, as water became more salty, and therefore heavier, it would sink and that, as the depths of the Mediterranean filled with this more saline water, it would flow out into the Atlantic (see text). When more detailed physical surveys of the Mediterranean and the Atlantic came to be made in the late nineteenth and twentieth centuries, these showed that saline water does, indeed, spill out over the lip of the Strait of Gibraltar and spread out into the Atlantic, where it has a significant effect on the wider oceanic circulation (see Chapter 11). Currents can now be measured by moored instruments or acoustic remote sensing (see Chapter 19). (Reproduced by kind permission of the Master and Fellows, Magdalene College, Cambridge, England.)

young man, he accompanied a diplomatic mission to Constantinople and, while there, investigated reports of a counter-current in the Bosphorus, beneath the surface current flowing out of the Black Sea. He showed that the depths of the strait were occupied by more saline, and therefore heavier, water of Mediterranean origin, and that this water must reach the Black Sea, which would otherwise be entirely fresh because of the rivers flowing into it (*Figure 1.3*). He demonstrated the way this could happen with an experiment showing how, when two liquids of differing specific gravity were introduced into a container, they would form layers with the heavier liquid below and the lighter one above.

Marsigli spent many years on military service in eastern Europe before returning to the science of the sea in later life. He wrote about his researches off the southern coast of France in his book, *Histoire Physique de la Mer*[15]. Earlier works on geography and navigation had contained sections on tides and currents, and occasionally other aspects of the sea, but this was the first book on a truly oceanographic theme. However, its title is misleading as it is largely about marine invertebrates. These interested Marsigli because they had so far attracted little attention – there were earlier works on fish. He himself was particularly interested in coral, valued for use in jewellery and decorative objects, but wrongly classed it as plant rather than an animal, because of what he concluded were its 'flowers'[17].

The Eighteenth and Early Nineteenth Centuries

In spite of this promising beginning, marine science did not develop as rapidly as one might have expected during the eighteenth century. An indication of why this was so is found in Marsigli's book[15]. where he points out that science at sea is beyond the resources available to individuals and that further progress would only be made with government aid. This is because oceanographic research demands expensive items, like ships, people, and apparatus, but state funding of science is a comparatively recent innovation. There were also technical obstacles, especially to the exploration of the deep ocean. Above all, in spite of the work that had already been done, there was at this time no recognised 'science of the sea'. Nevertheless, important advances in understanding were made during the eighteenth century. At the same time, a number of related developments contributed to laying the foundation of oceanography as we know it today. These included improvements in navigation and marine surveying, in particular the discovery of methods to measure longitude at sea, which made it possible for the first time to fix a ship's position accurately when out of sight of land – an essential prerequisite for studying the ocean. These improvements were exploited first on official voyages of exploration, expeditions despatched by governments with political and economic objectives in mind, but from the time of Captain James Cook onward (the circumnavigations he commanded spanned the years 1768–1780) they became increasingly scientific in nature. A considerable amount of oceanographic work was done, especially by French and Russian expeditions, in the early nineteenth century. Their observations of surface temperature and salinity were used by geographers like Humboldt in studies of the world climate. However, as deep-water observations were more difficult, few were made. Much depended on individual scientists being fortunate enough to have the opportunity of observing for themselves, or through interested laymen, especially naval officers.

Even so, sufficient new information was obtained to encourage the development of ideas about the interior of the ocean. Toward the middle of the eighteenth century Stephen Hales[10], who had already tried to improve Hooke's sounding machine (*Figure 1.4*), also produced an apparatus designed to measure temperature by raising water from ocean depths. Such attempts were not entirely new. Hooke had proposed a design for a deep-sea

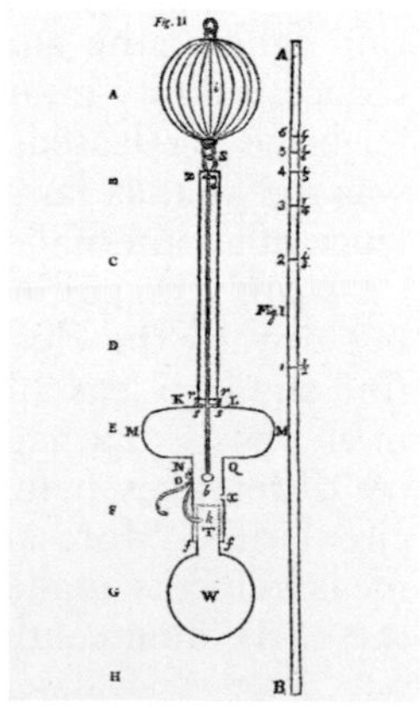

Figure 1.4 Deep-sea sounding machine devised by Stephen Hales[10]. A clergyman and scientist, Stephen Hales (1677–1761) befriended the naturalist Gilbert White. In the early years of the eighteenth century, he and J.T. Desaguliers, the Royal Society's curator, tried to develop new ways of measuring the depth of the sea which would be more reliable than those suggested by Hooke half-a-century earlier. They relied on the compression of air inside the apparatus, but most were made of glass and would not have been sufficiently robust for use in the marine environment. An example of this type is held in the George III collection at the Science Museum in London. The more durable apparatus illustrated here employed a rifle barrel (Fig. II, K–Z), with a removable rod (Fig. I, A–B) inside. Hales intended using coloured oil to mark the height reached by the water inside, which would enable the pressure, and therefore the depth of water, to be calculated. He did not appreciate the weakness of this design in that it still relied on a float (Fig. II, i) made of wood to bring it to the surface. In the sea's depths, pressure would force water into the pores of the wood so that it would lose its buoyancy and not bob up again. When this machine was tried in mid-ocean it never returned to the surface.

thermometer and Marsigli[15] had measured sea temperatures – until his only thermometer was broken in a raid by pirates. Hales' apparatus[10] consisted of an enclosed bucket with hinged flaps, opening upward only, in both top and bottom. This allowed water to flow freely through the device on the way down, but trapped a sample inside as soon as the observer began to haul it up. Its temperature was measured at the surface. However, the objection could be made that the water's temperature might have altered on the way up. To overcome this problem, the Swiss scientist H.B. de Saussure insulated the thermometer itself and left it down long enough to take on the temperature of the water at that depth. This took such a long time that the method was not much used, though it proved the most reliable at the time. Self-registering thermometers, pioneered by James Six in the 1780s, were more popular, but had disadvantages that did not become apparent until later.

The reason for this growing interest in temperature measurement was initially connected with geological debate – was the interior of the Earth hot or cold? It was some time before people began to speculate what this work was telling them about the sea. In the mid-eighteenth century a German scientist, J.S. von Waitz[6] pointed out that Marsigli's arguments[15] about the Bosphorus could equally well apply to the Strait of Gibraltar. Beneath the surface current from the Atlantic there must be an outflow of more saline water (see *Figure 1.3*, caption), otherwise the Mediterranean's salinity would be far higher. Waitz then suggested that similar imbalances gave rise to currents in the ocean. These were due to differences in density between equatorial regions, where the Sun's heat caused evaporation and increased salinity, and high latitudes where rainfall would lessen it. Saline water in the tropics sank and spread toward the poles in the ocean depths, while lighter, fresher water at the surface flowed toward the equator to replace it. The situation is more complex than Waitz supposed. The density of sea water depends on temperature as well as salinity and one can counteract the other. Heavy rain falls at the equator, and though melting ice reduces salinity in high latitudes in summer, in winter brine is released as the sea freezes. However, he was the first, as far as we know, to suggest the existence of an internal circulation in the ocean.

Waitz's suggestion had little immediate impact, but toward the close of the eighteenth century we find similar ideas appearing, but with an alteration in emphasis that had temperature rather than salinity differences being responsible for maintaining circulation. More ambitious deep-sea temperature measurements made on voyages of exploration in the early nineteenth century provided supporting evidence for this view, in particular those made by the French scientist François Péron. He used de Saussure's method to reveal the existence of low temperatures in the depths of seas in warm latitudes. Since it could not have formed there, it was argued that this colder water must have originated in polar regions.

This idea was widely accepted on the continent, but, in Britain in particular, the supposition that sea water, like fresh water, expands before freezing led to the widely held belief that water in the depths of the sea could not fall below 4°C, the temperature of maximum density of fresh water. In 1819 Alexander Marcet, a Swiss physician living in London, published an important paper on the salinity of sea water in different parts of the world[5]. In this he showed that sea water of average salinity behaves differently from fresh water, and that its density increases with cold until it freezes. This meant that in a theoretical ocean, where salinity was uniform and density was a function of temperature only, the coldest water would always sink to the bottom. The delay in accepting Marcet's findings, which were later confirmed by other scientists, was due principally to poor communication.

There was no recognised science of oceanography during the first half of the nineteenth century and individuals interested in marine science at that time came from a variety of backgrounds. Another reason for the continuing confusion over ocean circulation was that the Six self-registering thermometers (see earlier), then widely used to measure deep-sea temperatures, were not sufficiently protected and gave readings distorted by the effect of pressure. On his voyage of discovery in the Southern Ocean between 1839 and 1843, Sir James Clark Ross[30] measured deep-sea temperatures assiduously, but was not surprised that they never apparently fell below 4°C. As a naval officer, although one who was active in scientific research, Ross could not be expected to be fully up to date in all the branches of science represented on the expedition (its primary task was observations on terrestrial magnetism in the southern hemisphere). He should have been better advised by scientific colleagues at home, but the information he needed in this instance was possessed by chemists and physicists with whom he had no direct contact. It was not until the events leading up to the *Challenger* expedition (see later) that the misunderstanding was exposed, but such widely held misapprehensions are hard to eradicate and the 4°C error can be found in some twentieth century publications, including *Hansard*[12] (1961), in a reply to a question about the operation of Royal Navy submarines in the Arctic!

The situation just described had partly arisen because naval surveyors, geographers, and some

Figure 1.5 Chart of Atlantic currents by James Rennell, from Rennell's book, *An Investigation of the Currents of the Atlantic Ocean,* 1832[23]. Rennell was the first to produce charts of winds and currents, based on observation, for an entire ocean and to show how the course of ocean currents was largely shaped by prevailing winds. A former naval officer and surveyor for the East India Company, he was already interested in currents before his return to England in 1778. The ship in which he was travelling narrowly escaped loss on the Scilly Islands, notorious for shipwrecks. He suggested that part of the problem might be an unidentified current, since known as Rennell's Current, flowing out of the Bay of Biscay. Later literature has generally discounted this, but some recent models of North Atlantic circulation provide a possible explanation for such a feature. Rennell devoted the last 50 years of his life to geographical research, including work on ocean currents throughout the world. Only part of his work, a volume of charts and accompanying memoir on the Atlantic, was published in 1832, two years after his death. These charts were based either on observations made by seafaring friends or derived from ships' log-books. Chapter 4 describes recent discoveries about ocean currents. (Courtesy of the SOC, Southampton, England.)

scientists were working in a different tradition, linked to hydrography and the interests and needs of seafarers, rather than to the physical sciences. Though charts of the Gulf Stream had been published somewhat earlier by Benjamin Franklin[27] and his less well-known predecessor, W.G. de Brahm[7], it was the introduction of chronometers toward the end of the eighteenth century which made it possible to collect information on ocean currents on a wider scale. This was because, once ships could fix their position when out of sight of land, the information contained in their log-books enabled the effect of currents upon them to be calculated. The first to take advantage of this was James Rennell[23] (*Figure 1.5*). Other hydrographers followed his example, so mid-nineteenth century sailing directions contained good accounts of the surface currents of the major oceans. Rennell had shown how closely such water movements were allied to the direction of winds blowing over the sea surface, something that had probably always been self-evident to seafarers, but which was slow to take root in the scientific literature. During the 1840s, Matthew Fontaine Maury, Superintendent of the United States Navy's Depot of Charts and Instruments, produced seasonal wind and current charts[20], based on averaging data from log-books and designed to speed the passages of sailing ships. He believed that these charts could be further improved if ships from all nations systematically collected and recorded details of wind and weather. As a result of his efforts, an international conference was held in Brussels in 1853 at which governments agreed to adopt a standardised scheme of observations. Scientists had always recognised the importance of co-operation and exchange of information and ideas with colleagues in other countries. The Brussels meeting was a milestone in the development of maritime meteorology, which is closely linked to several branches of modern oceanography (in particular the study of air–sea interaction, see Chapter 2). It also introduced the idea of scientific co-operation between governments, which has been of particular significance in the development of modern oceanography.

Maury's initiative came at a time of rapid change – sail was already giving way to steam for naval and commercial purposes. The technological developments of the nineteenth century revolutionised the opportunities for scientific research, and the study of the oceans in particular. It was as a result of this that the establishment of oceanography as a separate discipline took place, and much of the impetus came from increasing maritime activity. The construction of more and larger ships, of ports and harbours to receive them, and of lighthouses to guide their passage made it necessary, for example, to obtain better knowledge of tides

1.6

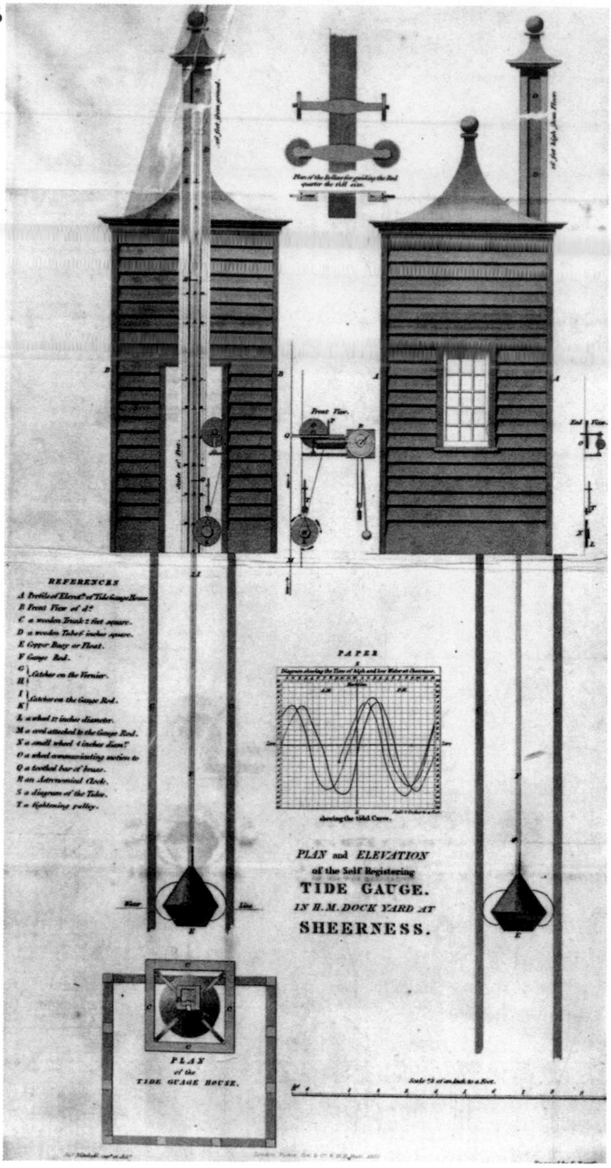

Figure 1.6 A tide gauge (from *Nautical Magazine*, 1832, **1**, 401–404), an apparatus for measuring the rise and fall of the tide devised by an engineer named Mitchell at Sheerness Dockyard. Similar gauges were being installed elsewhere at the time, but Mitchell's gauge incorporated an important innovation – it was self-registering, so did not require the presence of an observer. Tides are caused by the gravitational pull of the Moon (and to a lesser extent of the Sun), which varies with distance. Monthly and annual cycles can be detected, but tidal heights are also affected by weather. By the 1830s the Industrial Revolution was in full swing in Britain, and engineering projects needed precise information of such phenomena. The information was also welcomed by scientists who were trying to work out how the gravitational forces, well-known from the work of Newton and his successors in the eighteenth century, were translated into actual movement in the ocean. Tides can now be measured by instruments set on the sea bed in mid-ocean (Chapter 19).

(*Figure 1.6*) and waves. But perhaps the most important development for marine science came through the technology developed in response to the challenge of laying deep-sea telegraph cables, as this made scientific investigation of the ocean depths possible for the first time.

The Origins of Deep-Sea Exploration

The first functioning submarine telegraph cable was laid across the Straits of Dover in November 1851. From that time the prospect of extending this new means of rapid communication between continents was a powerful incentive to governments and industry alike. New technology had to be developed, not only to protect and lower cables to the sea bed and to raise and repair them if the need arose, but also to find out about the deep-sea environment – knowledge essential for routing and operating the cables[16]. The nature and contours of the sea bed had to be established, and also the temperature of the water. Up to this time, deep soundings had rarely been attempted because of the great effort involved, particularly if line and instruments were to be retrieved (*Figure 1.7*). The introduction of steam power made such operations possible on a more routine basis for the first time, although they were still laborious and time-consuming. By the mid-nineteenth century, hydrographic surveying was already a specialised activity in most navies, so new techniques and apparatus were developed rapidly for use in the deep sea. It was the combination of this new technology, and the accompanying professional expertise, with scientific thought that made possible further advance. Yet there was some delay before this happened – it was marine biologists rather than physical scientists who were the first to make use of the new opportunities that had been created.

Familiarity with the marine life of coastal and surface waters had been greatly extended during the latter part of the eighteenth and early nineteenth centuries, as biologists sought to expand their knowledge of living creatures and establish their affinities through schemes of classification[38]. By the middle of the nineteenth century many European and American zoologists specialised in marine work and their interest was further stimulated by the publication of Darwin's theory of evolution in 1859. Thus, the sea-shore collecting which became a popular craze among Victorian holiday-makers (*Figure 1.8*) served a more serious purpose among the scientific fraternity. They were not only interested in discovering new species, but also in learning more about the physiology and life history of individual organisms. This required working space and equipment, so seasonal laboratories were set up by the sea-shore, from which

Figure 1.7 *Deep Soundings; or, no Bottom with 4600 Fathoms*, a woodcut from Sir James Clark Ross[29]. This shows how laborious a task making deep-sea soundings was in sailing ships. Ross was an experienced polar explorer – before going to the Antarctic he had accompanied his uncle Sir John Ross on naval and private expeditions in search of the North West Passage and the North Magnetic Pole. The primary aim in this expedition, in H.M.S. *Erebus* and H.M.S. *Terror*, was to measure magnetic variation in the southern hemisphere and locate the South Magnetic Pole, in conjunction with a survey of terrestrial magnetism being made by scientists from many nations. He was also interested in ocean science and made several deep soundings and many sea-temperature measurements. In fact, his soundings greatly exaggerated the depth of water in the areas he covered. During the next few decades surveys of routes for submarine cables using steam vessels resulted in the development of new techniques and greater precision in such observations, and made the scientific study of the deep sea practical, 200 years after it had been proposed by the Royal Society.

more permanent institutions, the marine biological laboratories, emerged. One of the most famous and influential of these, though not the first to be established, was the Stazione Zoologica at Naples, founded by a German zoologist and follower of Darwin, Anton Dorhn, in the early 1870s. Others followed – on both sides of the Atlantic. Both the Woods Hole Biological Laboratory on Cape Cod and the Scripps Oceanographic Institution, part of the University of California, started life in this way.

When these developments took place the leading figures were, increasingly, professional scientists and academics, but in the mid-nineteenth century amateur collectors were still numerous and made an important contribution. Some were keen yachtsmen and began to extend their interests to the waters of the continental shelf. Dredging was also supported by the newly formed British Association for the Advancement of Science[21]. Some of the impetus for this work came from geological discoveries. When modern relatives of fossil remains were discovered in seas further to the north or south, this suggested that climate might have undergone considerable shifts (see Chapter 3). This finding was of much interest in the light of

Figure 1.8 *Pegwell Bay, Kent – a Recollection of October 5th 1858*, a painting by William Dyce, shows a scene that would have been familiar at the time. The growing interest of nineteenth-century biologists in marine life-forms was for a short while reflected in a more widespread enthusiasm for sea-shore collecting of natural history specimens. This became a feature of seaside holidays, taken increasingly by the expanding middle class as rail transport made travel cheaper. From the 1850s onward many semi-scientific books and guides were written to cater for this market. (Courtesy of the Tate Gallery, London, UK.)

evidence being put forward, particularly by Swiss geologists, for periods of extensive glaciation in the past.

Almost without exception, those engaged in this work failed at first to appreciate the new opportunities which deep-sea surveying work was creating. This was because it was widely believed that life could not exist in the conditions of darkness, cold, and immense pressure that exist in the depths of the sea. There was then no conception of the range of adaptations (which modern biologists are still discovering) that enable creatures to live in such environments (see Chapters 13 and 15). The findings of Edward Forbes, who had worked in the Mediterranean in the 1840s, were often quoted to support the idea of an azoic (without life) zone below 400 fathoms (780 m). There was, it was true, some evidence from elsewhere that seemed contrary to this view, but for some years most people regarded it as unconvincing[25]. However, in the late 1860s accumulating observations from a number of sources suggested to workers in several countries that the supposed limit was erroneous. In 1868, two British biologists, W.B. Carpenter and C. Wyville Thomson, backed by the Royal Society, persuaded the Admiralty to allow them the use of one of its survey ships so that they could dredge in deep water. In three voyages, first in the *Lightning* and then in the *Porcupine*[24], they obtained incontrovertible evidence that the deep sea was populated by a thriving community of creatures previously unknown to science.

The Voyage of H.M.S. Challenger

As the work went on, Carpenter's attention increasingly turned to the physical observations being made by the naval personnel. Improved thermometers gave a more accurate picture of the distribution of temperature with depth, and more was known about the behaviour of sea water at low temperatures. Carpenter adopted the idea that density differences between equatorial and polar regions cause internal circulation in the ocean, with warm, light water moving poleward at the surface to compensate for colder, denser water spreading toward the equator in the depths[5]. Such an idea was largely unfamiliar to a British audience brought up on Rennell and his successors, so it had a mixed reception. Carpenter's fiercest critic was James Croll, who had recently put forward a theory to account for the ice ages which held that shifts in the pattern of trade winds, and the ocean currents which they generate, were responsible for climate change. Carpenter believed that if he could obtain information from the other oceans he would have irrefutable proof of his theory. A respected elder scientist, he used his contacts with other scientists and politicians to win support for a large-scale expedition – the first to have marine science as its primary objective. This resulted in the round-the-world voyage of H.M.S. *Challenger* between 1872 and 1876 (*Figure 1.9*), with a naval crew and a team of civilian scientists led by Wyville Thomson.

The voyage of the *Challenger* (*Figure 1.10*) was a major landmark in the development of oceanography, both in essence and in its findings. The observations of temperature and salinity (*Figure 1.11*) showed hitherto unsuspected features, like the spread of saline water from the Mediterranean into the North Atlantic. Neither instrumentation[16] nor theory were yet good enough to enable a detailed picture of ocean circulation to be made, but its existence in some form could no longer be doubted (*Figure 1.12*).

1.9

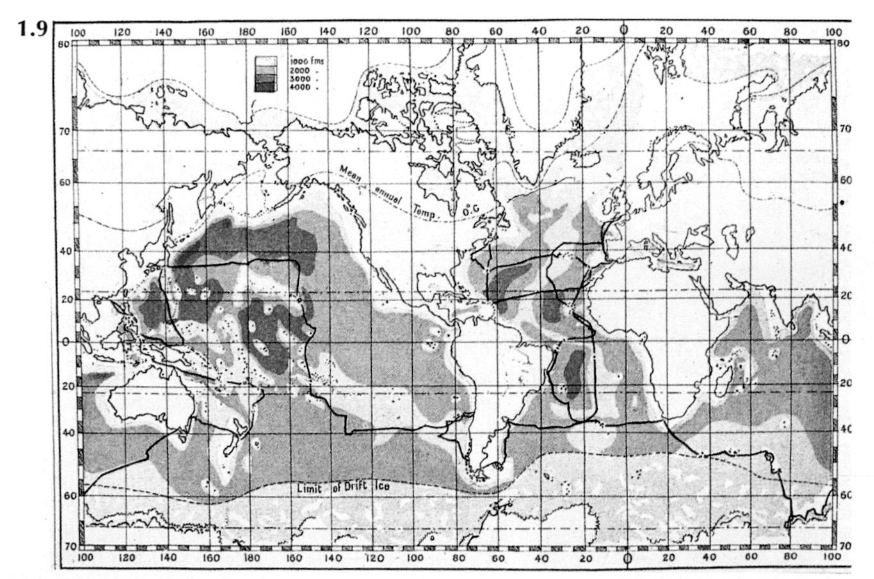

Figure 1.9 The route followed by H.M.S. *Challenger* during her oceanographic voyage round the world of 1872–1876, taken from Wild[36]. Wild was the expedition's artist and secretary to the scientific leader, C. Wyville Thomson. Other circumnavigations, including the voyages of Captain Cook in the eighteenth century, French expeditions, and Russian voyages in the early nineteenth century, had added much to geographical and scientific knowledge, but the *Challenger* expedition was the first large-scale expedition devoted primarily to the science of oceanography. For a non-scientist's view of the voyage, see the recently published letters of Joseph Matkin[22].

1.10

1.11

Figure 1.10 Challenger in the ice (reproduced from Wild[37]). (Courtesy of the Southampton Oceanography Centre, Southampton, England.)

Figure 1.11 This drawing, by Elizabeth Gulland, was one of a series commissioned as illustrations for the narrative volumes of the *Challenger Report*. It shows a scientist and members of the crew taking readings from deep-sea thermometers. [Courtesy of Edinburgh University Library (Special Collections), Edinburgh, Scotland.]

1.12

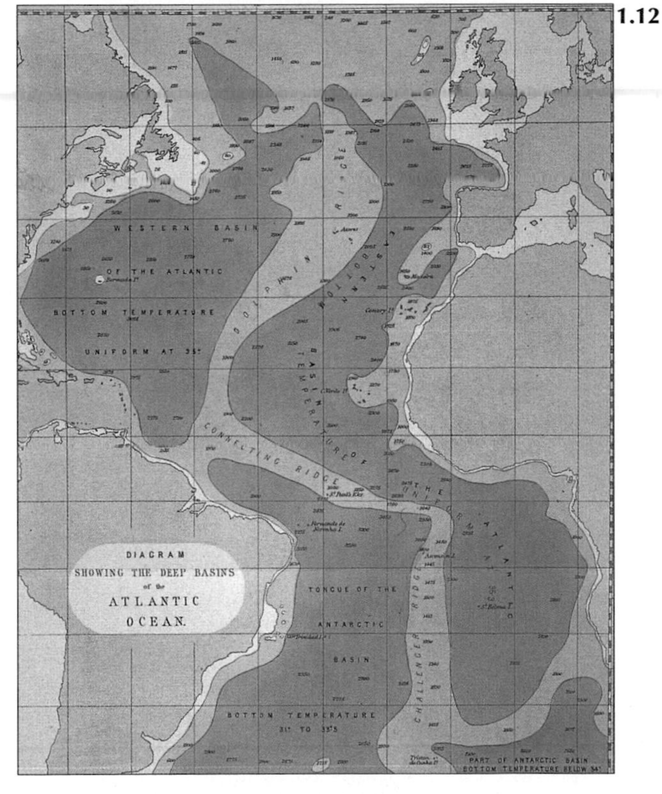

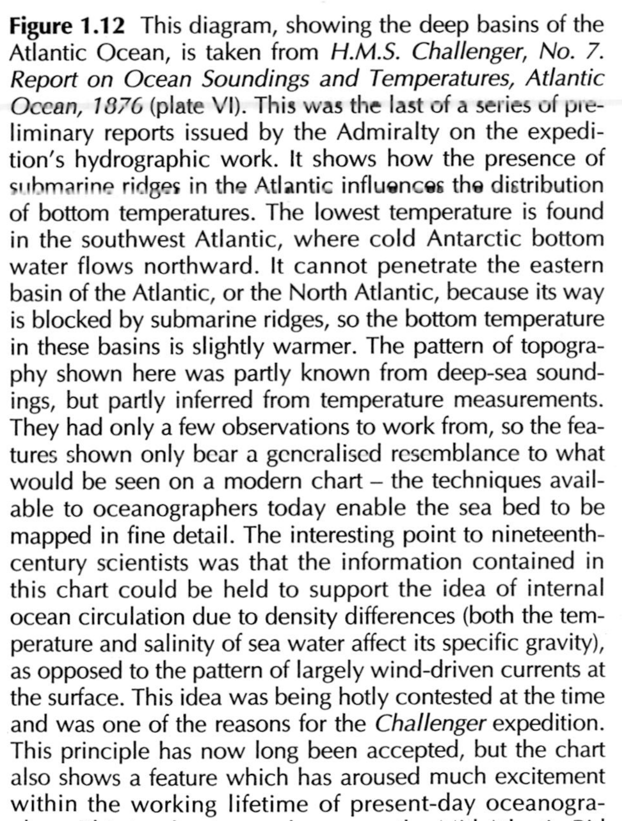

Figure 1.12 This diagram, showing the deep basins of the Atlantic Ocean, is taken from *H.M.S. Challenger, No. 7. Report on Ocean Soundings and Temperatures, Atlantic Ocean, 1876* (plate VI). This was the last of a series of preliminary reports issued by the Admiralty on the expedition's hydrographic work. It shows how the presence of submarine ridges in the Atlantic influences the distribution of bottom temperatures. The lowest temperature is found in the southwest Atlantic, where cold Antarctic bottom water flows northward. It cannot penetrate the eastern basin of the Atlantic, or the North Atlantic, because its way is blocked by submarine ridges, so the bottom temperature in these basins is slightly warmer. The pattern of topography shown here was partly known from deep-sea soundings, but partly inferred from temperature measurements. They had only a few observations to work from, so the features shown only bear a generalised resemblance to what would be seen on a modern chart – the techniques available to oceanographers today enable the sea bed to be mapped in fine detail. The interesting point to nineteenth-century scientists was that the information contained in this chart could be held to support the idea of internal ocean circulation due to density differences (both the temperature and salinity of sea water affect its specific gravity), as opposed to the pattern of largely wind-driven currents at the surface. This idea was being hotly contested at the time and was one of the reasons for the *Challenger* expedition. This principle has now long been accepted, but the chart also shows a feature which has aroused much excitement within the working lifetime of present-day oceanographers. This is what is now known as the Mid-Atlantic Ridge. Nineteenth-century cable surveyors were surprised to find that the greatest depths in the North Atlantic Ocean were not in mid-ocean, but to either side. A combination of soundings and deep-sea temperature observations suggested to the *Challenger* staff that the ridge might continue into the South Atlantic, and on the voyage home in 1876 they carried out soundings that showed this was so. Geological theory at the time could not easily explain such a feature and it was not until the 1960s that it became widely accepted that it is, in fact, a spreading centre at which new ocean floor is being created, and part of a world-wide system whose dynamics are explained by the theory of plate tectonics (see Chapter 8). Connecting ridges, like the Walvis Ridge, whose approximate course is shown here linking to southern Africa, are now thought to be due to hot-spot activity.

1.13a

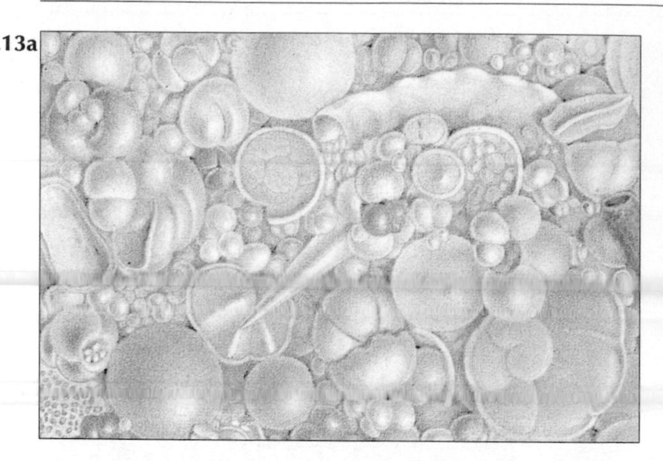

1.13b

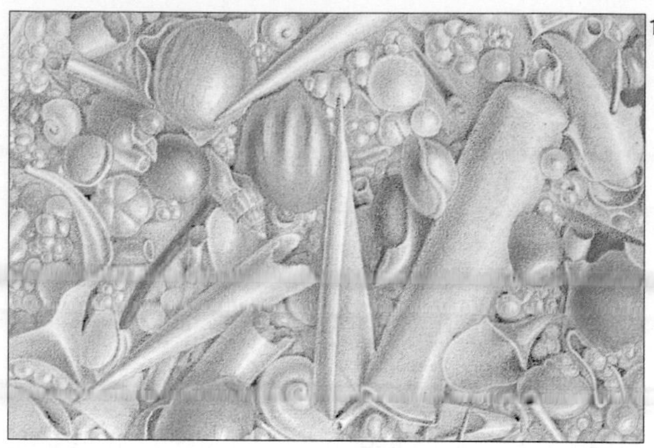

Figure 1.13 (a) Globigerina ooze and (b) Pteropod ooze, from Murray and Renard[19], plate XI, *Figures 5* and *6*, respectively. One of the principal scientific achievements of the *Challenger* expedition was to produce the first global map of what the sea bed is made of. John Murray, who edited the *Challenger Report* after the death of Wyville Thomson in 1882, had made observations during the voyage which enabled him to show that, in most parts of the ocean, sediments reflect the composition of marine life in the surface layers (plankton). Globigerina ooze was named after the remains of microscopic calcareous species of Foraminifera which form it. In the Southern Ocean, siliceous remains of single-celled phytoplankton (plants) predominate in diatom ooze. In deep water, far from land, where calcium carbonate dissolves and there are no terrigenous sediments (derived from land), with the exception of volcanic pumice which can float long distances, the *Challenger* found slowly accumulating 'red clay' and manganese nodules (see Chapters 6 and 7). (Courtesy of the Southampton Oceanography Centre, Southampton, England.)

Sediment samples (*Figure 1.13*) collected during the expedition formed the basis for the first world chart of sea-floor deposits. However, the emphasis both during the voyage and afterward was on marine biology. Of the 50 volumes of the *Challenger Report*, edited by John Murray after Thomson's early death, 33 were reports on the zoological collections, contributed by specialists from both Europe and America. Most of the species described were new to science and the reports are still a basic work of reference for oceanographers. The expedition also aided the growth of marine science in a more general sense. Scientists in other countries used the example of the *Challenger* to obtain government support for their work. As this progressed, the common ground between their researches gave them a sense of identity strong enough to override links with their various parent sciences, such as physics, chemistry, or biology. They called the new science 'Oceanography'.

Oceanography from the 1880s to the 1930s

While single-ship expeditions continued to make important contributions to oceanography[40], unlike the *Challenger*, which had been exploratory in nature, they tended to concentrate more on a particular area or problem. The *Challenger* expedition had been organised without much time to prepare. On the whole, it made use of well-tried techniques which were already a little old-fashioned. It was during the post-*Challenger* period that many basic types of oceanographic equipment, such as current meters, reversing thermometers, and self-closing nets, became standard[13,16]. There was considerable variation in the provision for oceanographic work from one country to another, depending on local customs and arrangements for supporting science. In some countries, as, for example, Germany, which had only recently been unified, state funding was relatively generous. The German government supported the Stazione Zoologica at Naples (whose founder, Anton Dohrn, was German), as well as oceanographic expeditions. A number of government bodies were involved in various aspects of marine science and a research institute (the Institut für Meereskunde), attached to Berlin University, was set up in 1900. In America, Maury's methods and ideas had been criticised by more orthodox scientists, but when he supported the South in the American Civil War, marine science suffered through the loss of what Schlee[32] has described as 'his stubborn and passionate interest in the sea and his ability to channel funds toward its exploration'. Schlee[32] shows how oceanographic work carried out by US government agencies actually declined in the latter part of the nineteenth century, with the exception of Pillsbury's survey of the Gulf Stream (in the Coast Guard steamer *Blake*) in the early 1880s.

Prince Albert I of Monaco, a wealthy patron of science and himself an active oceanographer[13,24] (*Figures 1.14–1.17*), established an oceanographic institute in Paris and the Musée Océanographique

1.14

1.15

1.16

1.17

Figures 1.14–1.17 Prince Albert I of Monaco – his statue, by Françoise Cogné, stands in the gardens adjacent to the Musée Océanographique, Monaco (**1.14**) – used his wealth to further a number of sciences, but oceanography was his principal interest[13]. He was an enthusiastic yachtsman and made almost annual voyages in his own research ships[24] from the 1880s until the outbreak of war in 1914 (he died in 1922). **1.15** shows one of his vessels, the Princesse Alice I (courtesy of Musée Océanographique, Monaco). With the help of specialist assistants, and visiting colleagues from France and other countries, he investigated a wide range of physical and biological problems during these cruises. Much attention was paid to the improvement of existing apparatus and to the development of new methods[16]. One of the Monaco inventions was the 'nasse triédrique'[26], (**1.16**) a baited trap which could be lowered to predetermined depths to catch creatures that evaded traditional nets and trawls. Prince Albert benefited the French oceanographic community by founding an institute for research and teaching in Paris. His superbly situated Musée Océanographique at Monaco (**1.17**), inaugurated in 1910, continues to provide a valuable resource for visitors and students interested in oceanography, and its history (courtesy of Musée Océanographique, Monaco).

1.18

Figure 1.18 Alexander Agassiz on board the US Fisheries vessel *Albatross*. Agassiz was a mining engineer, and one of the leading marine zoologists of his day (the Swiss zoologist Jean Louis Agassiz was his father[13]). He used his wealth to build up his father's Museum of Comparative Zoology at Harvard and undertake research expeditions, many of them to study coral reefs, in the hope of throwing new light on the origin of atolls. In the late 1870s he made three dredging cruises to the Caribbean in the US Coast Survey Ship *Blake*. The US Fish Commission was founded in the early 1870s, one of a number of such national organisations to come into being at that time (see text). Its ocean-going research vessel, the *Albatross*, built in 1882, was used by Agassiz on several expeditions, for which he paid part of the expenses. The first cruise, in 1891, was undertaken to compare deep-sea fauna on the Pacific side of the Isthmus of Panama with Caribbean forms and perhaps arrive at an approximate date for the closure of the sea-way between North and South America, which his earlier researches on the Atlantic side had suggested must have persisted until comparatively recent geological times. (Courtesy of the Southampton Oceanography Centre, Southampton, England.)

at Monaco in the first decade of the twentieth century. In the US the situation outlined above had led to a greater reliance on private funding. One of the outstanding marine scientists of the late nineteenth century was Alexander Agassiz (*Figure 1.18*), who financed expeditions in the Pacific[13,32]. The Woods Hole Oceanographic Institution was set up in 1930 through the agency of the National Academy of Sciences, with an endowment from the Rockefeller Foundation[32]. Only a few such specialised research institutes for oceanography were founded before World War II, though some others, such as the Geophysics Institute at Bergen in Norway, established in 1917, also did important marine work.

Existing educational and administrative structures did not generally lend themselves to such developments, and private wealth was not usually available on the scale required for such enterprises. In spite of this and the somewhat differing attitudes to science in different countries, marine science developed through a variety of local, national, and international agencies during the period from the 1880s to the 1930s. This happened partly through the expansion of higher education. University departments of oceanography were created – in the UK there was one at Liverpool, established in 1919, and one at Hull, established in 1928. Other scientific departments also did marine work. A considerable number of marine biological laboratories were also attached to universities, although some, like the Marine Biological Association's laboratory at Plymouth, were maintained by private bodies. John Murray's Scottish Marine Station for Scientific Research in the 1880s had been a short-lived attempt to create a more diversified institution.

In the early twentieth century, with more attention being paid to the importance of physical oceanography than to marine biology, laboratories began to widen their interests. The Scripps Institution of Oceanography in the US took this step to its logical conclusion when it transformed itself from a marine station in 1925[32].

During this period, whatever the prevailing attitude to support for science, most technologically advanced nations began paying more attention to marine research because of a variety of economic and political needs. They also discovered the benefits of international co-operation. By 1900 much of the marine survey work connected with submarine cables was being done by the cable companies themselves, but naval hydrographers also continued their interest in deep-sea work[28]. At the International Geographical Congress held in Berlin in 1899, it was decided that details of all these soundings should be collected onto a continually updated General Bathymetric Chart of the Ocean. This work was undertaken by Prince Albert of Monaco, and after his death, by the International Hydrographic Bureau which had been established in Monaco in 1921.

Owing to the growing economic importance of fisheries, scientific research in this area developed rapidly during the late nineteenth century. National research organisations were set up in many countries[13] to study the life histories of food fish, in the hope of reviving dwindling fisheries and creating new ones. From the start it became apparent that it was not sufficient just to study individual species; more needed to be known about biological diversity in selected regions, and about the physical environment and its influence on populations. This was the starting point for much local activity, such as H.B. Bigelow's biological survey of the Gulf of Maine in the early 1900s[32]. However, the wider questions posed had a considerable impact on the development of oceanography in the broad sense. In the 1890s, Scandinavian oceanographers proposed that there should be a joint programme of observations and this resulted in the setting up of the first intergovernmental body for marine science, the International Council for the Exploration of the Sea (ICES), in 1902. For several years the Council maintained a Central Laboratory in Norway, where important work was done toward improving oceanographic apparatus. Surveys of important fishing grounds were carried out in North America, Australia, and by the European nations, both at home and in their colonies around the world. British scientists working for the Discovery Committee carried out research in the Southern Ocean (*Figure 1.19*) during the 1920s and 1930s, with the aim of putting the whaling industry on a sustainable basis.

Work undertaken by these various organisations, as well as by individual institutions and expeditions, contributed to the growing knowledge of many aspects of the oceans during the late nineteenth and early twentieth centuries. The two most important areas of research developed during these years were ocean circulation studies and biological oceanography. In the 1870s the American meteorologist, William Ferrel, drew attention to the deflecting effect of the Earth's rotation (the Coriolis force) on ocean currents[32], but it was the circulation theorem of Vilhelm Bjerknes and work by Bjorn Helland-Hansen and Fridtjof Nansen on currents in the Norwegian sea, carried out in the fishery steamer *Michael Sars* in the early 1900s, that formed the basis for modern dynamical oceanography. The behaviour of oceanic winds and currents had been of crucial importance for Nansen's famous attempt to drift to the North Pole in the *Fram* (1893–1896). Observations he made then also formed the basis of two important discoveries by V.W. Ekman[35]. One was the existence of internal waves, now known to occur naturally throughout the ocean, at the interface between layers of water of differing density. In northern seas such waves are generated when a ship moves through a shallow layer of fresh water, originating from rivers or melting ice, that overlays normal sea water and hampers the ship's progress, a phenomenon known to sailors as 'dead water'. The 'Ekman spiral' is the name now given to the discovery that the direction of near-surface currents is increasingly deflected with depth. This results from the Earth's rotation and from frictional forces, and causes a mean current drift to the right of the wind in the northern hemisphere (and to the left in the southern hemisphere).

Important contributions were also made at this time to our knowledge of general oceanic circulation. In his report on the work of the German research ship *Meteor* in the 1920s, Georg Wüst[39] incorporated data from earlier expeditions to show the origin and distribution of the main Atlantic

Figure 1.19 R.R.S. *Discovery II*, at Port Lockroy, Wiencke Island, Palmer Archipelago, off the Antarctic Peninsula, in January 1931. The *Discovery II* was built in 1929 for the Discovery Committee, to replace the sailing vessel *Discovery* (originally built for Captain Scott in 1901) which had been used for initial work in the seas around South Georgia from 1925–1927. *Discovery II* worked throughout the Southern Ocean during the 1930s, contributing to an understanding of ocean circulation and the marine environment. A central theme was the study of the distribution and life history of Antarctic krill, the principal food of southern hemisphere baleen whales. After World War II, *Discovery II* became the research vessel of the newly formed National Institute of Oceanography (later Institute of Oceanographic Sciences), until replaced by the modern R.R.S. *Discovery* (built 1962). (Courtesy of the Southampton Oceanography Centre, Southampton, England.)

1.20

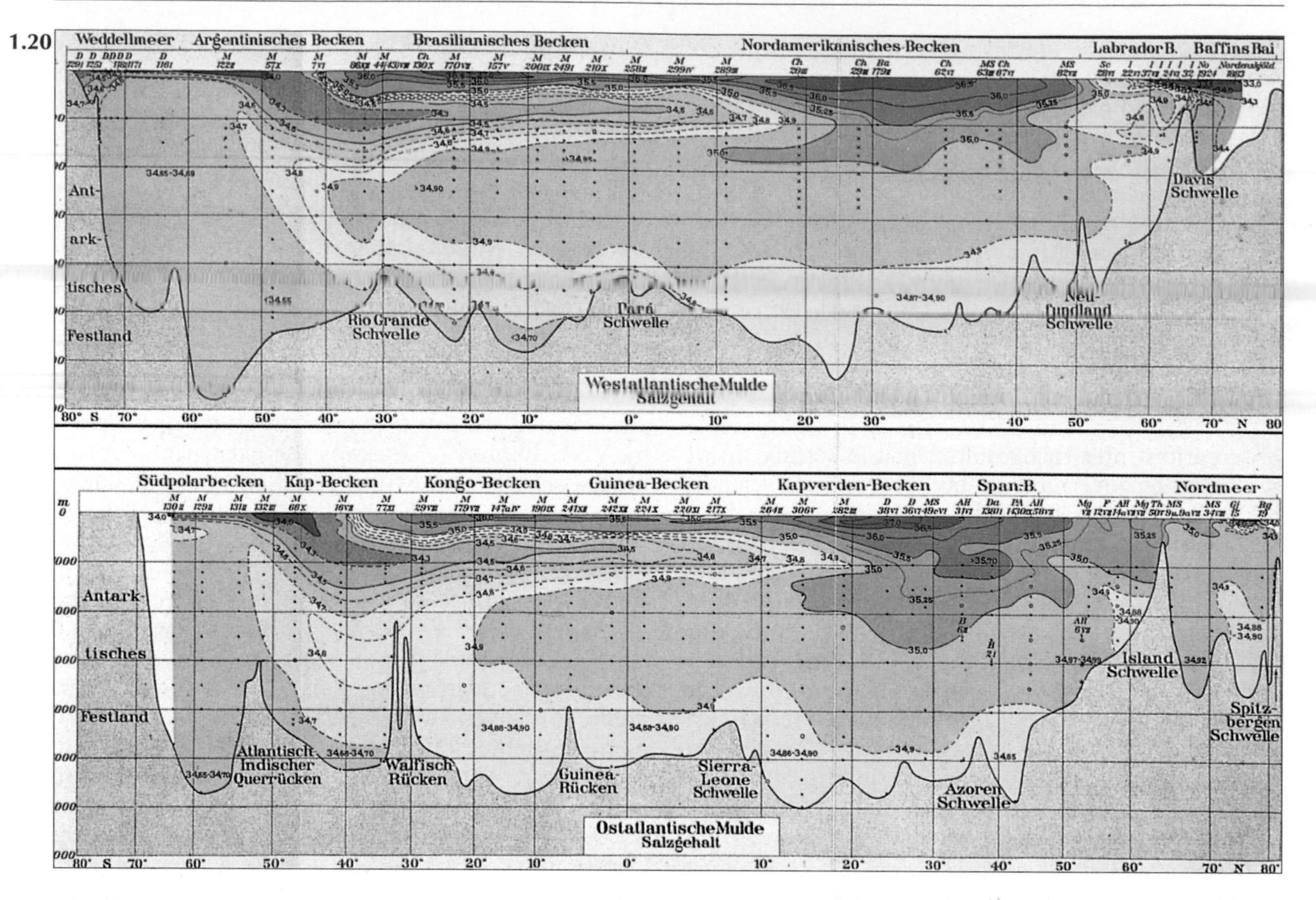

Figure 1.20 Diagram (Plate 33) by Georg Wüst[39] showing the longitudinal distribution of salinity in the Atlantic Ocean. Wüst was one of the leading physical oceanographers of the mid-twentieth century. As a young man in the 1920s he participated in the cruise of the German research vessel *Meteor* and used data collected during the voyage, together with observations made by other expeditions, from the *Challenger* onward, to show how density differences, due to salinity and temperature variations, are responsible for the movement of water masses within the body of the ocean.

water masses (*Figure 1.20*). His picture was extended by Discovery Committee scientists working in the Southern Ocean in the 1920s and 1930s[24].

Fisheries science was also responsible for the growth of interest in marine productivity. This field was developed by German scientists of the 'Kiel School'[18] in the late nineteenth century. Their researches showed that life in the sea depends on primary production – the phytoplankton, single-celled plants grazed by the zooplankton on which all other creatures in the sea depend, either directly or indirectly, for food (modern research also highlights the importance of bacteria in primary production – see Chapter 6). Victor Hensen[18] developed vertical nets for sampling plankton in order to obtain quantitative data on the productivity of the ocean. In the Plankton Expedition of 1889 he extended his work to the North Atlantic in the ship *National*. Karl Brandt[18] showed that the growth of phytoplankton is controlled by the supply of available nutrients.

Modern Oceanography

When Thomas Wayland Vaughan, the Director of the Scripps Institution of Oceanography, surveyed world oceanographic activity on behalf of the US National Academy of Sciences in the 1930s, he found nearly 250 institutions devoted to marine research throughout the world, from Russia to Japan and from Australia to Argentina. He had been asked to undertake this survey because of concern that not enough was being done to encourage the study of the sea, which, as we have seen, requires special conditions to make progress. The trend toward increasing state support for science was soon to be greatly accelerated by the demands of war, and of the Cold War which followed, and by the role of science in post-war economic growth.

Oceanography developed rapidly during and after World War II. It shared in the general expansion of science, as the impact of science and technology on almost every aspect of modern life has led to increased funding for research. National

1.21

Figure 1.21 Launching the clover-leaf buoy from R.R.S. *Discovery* in 1967. The 1950s and 1960s saw a dramatic expansion in oceanography, encompassing people, ideas, and methods. This apparatus was one of a number of different types designed and made at the National Institute of Oceanography for studying properties of waves. (Courtesy of Laurence and Pamela Draper, Rossshire, Scotland.)

defence needs and economic concerns, particularly the search for new energy sources (offshore oil and gas), as well as renewed concern about fisheries stocks, are among the factors that have led to a wider interest in marine research. Such needs have increased rather than diminished with time. While the political international situation has eased over all, environmental worries have come to the fore, in particular the problem of climate change, in which the oceans must play an important part.

However, it is interesting to see how many of the ideas and techniques that are important in present-day oceanography have their roots in the first half of the twentieth century. For example, work on underwater sound was begun before World War I[9], but then developed rapidly in the search to perfect a means of submarine detection by echo location (sonar). The invention of the hydrophone had many other scientific and peace-time applications. Echo-sounding by ships provided more detailed information about the topography of the sea floor. Seismic surveying, to investigate the internal structure of the sea bed, was first employed by Maurice Ewing in the 1930s. After 1945, this work played an important part in obtaining the information on which modern ideas about sea-floor spreading and plate tectonics are based[32]. Magnetic surveys also contributed to these developments, as did gravity measurements, first obtained at sea in submarines in the 1930s by the Dutch scientist, F.A. Vening-Meinesz. Modern oceanographers rely heavily on continuous-recording instruments, which in many areas have replaced the older single-observation measuring devices (see Chapter 19). Such devices were foreshadowed by the bathythermograph, invented by US scientists in the 1930s to measure the temperature of the upper layers of the ocean. This and similar ideas were taken up by scientists during World War II.

Other fields of study originated at that time, including wave research. Until then no way had been found to study sea waves that was not purely descriptive[37]. A major breakthrough occurred when a war-time research team based at the Admiralty Research Laboratory, in Teddington, England, developed a method of analysing wave spectra that enabled their components to be identified. This work continued after the war at the newly established National Institute of Oceanography, where new kinds of wave recorders were developed to measure waves at sea (*Figure 1.21*).

During the past 50 years such developments, and others described in the following chapters, have profoundly transformed our knowledge of the oceans. The main subject areas – marine physics, including ocean-circulation studies, knowledge of the sea floor (which has played a major role in the revolution of the earth sciences leading to modern theories of plate tectonics), marine chemistry, and biological oceanography – have all made important advances. The development and use of more sophisticated apparatus has been assisted by the introduction of computers and satellites, which permit the gathering, transmission, and analysis of data in quantities that would have been inconceivable a generation ago, let alone to the first scientific observers of the sea in the 1660s. Oceanography is still an expensive science, with the ship remaining a fundamental tool, although this too may change in the future. This expense has provided a strong incentive for co-operation and sharing on a more formal basis, so joint expeditions have become an important aspect of modern oceanography, from the International Indian Ocean Expedition of the early 1960s to the World Ocean Circulation Experiment (WOCE), designed to throw new light on the relation between the oceans and climate, in the 1990s.

General References

Deacon, M.B. (1971), *Scientists and the Sea, 1650–1900: A Study of Marine Science*, Academic Press, London and New York, 445 pp.

Herdman, W.A. (1923), *Founders of Oceanography and their Work. An Introduction to the Science of the Sea*, Edward Arnold, London, 340 pp.

McConnell, A. (1982), *No Sea Too Deep: The History of Oceanographic Instruments*, Adam Hilger, Bristol, 162 pp.

Rice, A.L. (1986), *British Oceanographic Vessels, 1800–1950*, The Ray Society and Natural History Museum, London, 193 pp.

Schlee, S. (1973), *The Edge of an Unfamiliar World: A History of Oceanography*, E.P. Dutton, New York, 398 pp.

References

1. Birch, T. (ed.) (1744), *The Works of the Honourable Robert Boyle*, A. Millar, London, Vol. 3, pp 105–113 and 378–388.
2. Bourne, W. (1578), *A Booke Called the Treasure for Traveilers*, Thomas Woodcocke, London, 269 pp.
3. Burstyn, H.L. (1966), Early explanations of the role of the Earth's rotation in the circulation of the atmosphere and the ocean, *Isis*, 57(2), 167–187.
4. Burstyn, H.L. (1971), Theories of winds and ocean currents from the discoveries to the end of the seventeenth century, *Terrae Incognitae*, **3**, 7–31.
5. Deacon, M.B. (1971) *Scientists and the Sea, 1650–1900: a Study of Marine Science*, Academic Press, London and New York, 445 pp.
6. Deacon, M.B. (1985), An early theory of ocean circulation: J.S. von Waitz and his explanation of the currents in the Strait of Gibraltar, *Progr. Oceanogr.*, **14**, 89–101.
7. De Vorsey, Jr, L. (1976), Pioneer charting of the Gulf Stream: the contributions of Benjamin Franklin and William Gerard de Brahm, *Imago Mundi*, **28**, 105–120.
8. Duhem, P. (1913–1959), *Le Système du Monde: Histoire des Doctrines Cosmologiques de Platon à Copernic*, 9 vols, Hermann, Paris.
9. Hackmann, W. (1984), *Seek and Strike: Sonar, Anti-Submarine Warfare and the Royal Navy, 1914–54*, HMSO, London, 487 pp.
10. Hales, S. (1754), A descripton of a sea gage, to measure unfathomable depths, *Gentleman's Magazine*, **24**, 215–219.
11. Halley, E. (1697), The true theory of the tides, extracted from that admired treatise of Mr Isaac Newton, entitled *Philosophiae Naturalis Principia Mathematica*; being a discourse presented with that book to the late King James, *Phil. Trans. Roy. Soc. Lond.*, **19**, 445–457.
12. Hansard (1961), *Parliamentary Debates*, Fifth Series, **638**, 235.
13. Herdman, W.A. (1923), *Founders of Oceanography and their Work: An Introduction to the Science of the Sea*, Edward Arnold, London, 340 pp.
14. Kircher, A. (1678), *Mundus Subterraneus*, Vol 1, 3rd edn, Apud Joannem Janssonium à Waesberge & Filios, Amstelodami, pp 134–135.
15. Marsigli, L.F. (1725), *Histoire Physique de la Mer*, Aux dépens de la Compagnie, Amsterdam, 195 pp.
16. McConnell, A. (1982), *No Sea Too Deep: The History of Oceanographic Instruments*, Adam Hilger, Bristol, 162 pp.
17. McConnell, A. (1990), The flowers of coral – some unpublished conflicts from Montepellier and Paris during the early 18th century, *Hist. Phil. Life Sci.*, **12**, 51–66.
18. Mills, E.L. (1989), *Biological Oceanography: An Early History, 1870–1900*, Cornell University Press, Ithaca and London, 378 pp.
19. Murray, J. and Renard, A.F. (1891), Deep-sea deposits, *Report on the Scientific Results of the Voyage of H.M.S.* Challenger *during the Years 1872–76*, HMSO, London, Plate XI.
20. Pinsel, M.I. (1981), The wind and current chart series produced by Matthew Fontaine Maury, *Navigation*, **28**(2), 123–137.
21. Rehbock, P.F. (1979), The early dredgers: 'naturalizing' in British seas, 1830–1850, *J. Hist. Biol.*, **12**(2), 293–368.
22. Rehbock, P.F. (ed.) (1992), *At Sea with the Scientifics: The Challenger Letters of Joseph Matkin*, University of Hawaii Press, Honolulu, 415 pp.
23. Rennell, J. (1832), *An Investigation of the Currents of the Atlantic Ocean*, J.G. and F. Rivington, 359 pp.
24. Rice, A.L. (1986), *British Oceanographic Vessels, 1800–1950*, The Ray Society and Natural History Museum, London, 193 pp.
25. Rice, A.L., Burstyn, H.L., and Jones, A.G.E. (1976), G.C. Wallich, M.D. – megalomaniac or mis-used oceanographic genius?' *J. Soc. Bibliogr. Natur. Hist.*, 7, 423–450.
26. Richard, J. (1910), *Les Campagnes Scientifiques de S.A.S. le Prince Albert Ier de Monaco*, Imprimerie de Monaco, p. 33.
27. Richardson, P.F. (1980), The Benjamin Franklin and Timothy Folger charts of the Gulf Stream, in *Oceanography: The Past*, Sears, M. and Merriman, D. (eds), Springer, New York, pp 703–717.
28. Ritchie, G.S. (1967), *The Admiralty Chart. British Naval Hydrography in the Nineteenth Century*, Hollis and Carter, London, 388 pp. (Reprinted 1995 by Pentland Press, Edinburgh.)
29. Ross, Sir J.C. (1847), *A Voyage of Discovery and Research in the Southern and Antarctic Regions, during the years 1839–43*, Vol. 2, John Murray, London, facing p. 354.
30. Ross, M.J. (1982), *Ross in the Antarctic: The Voyages of James Clark Ross in H.M. Ships* Erebus *and* Terror, *1839–43*, Caedmon of Whitby, Whitby, Yorkshire, 276 pp.
31. Royal Society (1667), Directions for observations and experiments to be made by masters of ships, pilots, and other fit persons in their sea voyages, *Phil. Trans. Roy. Soc., Lond.*, **2**, 433–448.
32. Schlee, S. (1973), *The Edge of an Unfamiliar World: A History of Oceanography*, E.P. Dutton, New York, 398 pp.
33. Stoye, J. (1994), *Marsigli's Europe, 1680–1730: The Life and Times of Luigi Ferdinando Marsigli, Soldier and Virtuoso*, Yale University Press, New Haven and London, 356 pp.
34. Vossius, I. (1993), *A Treatise Concerning the Motions of the Seas and the Winds*, together with *De Motu Marium et Ventorum Liber*, Deacon, M.B. (ed.), Scholars' Facsimiles and Reprints, Delmar, New York, for the John Carter Brown Library, 376 pp.
35. Walker, J.M. (1991), Farthest North: Dead water and the Ekman spiral, *Weather*, **46**(4), 103–107; **46**(6), 158–164.
36. Wild, J.J. (1877), *Thalassa: An Essay on the Depth, Temperature, and Currents of the Sea*, Marcus Ward, London, facing p. 14.
37. Wild, J.J. (1878), *At Anchor. A Narrative of Experiences Afloat and Ashore during the Voyage of H.M.S.* Challenger *from 1872 to 1876*, Marcus Ward, London, 198 pp.
38. Winsor, M.P. (1976), *Starfish, Jellyfish, and the Order of Life: Issues in Nineteenth Century Science*, Yale University Press, New Haven and London, 288 pp.
39. Wüst, G. (1928), Der Ursprung der Atlantischen Tiefenwässer, *Zeitschrift der Gesellschaft für Erdkunde zu Berlin Sonderband zur Hundertjahrfeier der Gesellschaft*, pp 506–534.
40. Wüst, G. (1964), The major deep-sea expeditions and research vessels, 1873–1960, *Progr. Oceanogr.*, **2**, 1–52.

CHAPTER 2:

The Atmosphere and the Ocean

H. Charnock

Introduction

The atmosphere and the ocean are held on the Earth by gravity and irradiated by the Sun. Both are shallow relative to the radius of the Earth and the motions within them are slow relative to that due to the Earth's rotation: they have similar dynamics. As they share a common boundary it is attractive to treat them as a single, coupled, system, but the physical and chemical properties of air and water are so very different that meteorology and oceanography have developed separately and at different rates. Electromagnetic radiation (light, microwaves, radar, radio ...) travels easily through the atmosphere and this, together with the commercial and economic benefit of weather-forecasting, has led to the existence of a global network of meteorological observing stations, making and transmitting regular routine surface and upper-air observations, as well as increasing information from sensors on satellites.

Most meteorological stations are on land, but winds, waves, currents, and weather affect ships, so have been observed by mariners from time immemorial: some selected merchant ships now report their observations as part of the global meteorological system.

Observation of the ocean away from the surface has developed more slowly (Chapter 4): water is almost opaque to electromagnetic radiation so oceanographers are essentially restricted to indirect observation of a fluid through which they cannot see; the period from the *Challenger* Expedition of 1872 has been described as a 'century of undersampling'.

More recent technological development has clarified some processes in restricted areas: it is now accepted that, like the atmosphere, the ocean is a three-dimensional turbulent fluid, with interacting motions and processes on all time- and space-scales. Understanding it requires an observing system which does not yet exist, one that may evolve if national governments perceive a need to forecast conditions in the ocean to the same extent as those in the atmosphere. In the meantime, we can attempt to use our knowledge of the atmosphere and of the underlying ocean surface, together with the limited but increasing observations of the deep ocean, to study the global transfers of heat, of fresh water, and of momentum in the coupled atmosphere–ocean system.

Energy and Water Exchanges in the Atmosphere–Ocean System

The general circulation of the atmosphere and ocean, regarded as a single system, is determined by the distribution of its sources and sinks of energy. Much the dominant external source is the absorption of energy from the Sun (solar radiation of wavelength between 0.2–4 μm). The near constancy of the temperature of the system requires an equal outgoing flux of energy in the form of terrestrial radiation (long-wave radiation of wavelength 4–100 μm).

The distribution and transformation, within the system, of the incoming solar radiation is complicated. Typical global mean values of the major components are shown in *Figure 2.1*, which indicates that nearly half the absorbed solar radiation reaches the Earth's surface, most of which is ocean. This sunlight does not penetrate far into the ocean (even in the clearest water 99% is absorbed in the upper 150 m): heat gains and losses take place at and close to the surface, so the ocean is relatively inefficient thermodynamically. Directly driven motions, due to the cooling of surface water at high latitudes (the thermohaline circulation) are slow. Indirectly driven motions (the wind-driven circulation) arise from the transfer of heat (and water vapour) from the ocean to the atmosphere, where it is converted by complicated processes into depressions and anticyclones, the winds of which provide energy to generate ocean waves and drive ocean currents. Their energy in turn is dissipated into heat by small-scale (viscous) processes and re-radiated. It is a complex, inefficient system with many interlocking components of different scale.

It can be seen from *Figure 2.1* that of the 153 W/m^2 received at the Earth's surface as solar radiation, 54 W/m^2 is lost as long-wave radiation

2.1

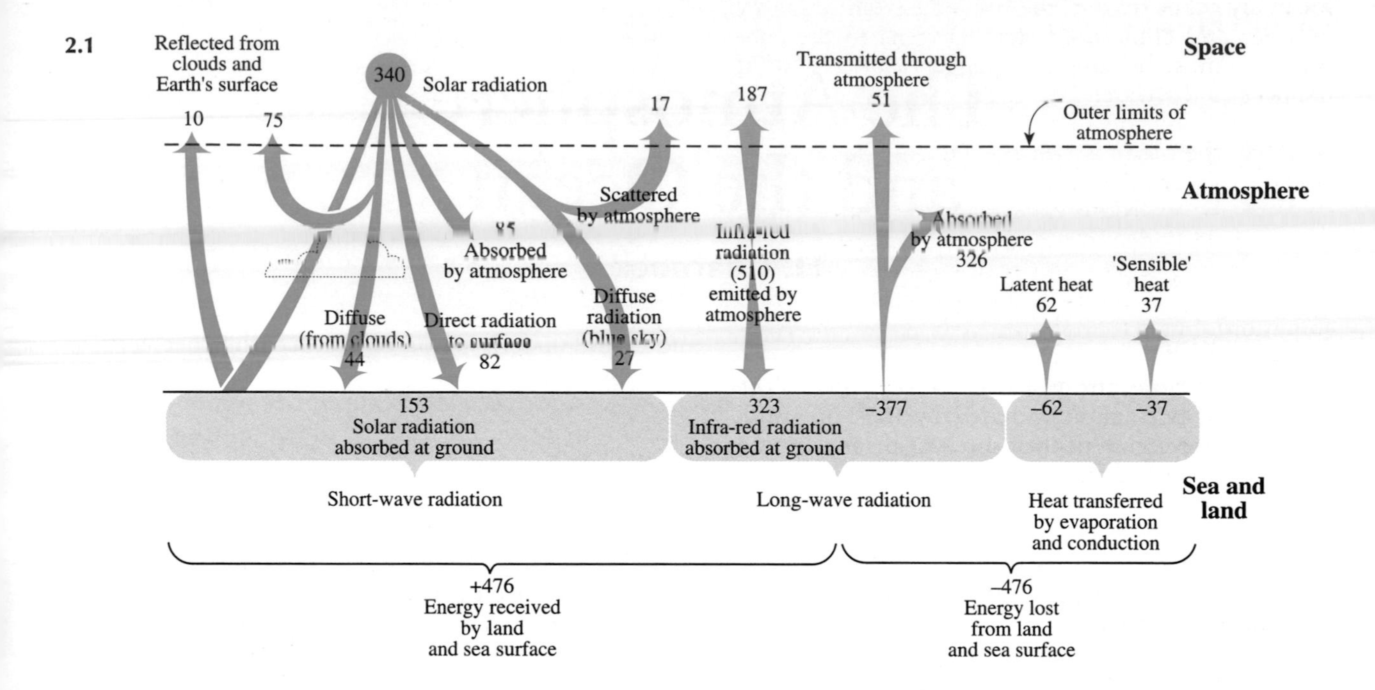

Figure 2.1 The average radiation balance (in W/m²) for the Earth as a whole (based on Nieburger *et al.*[11]).

and 37 W/m² is transferred by the conduction of heat (sensible heat) from a warmer sea to cooler air above. Energy is also used in evaporation of water vapour from the sea surface to the drier air above, heat (latent heat) being required to convert a liquid into a gas. Water has a high latent heat, 62 W/m² being used in evaporation. The latent heat is released to the atmosphere only when the water vapour condenses back into clouds, of liquid water or ice, often far from where it evaporated from the ocean into the atmosphere. Much of the cloud formation happens at relatively low altitudes, justifying the statement that the atmosphere can be regarded as heated from below, making it an active thermodynamic system with important vertical as well as horizontal motion.

Although both atmospheric and oceanic motions are ultimately powered by solar heating, the important working substance of the global heat-engine is water; as vapour, as liquid, and as solid ice. The infra-red characteristics of water vapour make it a major agent of long-wave radiative heat transfers (see *Figure 2.1*) The solar radiation absorbed at the surface is used to evaporate water, the large latent heat of which is released to the atmosphere when and where condensation occurs; the distribution of evaporation, precipitation, continental run-off, and ice are crucial to the determination of the salinity and so to the watermass structure and to the thermohaline circulation of the ocean. For these and

2.2

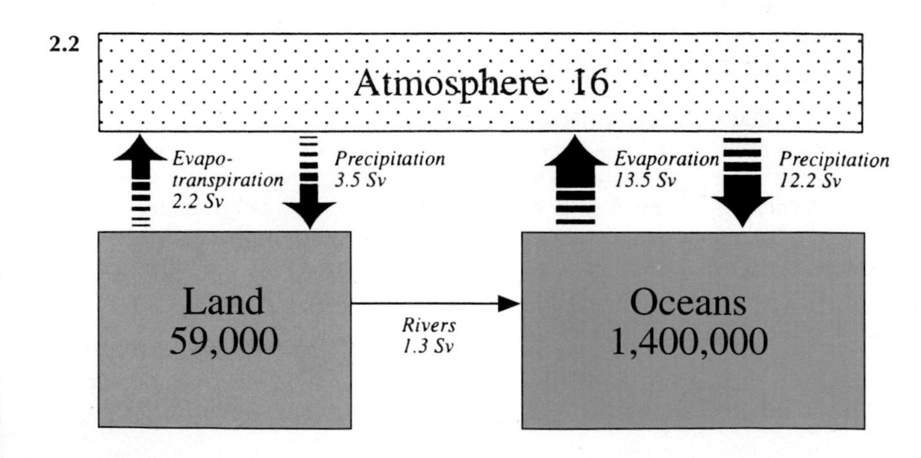

Figure 2.2 Estimates of the amount of water (in 10³ km³) in the atmosphere, the land, and the ocean, and of the fluxes between them (1 Sv = 10⁶ m³/s) (from Schmitt[15]; the figures are derived from Baumgartner and Reichel[2]).

for many other reasons the basic question posed by weather and climate – 'what happens to the sunshine?' – must be supplemented by asking – 'what happens to the water?'

The distribution of evaporation and precipitation over the ocean is clearly vital to understanding climate: unfortunately, this part of the hydrological cycle is not well known. Most treatments of the water cycle concentrate on exchanges over land (they are concerned with man's use of water, for agriculture and industry as well as human consumption), but it is estimated that 78% of the global precipitation goes into and 86% of the global evaporation comes from the ocean (see *Figure 2.2*). The corresponding transfer of (latent) heat represents a major component of the heat balance of the atmosphere and of the ocean.

Meridional Fluxes

The near-constancy of the global mean temperature implies a balance between the absorbed solar radiation and the outgoing long-wave radiation, as indicated in *Figure 2.1*, but their variation with latitude is significantly different. The solar radiation is absorbed mainly in the tropics: the long-wave radiation, determined mainly by the radiative properties of the atmosphere and the underlying surface, is observed to be much less dependent on latitude (*Figure 2.3*). It follows that there is a flux of heat from the tropics to the poles. Measuring this flux is important: it is fundamental to the maintenance of

2.3

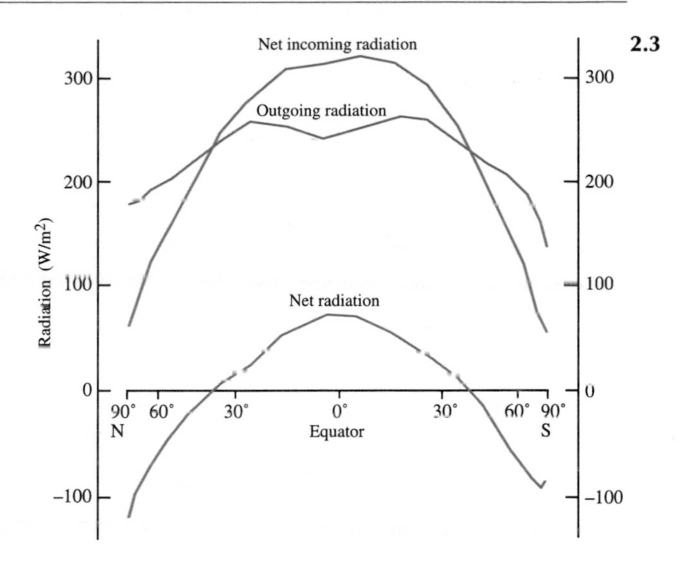

Figure 2.3 Zonal averages of radiation at the top of the atmosphere. Net incoming radiation peaks in the tropical regions; outgoing long-wave radiation varies less with latitude. To maintain a constant temperature, the excess of radiative heating within 35° of the equator is transferred poleward by atmospheric and oceanic motions to compensate for the deficit of radiative heating nearer the poles. A uniform bias of 9 W/m² has been subtracted from the net incoming radiation to ensure a balance between the total incoming and the total outgoing radiation (from Bryden[4]; the figures are derived from satellite observations reported by Stephens *et al.*[18]).

2.4

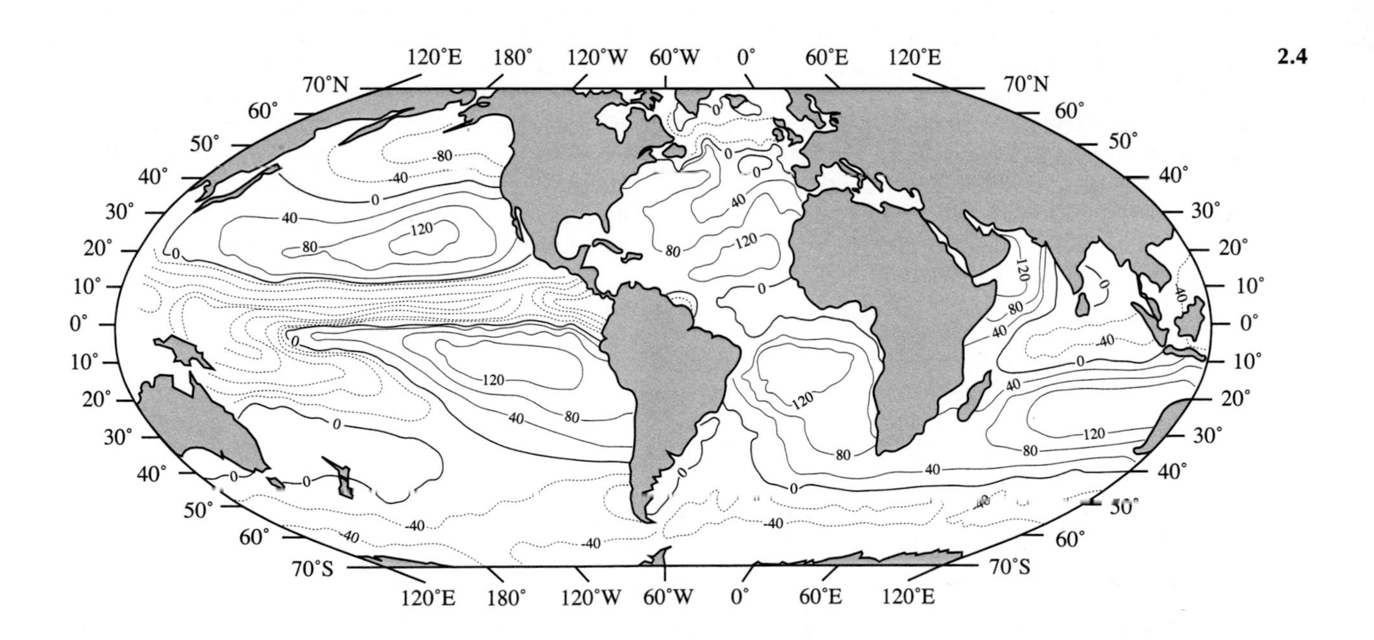

Figure 2.4 A chart of evaporation minus precipitation (*E-P*) over the ocean. Units are cm/yr; solid lines indicate *E>P*, dashed lines *E<P* (from Schmitt and Wijffels[17]; the figures are derived from Schmitt *et al.*[16]).

2.5

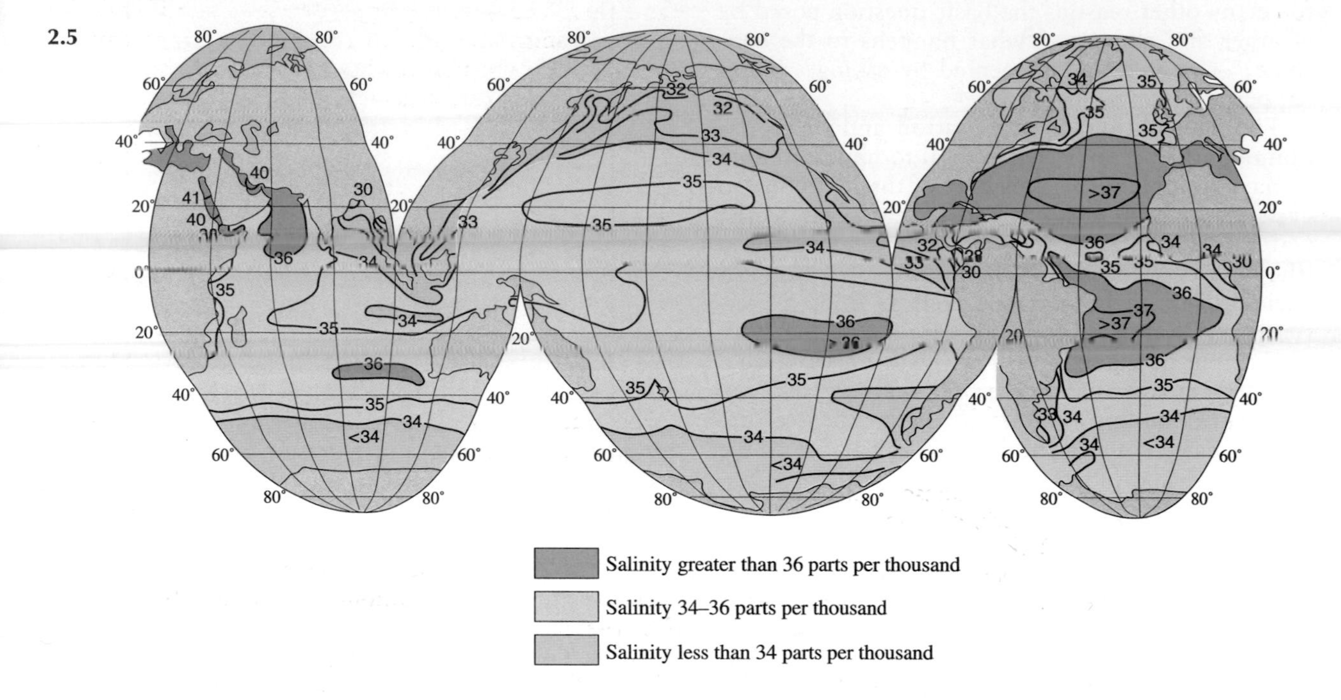

climate and its magnitude provides a significant constraint on atmospheric, oceanic, and coupled general circulation models. Understanding how the meridional heat flux is maintained requires a detailed knowledge of atmospheric and oceanic structures and processes.

Although the distribution of evaporation and (especially) of precipitation over the ocean is not well known, major features of the net water flux between atmosphere and ocean (evaporation minus precipitation, *E–P*) can be recognised (*Figure 2.4*). The distribution is roughly zonal (except for the North Indian Ocean); precipitation dominates in subpolar regions and especially in the Intertropical Convergence Zone (the ITCZ) at the thermal equator and in the South Pacific Convergence Zone to the northeast of Australia. Elsewhere, the subtropics have an excess of evaporation. There is a clear association with the surface salinity of the ocean (*Figure 2.5*); the latitudes where (*E–P*) is high are associated with high salinity at the ocean surface (and with deserts on land). The general pattern is of net annual precipitation at high and at low latitudes, with net annual evaporation between. That the mean structure is not changing implies meridional transports of fresh water by the ocean. As river transports are negligible in comparison, equal and opposite flows of water must occur in the atmosphere. These, in turn, transfer significant quantities of heat.

To assign numerical values to these important meridional fluxes presents difficulties. The radiation balance of the Earth as a whole has for many years been measured as part of a major programme (the Earth Radiation Budget Experiment, ERBE) using orbiting satellites fitted with radiometers sensitive to solar and to long-wave radiation. Although the measurements are technically demanding and present difficult problems of data analysis, recent results imply a near-balance between the global annual incoming solar radiation and the corresponding outgoing long-wave radiation. The imbalance was small (less than 10 W/m^2), but its distribution produces some uncertainty in the derived values of the meridional heat transfer by the atmosphere and ocean combined. *Figure 2.6* shows that it peaks near latitudes 30°N and 40°S where the flux amounts to almost 6 PW (1 petawatt, PW = 10^{15} W).

Estimates of the meridional flux of water and heat (and momentum) by the atmosphere can be made using the observations made daily from meteorological upper-air stations, where balloon-borne instruments measure the wind, the temperature, and the humidity in relation to pressure (height). The local meridional flux of, say, water vapour is carried by the northward component of the wind, V, and is measured by ρVq where q is the specific humidity and ρ the air density. The total meridional flux is given by $[\rho Vq]$ where the square brackets represent a mean value over the depth of the atmosphere and around a latitude circle, over a

Figure 2.5 Salinity of surface waters during the northern summer (from Gross[6]; the figures are derived from Sverdrup *et al.*[20] and from later sources).

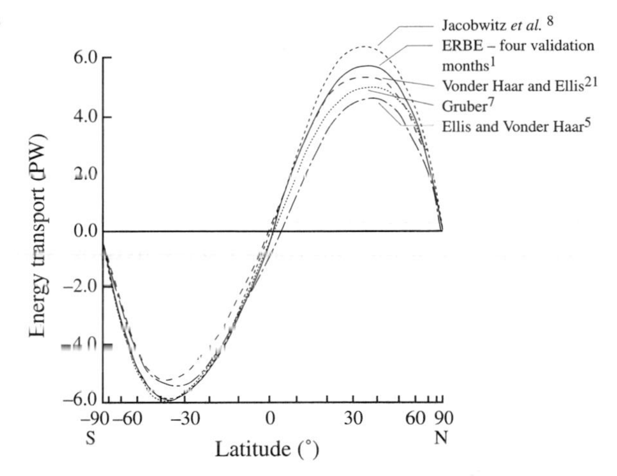

Figure 2.6 The total meridional transport of energy estimated from ERBE data for April, July, and October 1975 and January 1986 (from Barkstrom *et al.*[1]; other satellite-based estimates are from Vonder Haar and Ellis[21]; Ellis and Vonder Haar[5]; Gruber[7]; Jacobwitz *et al.*[8]).

time for which $[V]$ vanishes. The calculations can be made by using the upper-air observations directly or by assimilating them together with other observations into a suitable atmospheric computer model and using interpolated values.

The meridional flux of water vapour in the atmosphere (which is discussed later) implies a corresponding flux of latent heat. To estimate the total heat flux this must be supplemented by the sensible heat flux given by $[\rho C_p Vq]$, where C_p is the specific heat of air at constant pressure and q the temperature. Some results are shown in *Figure 2.7* for the total heat flux (compare with *Figure 2.2*). The atmospheric fluxes peak at about 40°N and 40°S where the flux amounts to about 4 PW.

There is no oceanic equivalent of the meteorological upper-air observing network, but it is possible to estimate the meridional flux of heat (and that of salt) by an analogous method: the heat transport is estimated by calculating the covariance (Vq) between the temperature and the inferred, as distinct from the directly measured, northward velocities. In the deep ocean away from the surface and the sea floor, the currents can be treated as frictionless and unaccelerated, so can be calculated as if they are in geostrophic balance: that is, the force due to the horizontal variation in pressure is balanced by the force due to the Earth's rotation, which is proportional to the speed of the current. In this case the currents flow along the isobars like winds on a weather map. The observations used are from a high-quality east–west hydrographic section between two continental land masses, giving accurate measurements of temperature and salinity (and therefore density), at all depths. The difficulty in determining deep-ocean currents geostrophically is that although the density field is known at all depths the pressure field is not known at any one, so the calculation of currents using the geostrophic balance requires a knowledge of the total transport. Given measurements or reliable estimates of the transport of western boundary currents, and making allowance for the near-surface currents due to the frictional drag of the wind, convincing estimates of the meridional heat flux by the ocean can be obtained. Estimates of the heat flux northward across latitude 24°N in the Atlantic have been made three times over the last 35 years and found to be consistently close to 1.2 PW to the north. A transoceanic hydrographic section made in 1985 has provided observations along 24°N in the Pacific from which an oceanic heat flux of 0.8 PW was obtained. Since there is virtually no heat flux across 24°N in the Indian Ocean the total ocean heat-flux across 24°N amounts to some 2.0 PW to the north.

If the atmospheric flux (*Figure 2.7*) across 24°N is taken to be 2.3 PW, the total (atmosphere + ocean) flux is 4.3 PW, significantly less than the 5.7 PW required from the most recent analysis of the ERBE results. Further transoceanic sections are being made as part of the World Ocean Circulation Experiment and should serve to clarify the climatically important meridional heat flux. They will be particularly important in the southern hemisphere,

2.7

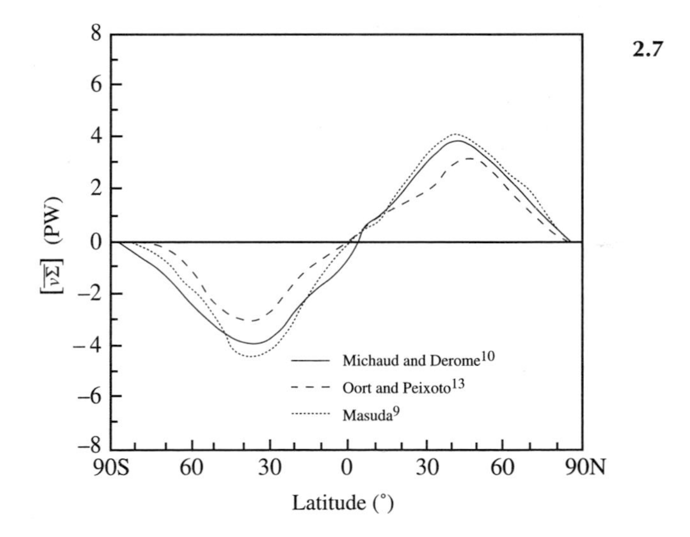

Figure 2.7 Annual mean northward flux of energy by the atmosphere, December 1985 to November 1986 (from Michaud and Derome[10]). The estimates of Michaud and Derome[10] are compared with those of Oort and Peixoto[13] (observed values from upper-air observations) and of Masuda[9] (ECMWF and GFD model assimilations for 1978–1979).

2.9

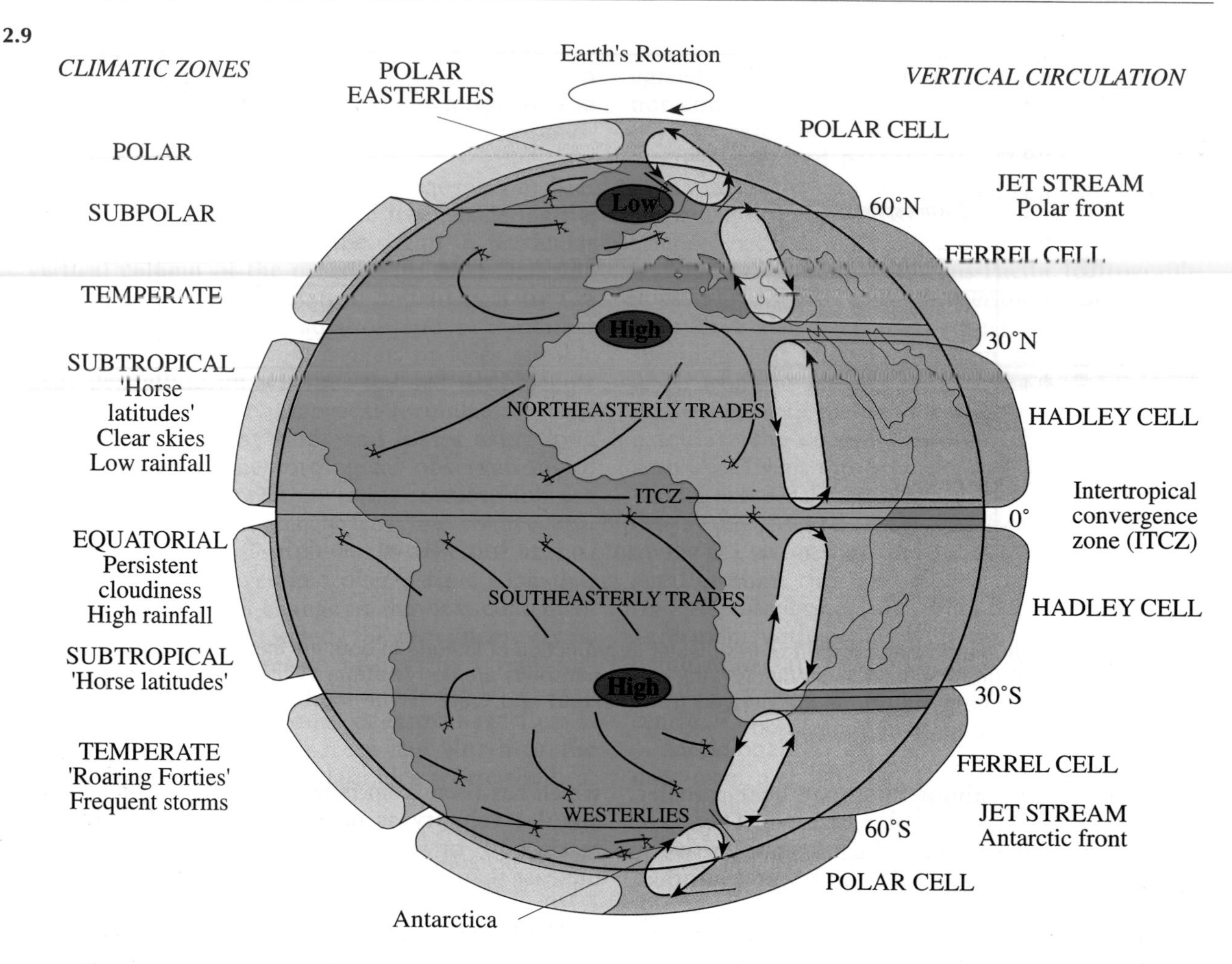

Figure 2.9 Schematic representation of features of the general circulation of the atmosphere (from Gross[6]).

2.10

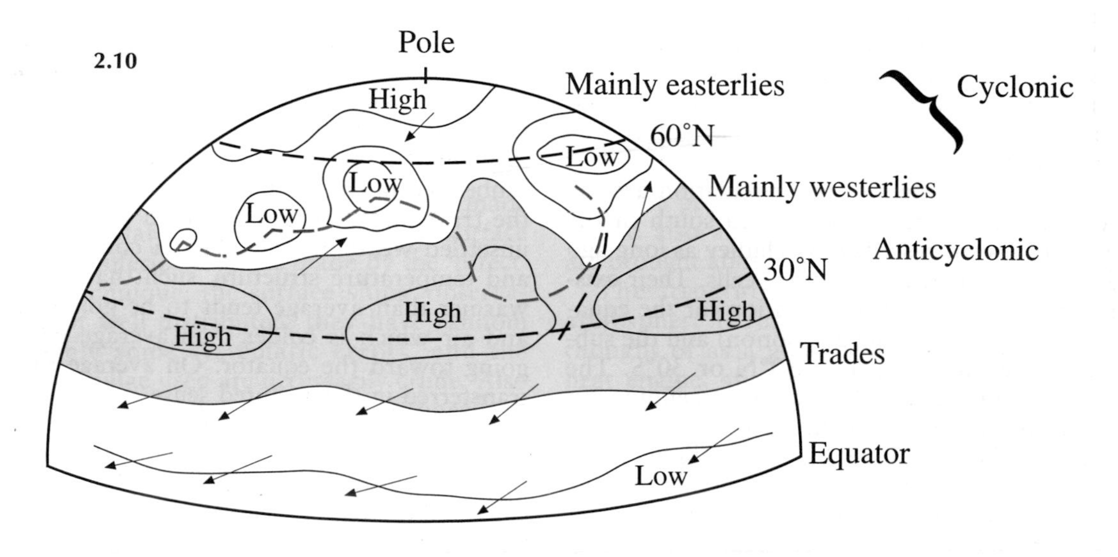

Figure 2.10 Atmospheric pressure at the surface of an idealised Earth as it might be on a particular day in comparison with the long-period average of *Figure 2.9* (from Sutcliffe[19]).

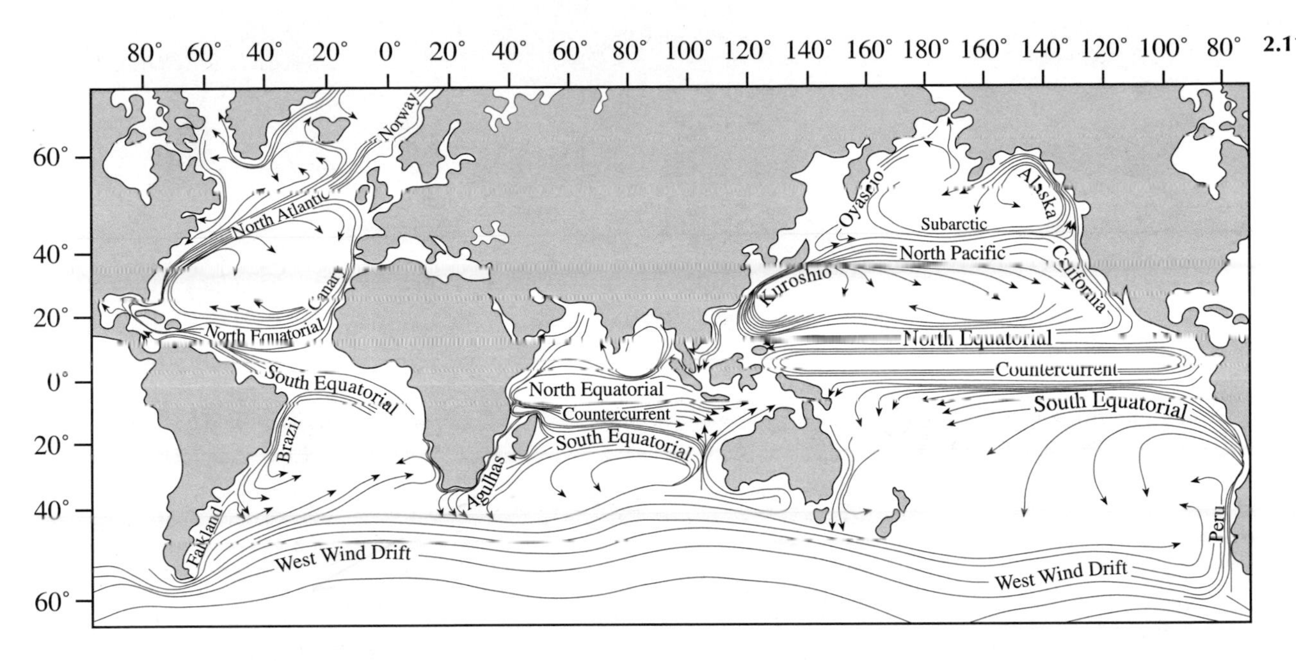

Figure 2.11 Schematic chart of the climatological average of the surface currents of the ocean (based on NRDC[12]).

sufficiently life-like to make one feel that the essential physics of the situation is correct.

The winds that bring about transfer of heat and water in the atmosphere have to be such as to conserve angular momentum about the Earth's axis, and their frictional drag at the Earth's surface such as to maintain the near-constancy of the rate of rotation of the Earth. These considerations limit the latitudinal extent of the Hadley circulation, from its ascent at the ITCZ to latitudes between 30°N and 30°S (*Figure 2.9*). Poleward of this tradewind belt the transfer of momentum, as well as that of heat and water, is brought about by smaller scale features such as depressions and anticyclones.

Calculations of momentum transfer using upper air observations are analogous to those of heat and water vapour transfer: they involve the covariance $[\rho Vq]$ of the northerly and easterly components of the wind velocity. They demonstrate that the large-scale eddies bring about most of the required transports of momentum, except at low latitudes, where meridional cells play an important role, especially in the vertical exchange of momentum. The middle latitude westerlies and the low latitude North-East and South-East Trades are a consequence of the conservation of angular momentum (*Figure 2.10*): the westerlies have to be the stronger to keep the rotation of the Earth constant.

Although the atmosphere and the ocean have certain basic similarities – both are vast bodies of fluid on a rotating Earth – their differences must be recognised. They have marked differences in physical properties, especially those controlling the transmission of radiant energy, and it must be recognised that their geometry also plays an important role. There are no barriers in the atmosphere which correspond to the continental barriers to the oceans.

Momentum transfer in the ocean is less well known but again it is found that large-scale eddies in the ocean are important in transferring angular momentum from strong surface currents into ocean depths. The distribution of surface currents, thought to be mainly wind-driven, has been compiled from ships' reports of their drift from their calculated course. The general features are shown in *Figure 2.11*: although the North Pacific, and the North Atlantic are quite different in shape they have a rather similar current pattern, or general circulation. There is an anticlockwise circulation (or gyre) in their northern parts and a huge clockwise one in the south. This is conspicuously asymmetric, the currents being much stronger in a narrow region near the western boundary of the North Pacific and the North Atlantic (the situation in the Indian Ocean is complicated by the seasonal variation due to the monsoon). These strong boundary currents (the Atlantic Gulf Stream and the Pacific Kuroshio) are the best known currents of the ocean.

Near the equator in all three oceans there are two west-flowing Equatorial Currents. The South Equatorial Current lies at or south of the equator and the North Equatorial Current to the north of it. In the Pacific and Indian Oceans, and in part of the Atlantic, the two west-flowing Equatorial

2.12

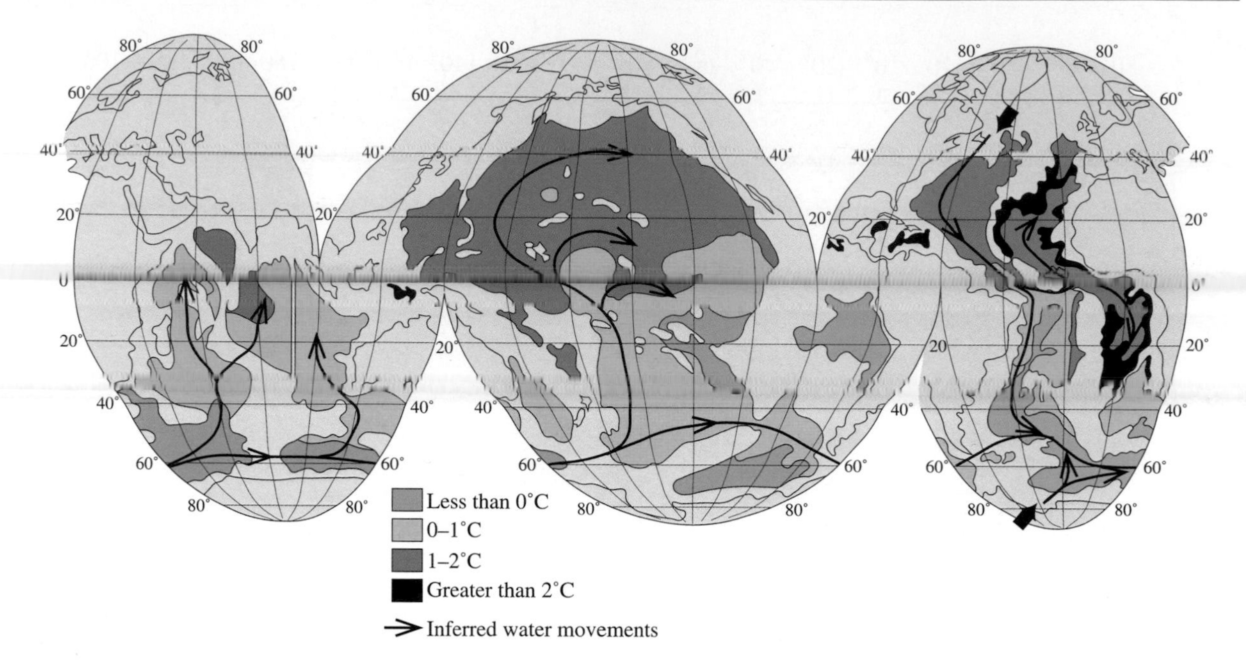

Figure 2.12 An impression of the flow pattern at 4000 m in the ocean. The major inputs are the North Atlantic Deep Water (NADW), which enters at the northern end of the Western Basin of the Atlantic, and the Weddell Sea Bottom Water, which enters from the margin of the Antarctic continent adjacent to the South Atlantic (based on Gross[6] and Broecker and Peng[3]).

Currents are separated by an Equatorial Countercurrent flowing toward the east.

In the Southern Ocean, surrounding the Antarctic, there is no continental barrier (though the relatively narrow Drake Passage may have a similar effect) and the main surface current flows round the Earth as an east-going flow referred to as the Circumpolar Current or as the West Wind Drift.

It must be emphasised that the charts of ocean currents are climatological; that is, they are based on averages of observations made over a long time. On any particular occasion a ship may find a current very different from the average current portrayed on the Pilot Chart. This is especially noticeable in the region of a fast western boundary current, such as the Gulf Stream, which meanders and changes the position of its axis in an unpredictable way (see *Figure 4.8*). In such a case the climatological chart can be misleading, for the observations at a particular place are averaged over a long period, irrespective of whether the current is present or not. It can be seen that a strong narrow current which varies in position is represented on a climatological chart as a broader but slower current. In this way, what may be called the 'climatological Gulf Stream' (as represented on a time-averaged climatological chart) is perhaps ten times wider, and considerably weaker, than the Gulf Stream on any particular occasion.

Currents at greater depth are much less well known; the mean currents are small and are obscured by the variability of the large-scale eddies that have been found to be ubiquitous in the deep ocean. A rough impression of the currents at 4000 m depth is given in *Figure 2.12*.

The estimation of meridional fluxes from trans-oceanic hydrographic sections has provided information on major differences between the mechanisms in the North Atlantic and the North Pacific. Study of the hydrographic section at 24°N in the North Atlantic shows that the heat flux is mainly due to a deep vertical-meridional cell. Warm and relatively saline water flows north near the surface, ultimately losing enough heat in winter to sink to great depth and return southward. The warm Gulf Stream water flows north near the surface but does not return south at similar depth; only after a high-latitude cooling process does it return to the south as deep water. The smaller northward heat transfer at 24°N in the Pacific is due to a nearly horizontal circulation: relatively warm water flows north in the Kuroshio on the western side and in the near surface layer, loses heat in the subtropical and subpolar North Pacific and returns southward in the central and eastern Pacific, at colder temperatures but still at depths less than 800 m. Unlike the North Atlantic, the North Pacific has no source of deep water and its deep circulation is correspondingly slower. These ocean circulation differences

2.13

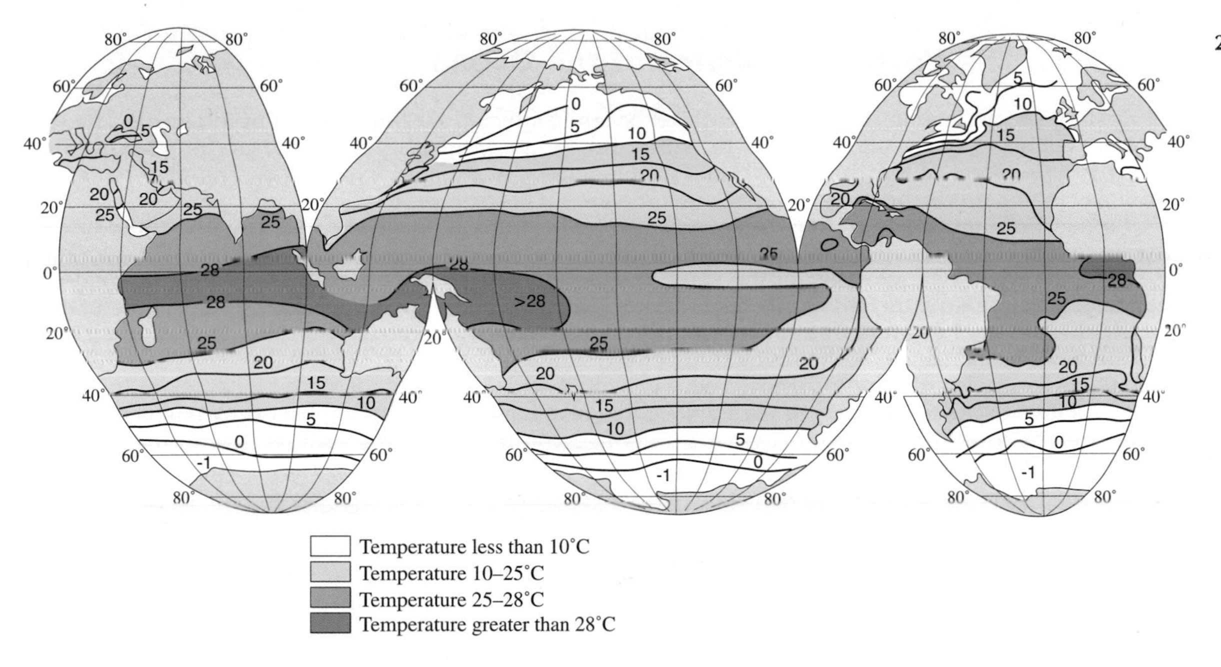

Figure 2.13 Average temperature at the sea surface in February (from Gross[6]; the values are derived from Sverdrup *et al.*[20] and from later sources).

are consistent with the marked climatic differences in the climate of the North Atlantic and the North Pacific – the Atlantic is much warmer, especially at subpolar and polar latitudes (*Figure 2.13*).

The climate of the ocean–atmosphere system depends on complicated interactions between the meridional fluxes, brought about by motions of relatively large scale, and the near-surface vertical fluxes, brought about by motions on a much smaller scale in the complicated turbulent interacting air–sea boundary layer. Processes that determine the properties of the ocean take place at its surface: the resulting horizontal transfers within it are such as to maintain the surface sources and sinks. How this is done is the central problem for observational and theoretical oceanographers alike. It is the target for those making computer models of the ocean and especially for those attempting to model the coupled ocean and atmosphere: they have great difficulty in matching the horizontal motions in the two media to the flux of heat and water between them.

The Coupled Atmosphere–Ocean Boundary Layer

Most of the atmosphere and most of the ocean, most of the time, can be treated as frictionless and adiabatic. But in some places there are vital processes that are more complicated: clouds, fronts in the atmosphere and the ocean, and especially the turbulent boundary layers which exist near the Earth's surface and at the sea floor. Most important is the coupled boundary layer of the atmosphere and ocean which occupies a layer typically 1 km in height above and 100 m in depth below the sea surface. This is a region in which many energy exchanges and transformations take place, processes which determine the properties of the ocean and many of those in the atmosphere. Two processes of the air–sea boundary layer are of particular importance: the production of vertical velocities and the transfer of boundary layer air to the less turbulent free atmosphere above, and of boundary water to the less turbulent deep ocean below.

The frictional stress of the wind on the sea surface, on a rotating Earth, drives a mass transport to the right (left) of stress direction in the northern (southern) hemisphere. If the stress varies from place to place it produces convergences or divergences that lead to vertical motion (*Figure 2.14*); a similar effect happens in the atmosphere. The resulting vertical motions are fundamental to the general circulation of the atmosphere and ocean through their effect on the vorticity balance (see *Box 2.1*). In the atmosphere they lead to the formation of cloud and rain, complications that are not present in the ocean. In the ocean, however, vertical motions (usually smaller than those due to wind stress) are also produced by the difference between Evaporation and Precipitation (*E–P*). Evaporation and precipitation are also fundamental to the near-surface energy exchanges and so to establishing the properties of the lower atmosphere and the upper ocean. These properties are communicated, by complicated and little understood processes, to the

Box 2.1. The Vorticity Balance of the Ocean

All the large-scale flow in the atmosphere and ocean is affected by the rotation of the Earth. The relatively small-scale flows encountered in bathroom, kitchen, or laboratory are dominated by other forces, so the effect of the rotation of the Earth is beyond our usual experience. The flow patterns it produces in the atmosphere and ocean sometimes seem bizarre.

The basic notion is one of the vorticity, or spin, which anything on a rotating globe must have. If one imagines a man standing astride the North Pole, for example, it is obvious that he will be rotating, or spinning, about his own axis, at the same rate as the Earth once per day. If the same man now stands astride the equator, the Earth continues to rotate about its axis but he no longer rotates about his. His local rate of rotation or spin is zero. So the spin which affects anything on the Earth is zero at the equator and increases to one revolution per day at the poles.

In these examples, our hypothetical man – we could equally have considered a parcel of fluid – was at rest relative to the Earth. He – or the fluid – could also have been rotating on his own axis relative to the Earth. The total spin is obviously made up of two components – the spin relative to the Earth and that due to the rotation of the Earth beneath it. It is a difficult but important concept – important because, in relatively shallow fluids like the atmosphere and the ocean, a body of water moves in such a way that its total spin, its vorticity, stays constant.

The vorticity relative to the Earth of a column of fluid increases if it is stretched: the stretching decreases the diameter and the rotation increases to conserve the column's angular momentum. Conversely shrinking the column increases the diameter and the vorticity decreases. The spinning of an ice skater provides a familiar example. To maintain its total vorticity constant, on the rotating Earth, a shrinking column must move equatorward and a stretched column poleward.

In the upper ocean shrinking and stretching are brought about by a vertical gradient of vertical velocity, produced near the surface by wind-stress convergence (*Figure 2.14*), and by *E–P*. Over much of the North Atlantic the wind distribution is such as to produce shrinking, so motion toward the equator. The *E–P* distribution is such as to produce stretching, so poleward motion of smaller magnitude. The resulting southward transport leads to the westward intensification of wind-driven currents, the necessary poleward return flow being accomplished in narrow western boundary regions – like the Gulf Stream – whose dynamics are more complicated.

In the deep ocean there is a very slow upward velocity to compensate for the sinking of deep water at high latitudes. This stretches the water column, and would be expected to produce generally northward transport with equatorial return flow being confined to a narrow region on the western boundary. Such western boundary currents are observed but the northward transport is obscured by the large-scale eddies and by the effects due to sea floor topography.

2.14

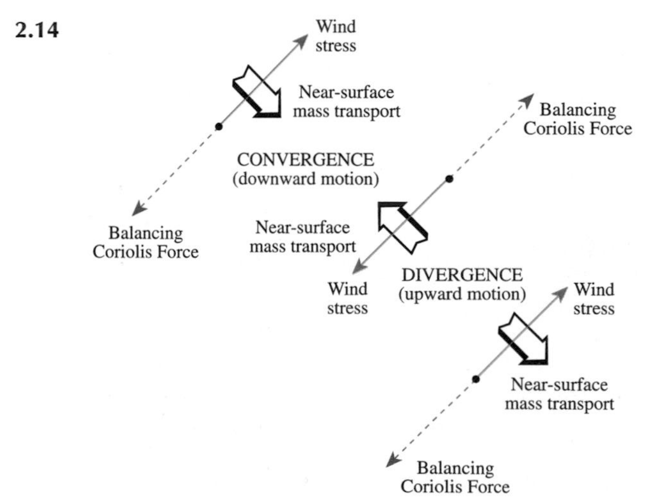

Figure 2.14 The balance of forces in the upper 50 m of the ocean produces a depth mean transport at right angles to the wind stress (to the right in the northern hemisphere). Variations of wind stress from place to place produce convergence or divergence in the surface layers and corresponding vertical velocities below the surface.

air above and to the water below the coupled boundary layer: they determine the properties of the atmospheric airmasses and the oceanic watermasses.

Unfortunately knowledge and understanding of the complicated boundary layer processes is very poor. There is some empirical information about the marine atmospheric boundary layer, especially its lowest 100 m or so (which allows the climatological estimates of heat transfer and evaporation) but uncertainty remains about the interaction of wind and waves. Processes near the top of the marine atmospheric boundary layer, where there is frequently cloud, are much less understood. In the upper layers of the ocean surface waves make observations difficult: there is a need for more information on the effect of spray and of bubbles, as well as on the living and non-living particulate matter that determines the transparency and the absorption of solar radiation. The transfer properties of the helical (Langmuir) circulation (*Figure 2.15*) are still uncertain. In both atmosphere and

2.15

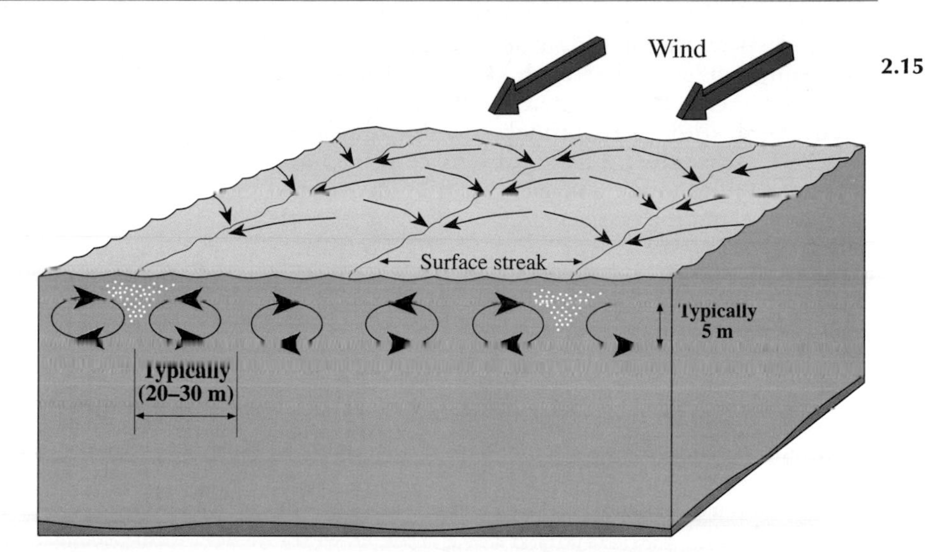

Figure 2.15 Helical (Langmuir) circulations in the upper layers of the ocean producing windrows of floating material on the surface.

ocean we need a reliable way of predicting the vertical extent of the boundary layer and a better understanding of how fluid is transferred from the boundary layer to the free atmosphere and the deep ocean. The problems are technically demanding, and the theory underlying them notoriously difficult, but an improved knowledge of the coupled air–sea boundary layer is vital to our need to understand and to model the atmosphere and the ocean.

Climate Studies

Numerical modelling of the atmosphere is already well advanced, both for weather forecasting and for simulating climate and climatic change. Ocean modelling is advancing rapidly as increasing computer power allows finer resolution to represent ocean eddies, which are smaller than atmospheric disturbances (see Chapter 4). The discrepancies between meridional flux estimates using different methods should soon be clarified: the ERBE project continues, more trans-oceanic hydrographic sections are being made, and there is the prospect of improved atmospheric flux estimates. The results will provide valuable constraints and tests of climate models simulating a coupled atmosphere and ocean.

Near-surface meteorological observations are vital as input to weather forecast models and for the verification of climate models. Their calculation relies on realistic simulation of the atmosphere–ocean boundary layer, but rapid progress in our knowledge of this complicated region is not to be expected. There is also a great need for better observation (preferably from space) and improved simulation of precipitation.

The difference between evaporation and precipitation (E–P) is important as providing a vertical velocity and affecting the vorticity balance, and it provides an input (a haline buoyancy flux) to the surface buoyancy flux which is sometimes comparable to that of the thermal buoyancy flux. It has been suggested that climatic freshening of high-latitude surface water could stop the formation of deep water in the North Atlantic: the overturning meridional cell is thought to be very sensitive to the freshwater flux. Recent estimates show that the thermal buoyancy flux dominates the haline buoyancy flux at high latitudes, suggesting that large changes in (E–P) would be needed to bring about what has been called the 'haline catastrophe'. However the effect of continental run-off, and especially of the freezing and melting of ice (*Figure 2.16*) could be significant, especially in the areas

2.16

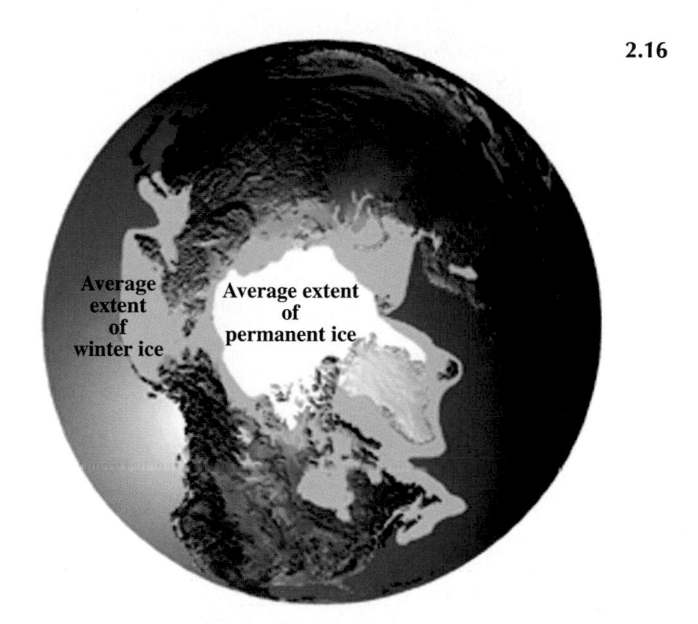

Figure 2.16 Chart to indicate the extent of permanent and winter ice in the Arctic (*from Oceanus*, **37**, 2, 1994).

where deep water formation now occurs. There is increasing evidence that both $(E-P)$ and thermal changes are closely connected to changes in ocean circulation, and so to climate on decadal and longer time-scales. The problem requires an improved understanding of how the motions of the atmosphere and the ocean maintain the sources and sinks of heat and water, and of the coupling between the large scale horizontal winds and currents and the small scale turbulent transfer processes of the coupled air–sea boundary layer.

General References

Gill, A.E. (1982), *Atmosphere–Ocean Dynamics*, Academic Press Inc., Orlando.

Gross, M. Grant (1992), *Oceanography, a View of Earth*, Simon and Schuster, Englewood Cliffs, New Jersey.

Peixoto, J.P. and Oort, A.H. (1992), *Physics of Climate*, American Institute of Physics (AID Press), Woodbury, New York.

References

1. Barkstrom, B.R., Harrison, E.F., and Lee, R.B. (1990), Earth Radiation Budget Experiment: preliminary seasonal results, *EOS*, **71**, 297–305.
2. Baumgartner, A. and Reichel, E. (1975), *The World Water Balance*, Elsevier, New York.
3. Broecker, W.S. and Peng, T.-H. (1982), *Tracers in the Sea*, Lamont–Doherty Geological Observatory, Columbia University, New York.
4. Bryden, H. (1993), Ocean heat transport across 24°N latitude, in *Interactions Between Global Climate Subsystems: the Legacy of Hann*, McBean, G.A. and Hantel, M. (eds), Geophysical Monographs, **75**, 65–75.
5. Ellis, J. and Vonder Haar, T.H. (1976), *Zonal Average Earth Radiation Budget Measurement from Satellites*, Atmos. Sci. Papers 240, Colorado State University, Fort Collins, Colorado.
6. Gross, M. Grant (1992), *Oceanography, a View of Earth*, Simon and Schuster, Englewood Cliffs, New Jersey.
7. Gruber, A. (1978), *Determination of the Earth–Atmosphere Radiation Budget from NOAA Satellite Data*, NOAA Tech. Rep. NESS 76, Washington DC.
8. Jacobwitz, H., Smith, W.L., Howell, H.B., and Hagle, F.W. (1979), The first 18 months of planetary radiation budget measurement from the Nimbus-6 ERB experiment, *J. Atmos. Sci.*, **36**, 501–507.
9. Masuda, K. (1988), Meridional heat transport by the atmosphere and the ocean: analysis of FGGE data, *Tellus*, **40A**, 285–302.
10. Michaud, R. and Derome, J. (1991), On the mean meridional transport of energy in the atmosphere and oceans as derived from six years of ECMWF analyses, *Tellus*, **43A**, 1–14.
11. Nieburger, M., Edinger, J.G., and Bonner, W.D. (1982), *Understanding our Atmospheric Environment*, W.H. Freeman and Company, San Francisco.
12. NRDC (1946), *Summary Technical Report*, Division 6, Office of Naval Research, Washington DC.
13. Oort, A.H. and Peixoto, J.P. (1983), Global angular momentum and energy balance requirement from observations, *Advances in Geophysics*, **25**, 355–490.
14. Peixoto, J.P. and Oort, A.H. (1983), The atmospheric branch of the hydrological cycle and climate, in *Variations in the Global Water Budget*, Street Perrott, A. (ed.), pp 5–65, Reidel, Dordrecht.
15. Schmitt, R.W. (1994), *The Ocean Freshwater Cycle*, JSC Ocean Observing System Development Panel, Texas A&M University, College Station, Texas.
16. Schmitt, R.W., Bogden, P.S., and Dorman, C.E. (1989), Evaporation minus precipitation and density fluxes for the North Atlantic, *J. Phys. Oceanogr.*, **19**, 1208–1221.
17. Schmitt, R.W. and Wijffels, S.E. (1993), The role of the ocean in the global water cycle, in *Interactions Between Global Climate Sub-systems: the Legacy of Hann*, McBean, G.A. and Hantel, M. (eds), Geophysical Monographs, **75**, 77–84.
18. Stephens, G.L., Campbell, G.C., and Vonder Haar, T.H. (1981), Earth radiation budgets, *J. Geophys. Res.*, **86**, 9739–9760.
19. Sutcliffe, R.C. (1966), *Weather and Climate*, Wiedenfield and Nicolson, London.
20. Sverdrup, H.U., Johnson, M.W., and Fleming, R.H. (1942), *The Oceans, their Physics, Chemistry and Biology*, Prentice Hall, New York.
21. Vonder Haar, T.H. and Ellis, J. (1974), *Atlas of Radiation Budget Measurements from Satellites*, Atmos. Sci. Papers 231, Colorado State University, Fort Collins, Colorado.
22. Wijffels, S.E., Schmitt, R.W., Bryden, H.L., and Stigebrandt, A. (1992), Transport of fresh water by the oceans, *J. Phys. Oceanogr.*, **22**, 155–162.

CHAPTER 3:

The Role of Ocean Circulation in the Changing Climate

N.C. Wells, W.J. Gould, and A.E.S. Kemp

Overview of Climate System

The relatively warm and stable climate which we have had for the past 10,000 years, since the end of the most recent glaciation, has been essential for the evolution of cultivated crops, which led to the development of human settlements rather than to a nomadic existence, and hence to the development of civilisation.

Recent research has shown that changes in ocean circulation played a key role in controlling climate change and regulating the glacial–interglacial cycles that have been a hallmark of the northern hemisphere climate for the past 2 million years. Understanding the nature of this link between ocean circulation and climate change is now a key goal of research in this area. In particular, the identification of periods of very rapid climate change in the recent past has given a new urgency to these studies.

The measurement of climate change by scientific instruments dates back to the invention of the thermometer by Galileo in the sixteenth century. It was, however, not until considerably later that systematic methods were applied to the measurement of temperature and rainfall, which allowed the concept of climatology to develop. These measurements have been and are unevenly distributed over the globe, with the majority covering the landmasses and, particularly, the well-populated areas of the northern hemisphere. The oceans, covering nearly 71% of the Earth's surface, have not been neglected; indeed, one of the first responsibilities of the UK Meteorological Office, when it was formed in 1854, was to implement the systematic recording of surface observations from commercial ships plying the trade routes of the world. Despite such efforts, measurements in the southern hemisphere, particularly over the Southern and Indian Oceans, remain rather sparse. We shall see demonstrated that these poorly observed oceans are a key element in the Earth's climate.

Recent interpretation of ice cores from Greenland[15] has shown that the climate system is not as stable as was once thought and may undergo extremely rapid changes (e.g., 5–7°C in a decade). As we concern ourselves with possible man-made influences on climate, one of our great challenges in the closing decade of the twentieth century is to measure and understand natural climate variability across the globe, together with the role played by the oceans.

Given the limitations of the observing network, where does our scientific evidence for climate change over longer periods come from?

Indirect evidence in ice cores from Antarctica and Greenland, and from sediment cores studied during the CLIMAP (Climate: Long Range Interpretation, Mapping, and Prediction) initiative and, more recently, from the Ocean Drilling Programme, have provided reliable indicators of past climate variations. The measurement of the air trapped in the Vostok ice core, from Antarctica, has revealed the levels of carbon dioxide from the most recent interglacial period (circa 125,000 years BP) to the present day[16], and thus has given a benchmark from which the more recent anthropogenic contribution can be determined. Measurements of oxygen isotope ratios, contained in bubbles trapped in the ice, are now known to be a proxy for air temperatures (see Chapter 8, *Box 8.3*). Recent analyses of ice cores from Greenland have thus provided time series of temperature from the most recent interglacial to the present day. This has shown that the climate over the last 5,000 years has been remarkable for its stability. The ice cores also provide evidence for rapid changes in climate during the previous interglacial, from temperatures similar to those of today to glacial conditions on the remarkably short time-scales of decades to centuries (see later).

Evidence of longer-term changes, those over millions of years, are found in marine sediments (see Chapter 8). The glacial and interglacial cycles that occurred during the last 2 million years have been linked to variations in solar radiation, associated with variations of the orbital parameters of the Earth around the Sun, known as Milankovitch cycles. Longer-term changes may be associated with the different configurations of the oceans and continents, due to continental drift, and with variations in continental uplift and mountain building.

In addition to these interpretations of past climate we now have new powerful methods which

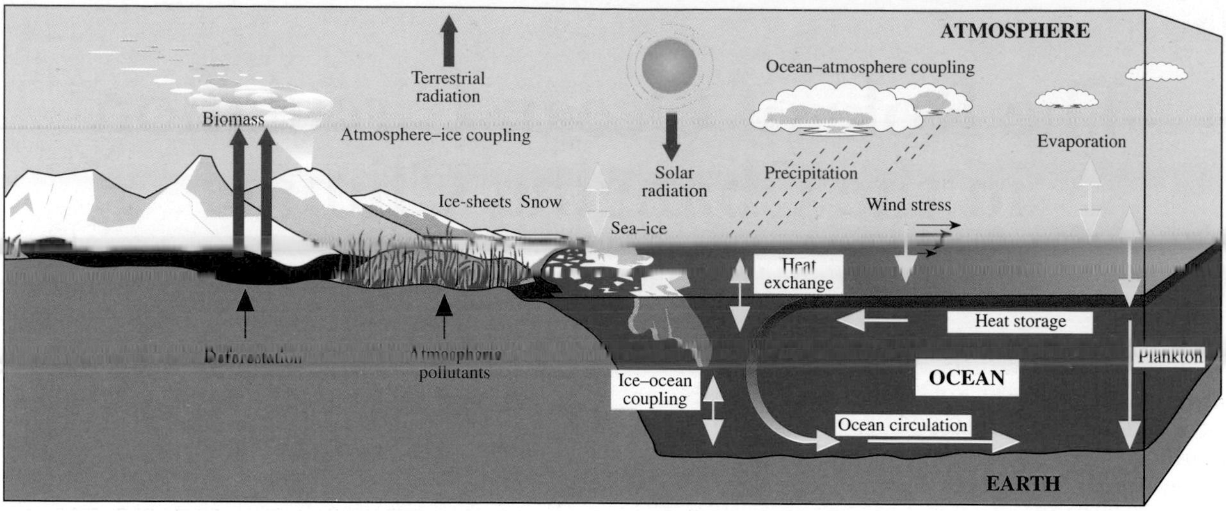

Figure 3.1 A schematic view of the Earth's climate system showing the roles of land, atmosphere, oceans, and sea ice.

are being applied to understanding how the Earth's climate works. These methods are based on an understanding of the different 'components' of the climate system (air, water, and ice), and the interactions between them (*Figure 3.1*).

Each component responds to change over different periods of time. The atmosphere takes the shortest time (of the order of a week to a month) to communicate changes throughout its mass, because winds are very much faster than ocean currents and mixing through turbulence is efficient. The dominant time-scale of the ocean is of the order of weeks to seasons in the surface layers and decades to centuries at abyssal depths. The ice caps of Antarctica and Greenland are excellent measures of longer-term change as they grow slowly on time-scales of centuries to millennia.

Mathematical computer models of each one of the components, bounded, constrained, or driven by interactions with others, allow a deeper understanding of both the components themselves and their interactions with one another. For example, mathematical models of the ocean (see *Boxes 3.1* and *3.2*) are based on the dynamical laws which govern the behaviour of the ocean. These ocean models can reproduce the large-scale, wind-driven circulations of the ocean basin gyres, and the deep vertical transports, or overturning, of the thermohaline circulation of the oceans driven by cooling at polar latitudes. The advent of faster, more powerful computers has allowed models to reproduce some of the smaller (*ca* 100 km), energetic ocean eddies (see Chapter 4), which are important for both the transport of heat and fresh water in some regions of the world's oceans and for the maintenance of the large ocean circulations. Observations are always needed to initialise and test the models, but while relatively abundant observations are available for the atmosphere, comprehensive, synoptic ocean measurements are few and far between. The World Ocean Circulation Experiment (WOCE; see later) will provide critical ocean measurements, such as heat and fresh water transports, on a global scale.

Improvements in our ability to model the various components of the climate system will allow the development of more comprehensive climate models in which the components are coupled together. These models will aid our interpretation and understanding of the complex climate system, as well as providing predictions of socioeconomic importance.

The Role of the Ocean in Climate

The ocean and atmosphere together transfer heat from the tropics to the polar regions, at a rate of the order of 5 PW at 30°N, (equivalent to the output of 5 million large power stations) in order to balance the deficit of incoming radiation in the mid-latitudes and polar regions with the excess of radiation in the tropics. The partition of this transfer between atmosphere and ocean is not well-determined, but recent ocean measurements at 24°N have shown transports within the ocean of the order of 2 PW; a value which is comparable with the atmospheric flux at this latitude (*Figure 3.2*)[7]; see Chapter 2 for further discussion.

The oceanic component of the heat flux is provided by ocean currents, which have much longer time-scales (of the order of 10 years for the wind-driven subtropical gyres and of many decades to centuries for the vertical overturning thermohaline circulation) than the atmosphere. When one considers that a 2.5 m deep layer of sea water covering the globe has the same thermal capacity as the entire atmosphere, it can be appreciated that the

Figure 3.2 Poleward transfer of heat by: (a) ocean and atmosphere together (T_A+T_O), (b) atmosphere alone (T_A), and (c) ocean alone (T_O). The total heat transfer (a) is derived from satellite measurements at the top of the atmosphere, that of the atmosphere alone (b) is obtained from measurements of the atmosphere, and (c) is calculated as the difference between (a) and (b) (1 PW = 10^{15} W). (Based on Carrissimo *et al.*[7]; results from other investigations are added for comparison.)

oceans make a very important contribution to the stabilisation of our climate system.

While, overall, the oceans transfer heat poleward, they also exchange heat between ocean basins in a more complex fashion. For instance, the North Atlantic Ocean loses more heat to the atmosphere than it gains from incoming radiation, so there has to be a net heat transfer from the Pacific and Indian Oceans into the South Atlantic Ocean and thence to the North Atlantic to compensate for this deficit. The thermohaline circulation, driven by the production of denser water at polar latitudes, is a mechanism by which heat is transferred within and between ocean basins. The oceans are interconnected by what has come to be known as The Global Thermohaline Conveyor Belt (*Figure 3.3*)[13].

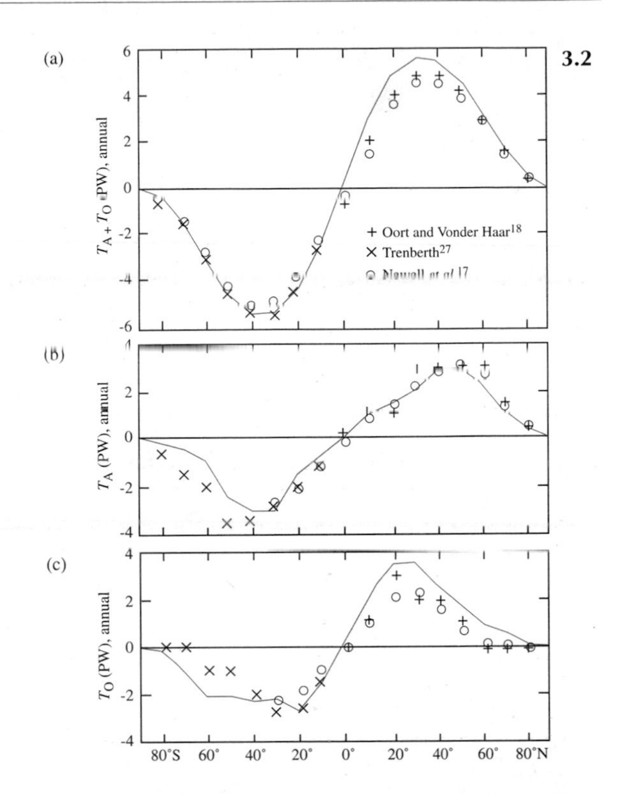

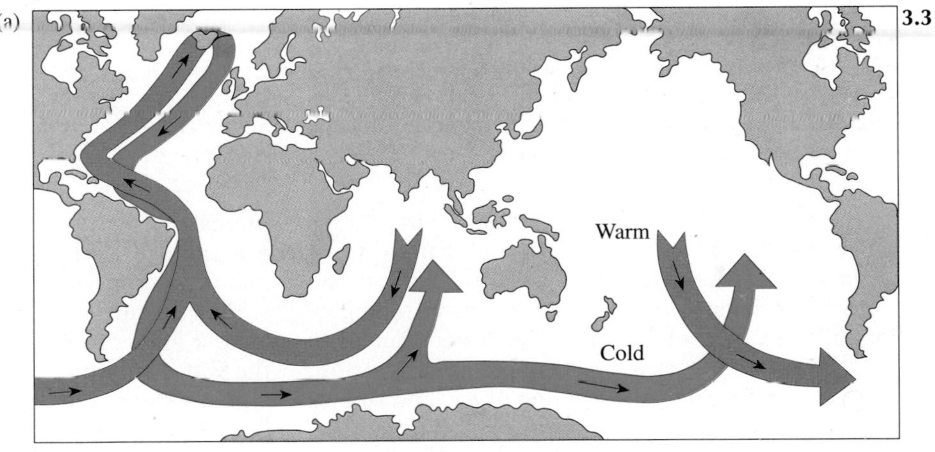

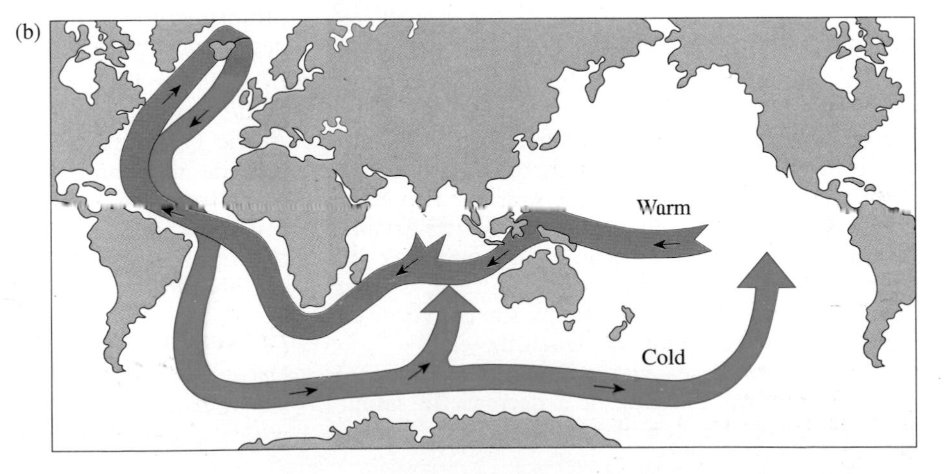

Figure 3.3 Schematic views of two versions of the 'conveyor' circulation of the oceans. Warm water associated with the surface and intermediate waters of the oceans (upper 1000 m) follows a pathway toward the northern North Atlantic Ocean, where it is subjected to intense winter cooling. This leads to the formation of cold North Atlantic deep water, which spreads southward into the Southern Ocean and returns to the Pacific Ocean. The conveyor is responsible for a northward transfer of heat throughout the whole of the Atlantic Ocean. (a) The upper water moving eastward from the Pacific into the Atlantic. (b) The upper warm water moving through the Indonesian Archipelago into the Indian Ocean and thence into the Atlantic Ocean.

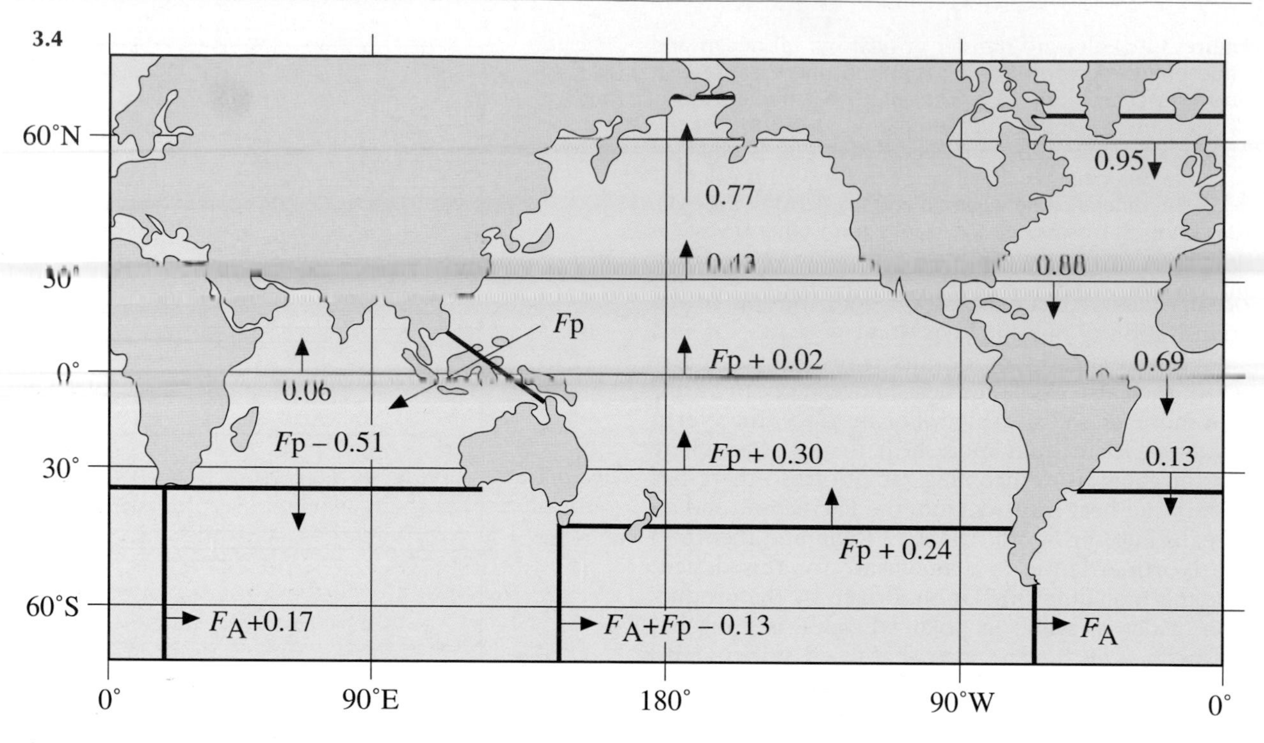

Figure 3.4 An estimate of the transfer of fresh water (x 10^9 kg/s) in the world oceans. In general, in polar and equatorial regions precipitation and river run-off exceed evaporation and hence there is an excess of fresh water, while in the subtropical regions there is a water deficit. A transfer of fresh water is required from the regions of surplus to the regions of deficit. For example, in the North Atlantic, there is southward flow of 950,000 t/s (t = metric tonne) of fresh water at 60°N, while at the equator the southward flow is 690,000 t/s – hence 26,000 tonnes of fresh water per second are evaporated in the North Atlantic. It can be seen that fresh water is exported from the North Pacific to the North Atlantic, through the Arctic Ocean. F_P and F_A refer to the fresh water fluxes of the Pacific–Indian throughflow and of the Antarctic Circumpolar Current in the Drake Passage, respectively.

The ocean also transports fresh water around the globe, another key element of the global conveyor (*Figure 3.4*). The total mass of salts in the ocean remains unchanged on time-scales shorter than geological time. In contrast, the fresh water content of the oceans changes in response to precipitation, evaporation, freezing and melting of ice, and run-off from the land. All these factors influence the dilution or concentration of ocean salt, and hence the salinity. Differences in the input of fresh water into the ocean, from one region to another, have to be balanced by a horizontal transport of fresh water by ocean currents and by sea ice in polar regions. Generally, in the subtropics there is a deficit of fresh water which is reflected in higher surface salinity, while in the higher latitudes there is an excess of fresh water and a lower surface salinity. The North Pacific Ocean is less saline than the North Atlantic, because of lower evaporation, and therefore is a source of fresh water. This water, in turn, is transported into the North Atlantic by the conveyor circulation, to make up for the deficit there.

The density of sea water depends on temperature and salinity. Heating or a decrease in salinity will decrease the density, while cooling or an increase in salinity will increase the density. The low salinity layer at the surface of the North Pacific maintains a stable stratification (low density at the surface), which cannot be destabilised (i.e., made to have a higher density at the surface than deeper in the water column) by surface cooling in the present climate state. Hence, there is no significant deep convection in the North Pacific and, consequently, little formation of cold deep water. By contrast, the northern North Atlantic, with a higher surface salinity, is destabilised by winter cooling, and produces deep cold watermasses by vertical convection. Polar watermasses, formed in the Greenland/Norwegian sea and, to a lesser extent, in the Arctic Ocean, also enter the North Atlantic and provide additional cold deep water. The transport of ice from the Arctic Ocean provides an additional source of fresh water to the North Atlantic, which through its variability can modulate the surface salinity on decadal time-scales[28].

Evidence of Climate Change in the Ocean

Despite its high thermal capacity, the ocean responds to exchanges of heat, fresh water, and momentum with the atmosphere on a range of time-scales from

3.5

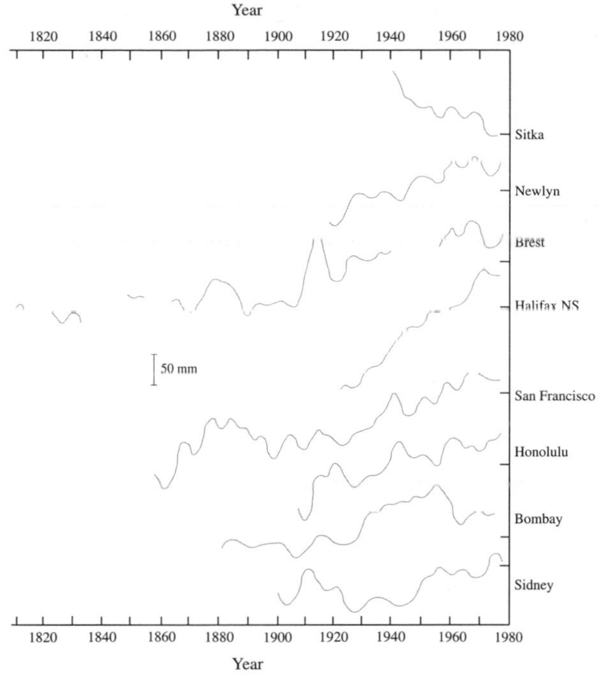

Figure 3.5 Long-term changes in annual mean sea level at selected ports. The general upward trend is seen at all but one station. The downward trend in sea level at Sitka, Alaska, is due to the vertical movements of the land at which sea level is measured (see text). The global mean sea level rise is estimated to be about 1–3 mm/year.

a day and shorter right up to geological scales (glacial and interglacial periods of 10,000–100,000 years; see Chapter 8). The thermal inertia of the ocean, however, means that oceanic changes are of much smaller amplitude than those seen in the atmosphere, but nevertheless they are easily measurable on daily to seasonal scales and, with care, on decadal scales. The state of the ocean at any time is predominantly a result of exchanges with the atmosphere over the previous 100 years or so.

By far the longest time series of 'oceanographic' measurement is that for sea level (*Figure 3.5*). Measurements made for the prediction of tides go back over 100 years and, after the tidal signal is removed, show long-term trends, rising in some places, in others falling. It has to be remembered that sea level is measured relative to a fixed point on land, so the changes seen are a summation of changes in true sea level (due primarily to thermal expansion of the water, and melting of glaciers and ice caps) and the not insignificant vertical movement of the land (isostasy). This movement can be caused by 'rebound' after release from the covering of ice during the most recent glacial period or by tectonic activity. Unravelling the changes due to each individual factor is difficult[21] and yet the prediction of sea level change is of immediate and practical importance to low-lying areas and, in particular, to the inhabited coral atolls of the Pacific and Indian Oceans.

Measurements of temperature and salinity with useful accuracy have only been available for the past 50 years or so, and there are few places where high-quality measurements have been made over several decades. This severely limits our ability to directly measure climate-scale change in the ocean.

At the surface of the ocean there is a great deal of temperature variability caused by daily and seasonal heating and cooling, so here salinity is a better indicator of decadal change. Salinity records made in the area west of Scotland show a remarkable decrease that lasted from 1973–1979, with the lowest values in 1975. The lowest value deviated from the average by almost four times the typical variability.

Subsequent analysis of other salinity data from around the Atlantic showed that this salinity change, now referred to as the 'Great Salinity Anomaly', was not confined to Scottish waters, but was a phenomenon that took over 10 years to propagate around the North Atlantic (*Figure 3.6*)[10,17]. There is still much conjecture about the cause of the 'Great Salinity Anomaly'. Recent evidence has suggested that in the 1960s an unusually large export of ice into the North Atlantic from the Arctic Ocean may have caused the low salinity

3.6

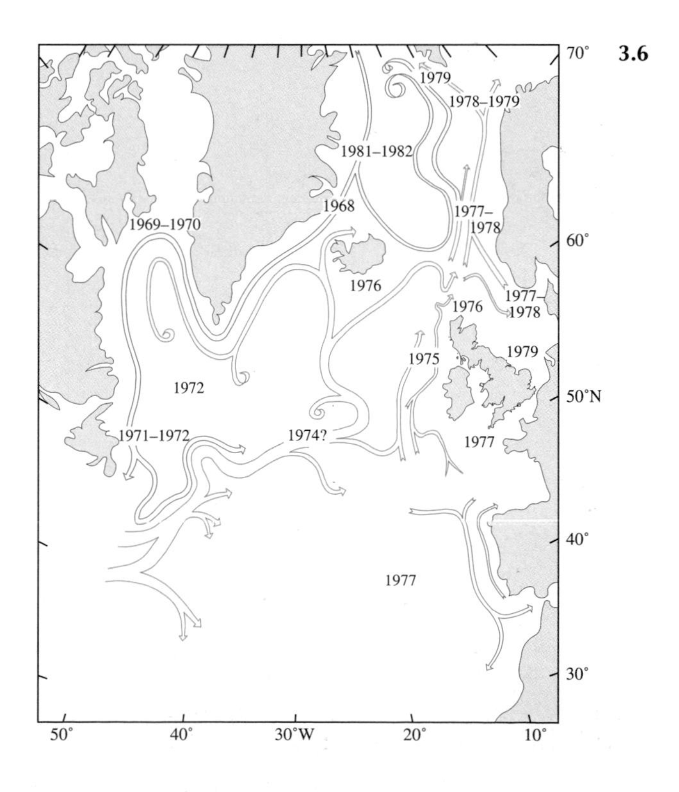

Figure 3.6 Timing of the propagation of the 'Great Salinity Anomaly' around the North Atlantic.

3.7

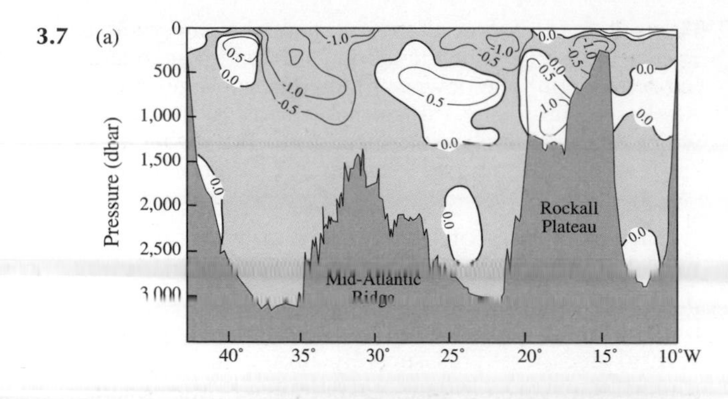

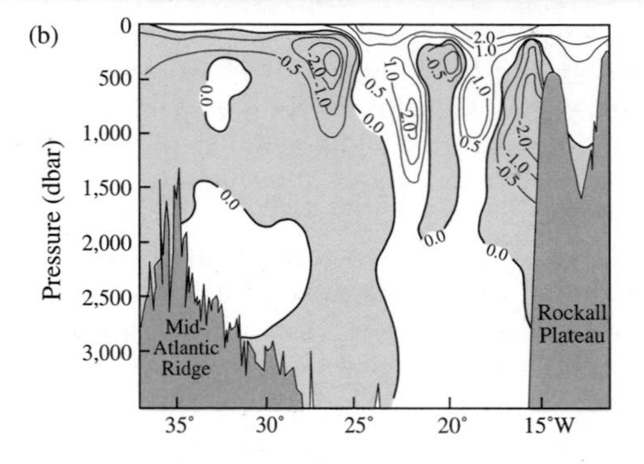

Figure 3.7 Cooling (blue areas, °C) along the section of the subpolar North Atlantic between northwest Europe and Greenland: (a) 1981–1991, (b) 1962–1991.

3.8

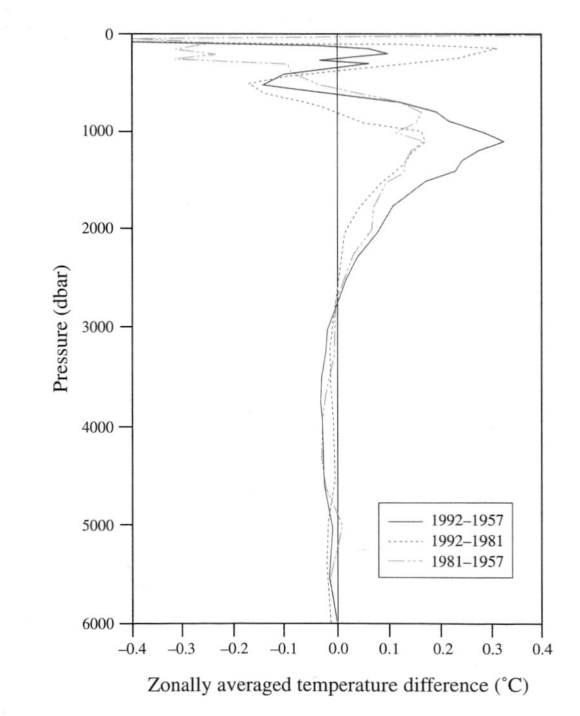

Figure 3.8 Temperature changes (°C) in the subtropical North Atlantic (24°N), 1957–1992. The measurements have been averaged across 24°N between North Africa and Florida.

anomaly, which suppressed deep convection for more than a decade[1].

Whatever the cause, the significance of such changes cannot be over-emphasised. The global thermohaline conveyor belt is driven by deep convection in high latitudes and salinity is, in many cases, the controlling factor that decides whether such deep convection is possible. In simplified terms, fresher water at the surface results in the formation of an insulating layer of ice in winter, rather than plumes of dense cold saline water. So it is possible that extreme salinity anomalies could be responsible for the transitions between glacial and interglacial episodes. We give an example of this in the next section.

In the interior of the ocean, patterns of temperature and salinity change have been determined in three ways, from time series at fixed stations, by the subtraction of averages of values accumulated, say, over one decade from those in another decade, and from the comparison of repeated sections (lines of measurements across an ocean). Each of these techniques has its drawbacks. There are very few time series stations (Bermuda, Hawaii, and off the West Coast of Canada), the decadal averages are subject to errors of non-uniform spatial sampling, and the sections are instantaneous pictures that are contaminated by the presence of energetic transient eddies (see Chapter 4).

Despite these limitations, results from each of these lead to the conclusion that temperatures in the ocean change by a few tenths of a degree over a typical 10-year time-scale. The changes are largest in the upper part of the water column and decrease with depth (see *Figures 3.7* and *3.8*). The horizontal areas of such changes are large, comparable to the width of an ocean basin[19,23].

Thermohaline Catastrophes and the Younger Dryas

During the most recent ice age, the North Atlantic component of the Global Thermohaline Conveyer was partially shut down and the ocean is thought to have operated in a different mode to that of the present day. The northern North Atlantic was considerably cooler and the transport of the North Atlantic current (the northward extension of the Gulf Stream) much reduced (*Figure 3.9*). The North Atlantic component of the conveyer was reactivated at the end of the most recent glaciation, at about 14,000 years BP.

Recent research on deep-sea sediment cores has shown that this reactivation of the conveyer was not without hiccups! Part of this conveyer stopped abruptly at about 11,000 BP – a period known as the Younger Dryas[29]. This led to a catastrophic cooling of the North Atlantic region and caused the build up of small glaciers in the British mountains in what geographers call the Loch Lomond glacial

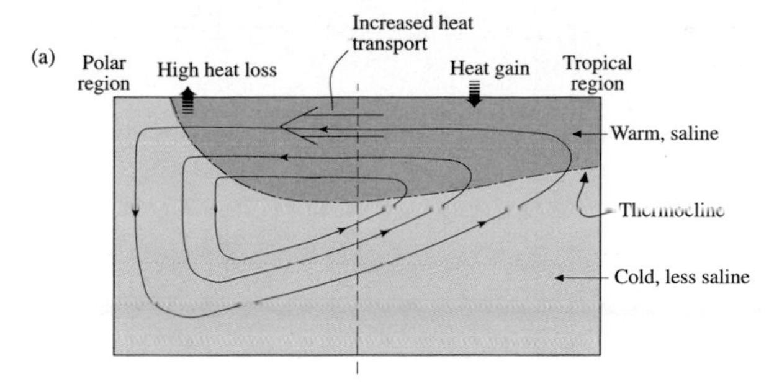

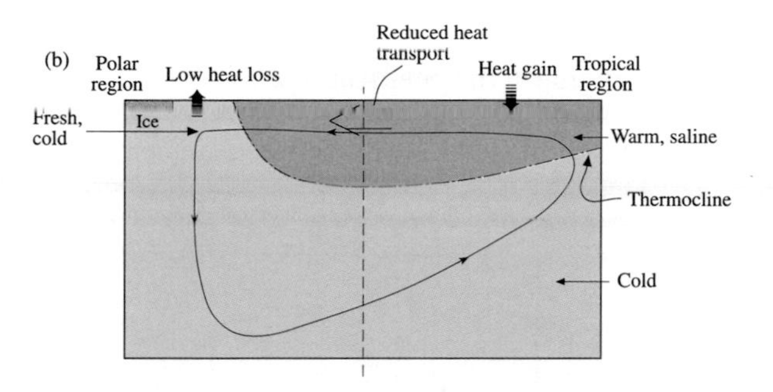

Figure 3.9 (a) The 'present-day' North Atlantic with a vigorous conveyor (thermohaline) circulation, and a large northward transfer of heat. (b) The 'ice-age' North Atlantic with a weak conveyor (thermohaline) circulation, and a reduced northward heat transfer. The northern North Atlantic was therefore cooler and fresher than the 'present-day' ocean. The extended sea-ice cover is associated with a fresher surface layer, which stabilises the water column and inhibits deep convection.

re-advance. This cooling only appears to have lasted a few centuries, but it developed very rapidly over decades. There are several theories about what exactly led to the shut down of the conveyer. One view is that a sudden influx of fresh melt-water from the Laurentide ice sheet into the North Atlantic could have stabilised the vertical stratification and reduced the rate of formation of North Atlantic deep cold water. This, in turn, could have shut down the North Atlantic conveyor circulation, resulting in a cooling of the surface waters of the northern North Atlantic[4].

The effect of changes in fresh water fluctuations on the thermohaline system is an example of positive feedback. The fresh water input weakens the thermohaline circulation, which makes the circulation more susceptible to further weakening. This process has been investigated in recent years using mathematical models of ocean circulation. The models show that there are a number of different states of the thermohaline circulation, some of which are stable (*Figure 3.10*). There are, however, transitions between stable states, which occur over periods as short as 40 years[9]. It has been speculated that the present North Atlantic Ocean may be close to one of these transitional states, of which the Younger Dryas is an example. One study has suggested that the transition between states is not necessarily symmetrical. The change from strong to weak thermohaline circulation may be more rapid (40 years) than the re-establishment of the strong circulation (500 years). Further study of the Younger Dryas and similar events in the palaeoclimate record may give us important clues to the

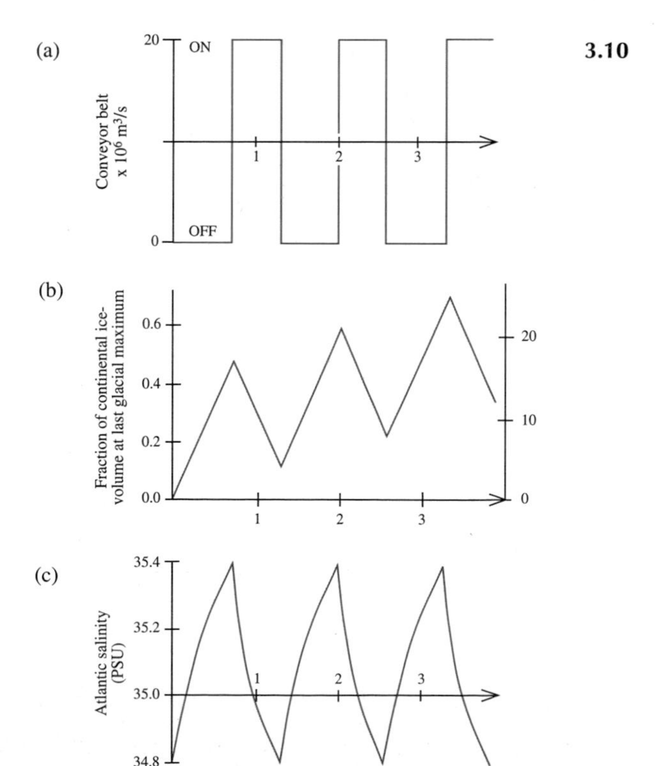

Figure 3.10 A theoretical model for oscillations of the ocean conveyor belt, continental ice volume, surface water flux, and Atlantic Ocean salinity as a function of time (1 unit = 1000 years). The oscillator model is based on bi-stable states for the conveyor circulation [(see *Figure 3.9(a)*]. When the conveyor circulation is turned off, northward heat transfer is reduced and cools (a), which results in a growth in ice volume (b). The reduced export of salt from the Atlantic, because of the 'turning off' of the conveyor circulation, causes an increase of salinity (c). The salinity increases to a threshold value, at which point the deep overturning of the northern ocean occurs, and the conveyor circulation is 'turned on' again. The continental ice volume (b) decreases, because of the increase in melting associated with the increased northward heat flux into the North Atlantic. With more melting and a greater export of salt (due to the 'turning on' of the thermohaline circulation), the Atlantic salinity decreases. These oscillations have a period of the order of 1000 years – the thermohaline circulation time-scale. (Based on Birchfield and Broecker[3].)

likely response of the present-day thermohaline circulation in the North Atlantic to global warming (remaining stable or 'flipping' to another state).

Monitoring Climate Change

So we know there are changes in the ocean that occur over periods of tens of years. Similarly, we suspect that effects of this type may be implicated in longer-period and more extreme (glacial and interglacial) changes. We have now to ask the question whether anthropogenic activities can be detected in the ocean and whether new technologies can help us to observe the ocean better.

There is at least one case in which repeated observations in the ocean seem to point to a man-made effect. The western Mediterranean Sea is an interesting 'laboratory' in which the process of deep water convection can be observed in conditions that are less hostile than those found in the high Arctic and Antarctic. The product of western Mediterranean convection is a homogeneous, warm, saline watermass that ultimately leaves the Mediterranean and enters the North Atlantic through the Straits of Gibraltar. Measurements of the properties of this water have been made since early in the twentieth century, since when they show that the temperature, salinity, and density of the watermass have increased slowly up to the mid-1950s and more quickly thereafter, *Figure 3.11*. This change has been attributed to a reduction of fresh water inflow into the Mediterranean caused by the damming of the Nile and those rivers flowing to the Black Sea[25].

The detection of anthropogenic effects in the open ocean is much more difficult. We know little about the inherent ocean variability, so detection requires a long-term commitment to systematic and careful monitoring of the state of the oceans. This has been embarked upon already in the WOCE. Over the period from 1990–1997, this experiment will provide a 'snapshot' of the state of the oceans (*Figure 3.12*), including the distribution of physical and chemical properties and an assessment of the role of ocean circulation in the transport of heat and water. The WOCE measurements, *Figure 3.13*, are being used to test and improve ocean models running on some of the largest computers now available[14].

WOCE will produce a baseline picture of the oceans against which future measurements can be compared. Such future measurements will be made by the Global Ocean Observing System (GOOS), the ocean element of a Global Climate Observing System (GCOS). WOCE observations are based on the use of expensive and sophisticated research ships, and these cannot be expected to provide much longer-term routine monitoring; to do this we must look to other techniques.

3.11

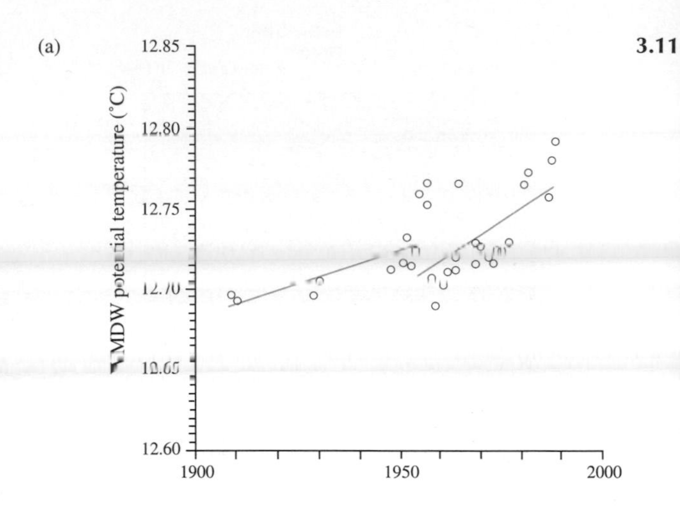

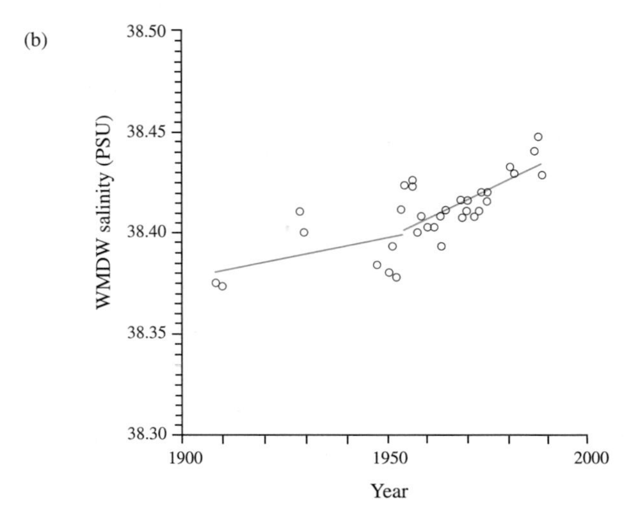

Figure 3.11 Changes in (a) temperature and (b) salinity of the western Mediterranean deep water (WMDW) during the twentieth century.

We already have some elements of a monitoring system in place. Sea level from coastal stations can be monitored centrally in real time and is used to analyse and predict the progress of El Niño events in the Pacific (see later). Satellite-tracked drifters as indicators of surface currents are becoming more and more reliable and can measure not just water movement, but also temperature, salinity (being developed), and meteorological parameters (see Chapter 19). Many merchant ships are equipped to measure temperature using expendable bathythermographs to provide a set of observations of subsurface temperature; but these cover only the major trade routes and contribute almost nothing over the remainder of the globe.

At present, satellites provide the only global rou-

3.12

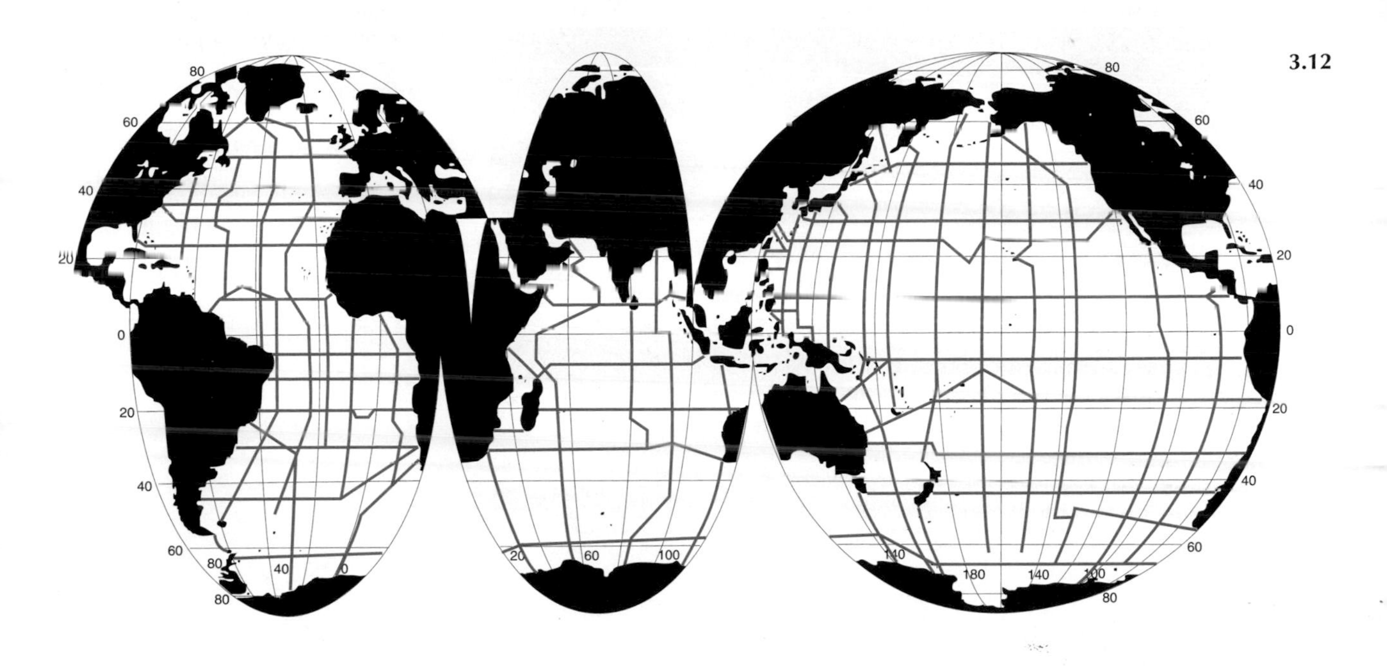

Figure 3.12 The WOCE hydrographic survey grid. Physical and chemical properties will be measured from surface to sea bed every 50 km along the red lines between 1990–1997.

3.13

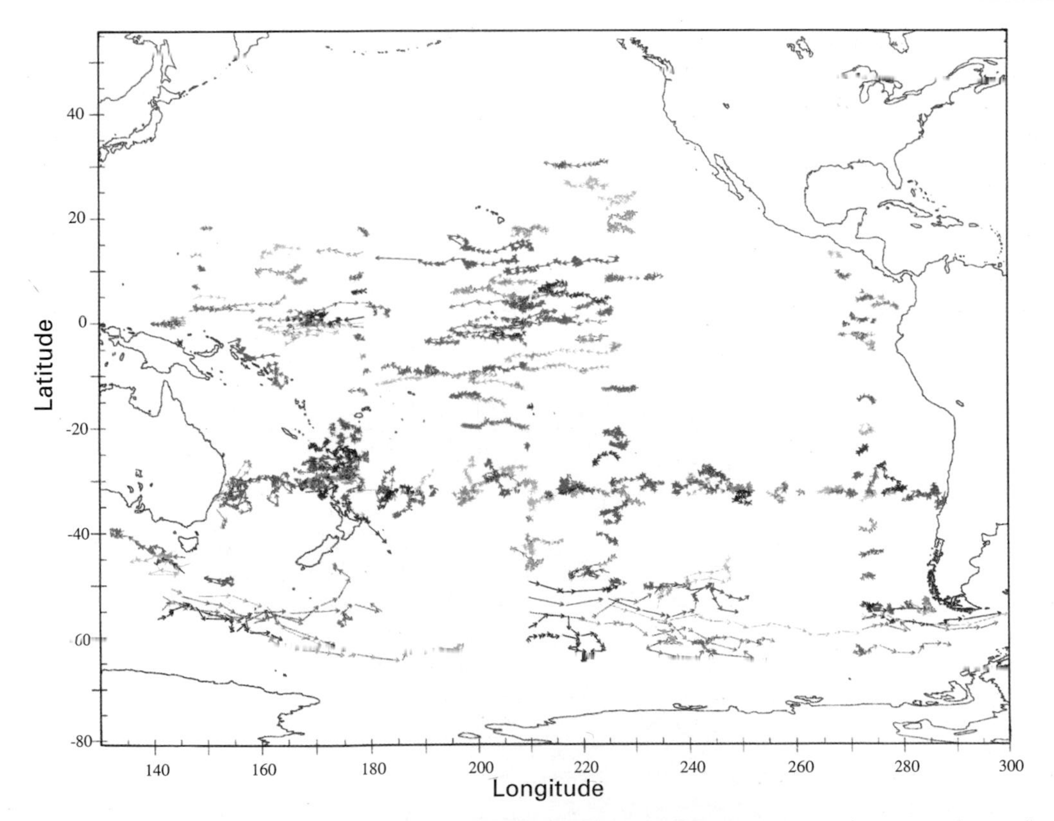

Figure 3.13 The tracks of subsurface ALACE (Autonomous LAgrangian Circulation Explorer) floats deployed in the Pacific Ocean during WOCE. The floats drift with the currents at a predetermined depth, surface every 20 days to transmit their position and data to a satellite, and then return to their programmed depth. This illustrates data points over 2.5 years.

3.14

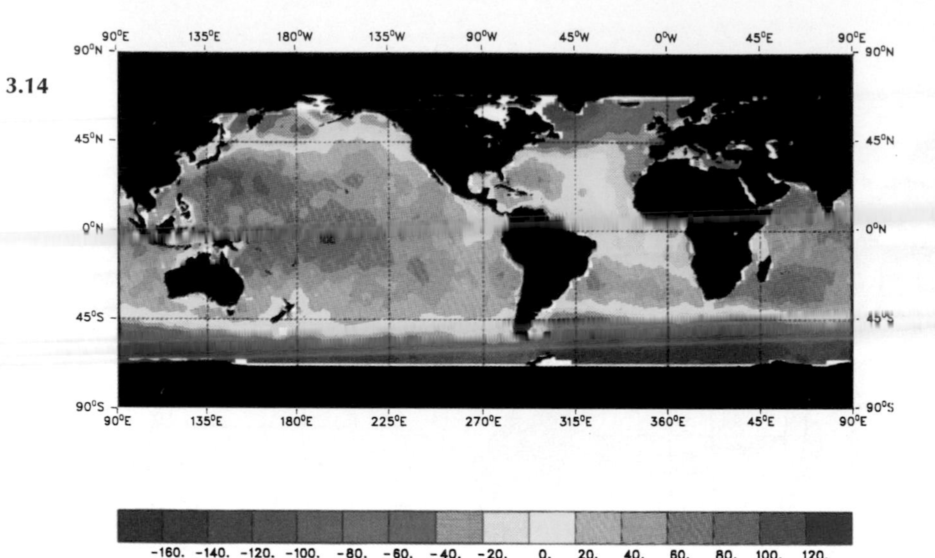

Figure 3.14 Sea surface topography from the Topex–Poseidon satellite altimeter. The range is from –180 cm to +140 cm. Most prominent is the large surface height change across the Antarctic Circumpolar Current and the clear shape of the North Atlantic and North Pacific ocean gyres.

tine monitoring capability (e.g., *Figure 3.14*), but what they can observe about the ocean is limited to surface temperatures and sea surface elevations. However, they do this well (to an accuracy of 0.5°C and less than 5 cm) and will continue to be part of any future ocean monitoring scheme.

For measurements of the interior of the ocean, new techniques will need to be developed (see Chapter 19). The UK is developing unmanned, autonomous submersibles, capable of carrying out many of the observations presently made from research vessels. Such vehicles would be capable of traversing entire ocean basins and making measurements from surface to sea bed on a regular basis (*Figure 3.15*).

Another novel technique is even now being deployed in the Pacific with a view to measuring ocean temperatures routinely. ATOC (Acoustic Thermometry of Ocean Climate; see *Figure 19.30*) relies on the fact that sound can be transmitted over vast distances in the ocean, and that the speed of sound in sea water is, at any given pressure, predominantly dependent on temperature. An experiment in 1991 transmitted sound from Heard Island in the Southern Ocean to receivers as far away as the east and west coasts of North America (16000 km). Based on these initial encouraging results, low frequency (70 Hz) sound sources off Hawaii and California and receivers around the rim of the Pacific were deployed in early 1994. This ATOC array, *Figure 3.16*, will show whether the kind of temperature anomalies that have been seen from repeated hydrographic station measurements can be reproduced and their evolution monitored.

3.15

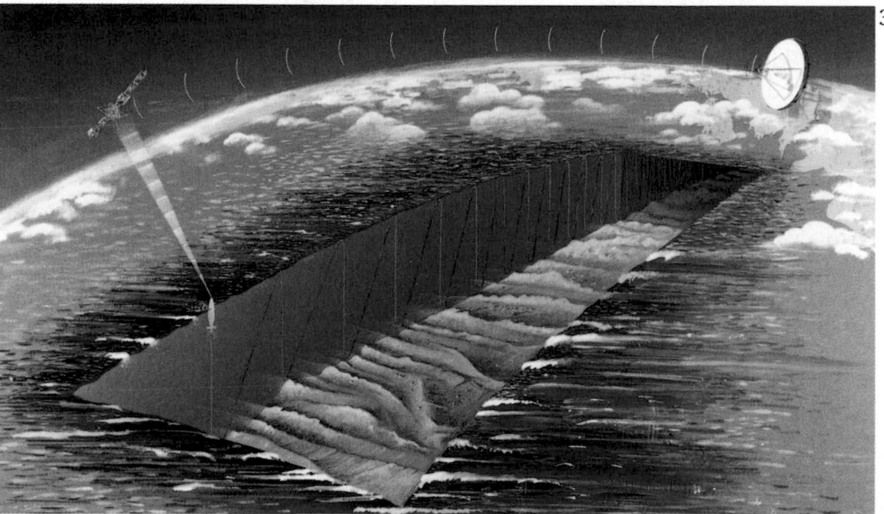

Figure 3.15 An artist's impression of the AUTOSUB vehicle carrying out a trans-oceanic hydrographic survey mission. The project involves the development of a very low drag body, an efficient propulsion system, sensors that can retain their accuracy over long missions, and deployment and recovery systems, as well as navigation and data telemetry schemes.

3.16

Figure 3.16 A proposed array of sources (north and south of Hawaii, off the Californian coast, and off Kamchatka, in the N.W. Pacific) and receivers (small circles, triangles and stars), to be deployed in the the ATOC project. The sound paths, for each source and receiver, are great circles.

Model studies based on coupled ocean–atmosphere models and a scenario of atmospheric increases in greenhouse gases suggest that ATOC measurements might even be able to detect signals due to global warming.

Modelling of Climate Change

The realisation that many of the large-scale processes involved in the ocean, atmosphere, and ice sheets may be described by a consistent set of dynamical equations has been with us since the early twentieth century. L.F. Richardson[24] described his experiment to predict the weather by numerical iteration of the equations of motion in 1922. He also discussed how the method could be extended to short-term climate prediction. Though his experiment was not a success, the methods were later successfully developed for numerical weather prediction in the 1950s. These methods were also used by N. Phillips[20], who developed the first general circulation model of the global atmosphere. The past 40 years have seen the further improvement of atmospheric general circulation models (AGCM) to include not only the dynamical processes, but also details of the radiations and their interaction with cloud, surface processes associated with hydrology and vegetation, and the ocean. Indeed, the atmospheric models have been at the forefront of estimating the response to the predicted changes in greenhouse gases.

In the late 1960s K. Bryan[5] published the first model of the general circulation of the ocean (see *Box 3.1*). This was developed further by M. Cox[8], and has been used extensively by research groups around the world. The Fine Resolution Antarctic Model (FRAM) project used this general circulation model as its basis (see *Box 3.2*), and it is now being used for other global ocean modelling projects. The development of the AGCM went hand-in-hand with improvements in the power of the computer – climate perturbation experiments for a few decades can be now run in a matter of a few weeks on a supercomputer.

Ocean circulation models, however, have been severely hampered by computer resources. First, the response time-scales in the ocean vary from days and weeks in the surface layers to the order of centuries in the deep ocean. This means that a global ocean model may have to be integrated for at least a 100-year period to reach equilibrium after a climate perturbation, compared with a period of a few years for an AGCM. Second, in some cases there is a need to model ocean eddies (scales of 50–200 km; see Chapter 4), which in some parts of the ocean are significant in the transport and mixing of ocean properties. A finer grid resolution of about 10 km is required to fully resolve these oceanic eddies in ocean models, compared with the lower resolutions needed in AGCMs. It is clear that computers of two orders of magnitude faster than

Box 3.1 An Ocean General Circulation Model

An ocean general circulation model is composed of a set of mathematical equations which describe the time dependent dynamical flows in an ocean basin. The basin is discretised into a set of boxes of uniform horizontal dimensions, but variable thickness in the vertical dimension. The horizontal flow (northward and eastward components) is predicted by the momentum equation, *Figure 3.17(a)*, at the corners of each box (*Figure 3.18*).

3.17

(a) **Momentum equation**

Rate of change of horizontal momentum in box = Transport of momentum across faces of box + Coriolis acceleration + Horizontal pressure gradient + Friction

- Coriolis acceleration ← (Rotation of earth)
- Horizontal pressure gradient ← (Gradients in sea level and in density)
- Friction ← (Internal dissipation and at ocean boundaries)

(b) **Temperature equation**

Rate of change of temperature in box = Transport of temperature across faces of box + Mixing of temperature between box and adjacent boxes

(c) **Salinity equation**

Rate of change of salinity in box = Transport of salinity across faces of box + Mixing of salinity between box and adjacent boxes

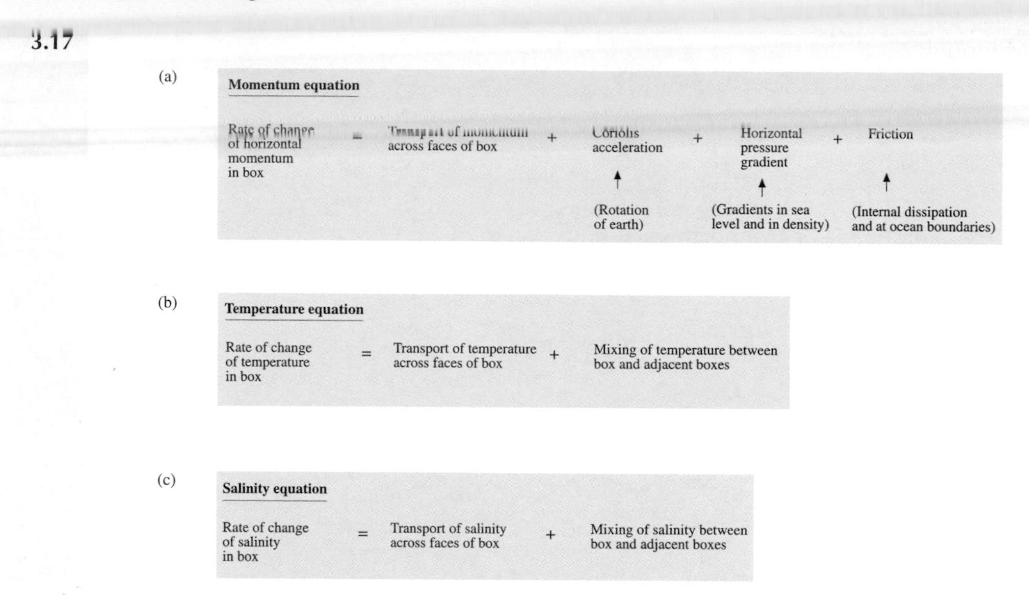

Figure 3.17 The basic equations for an ocean general circulation model.

The forcing for the flow may come from the surface wind stress (the frictional term in the momentum equation) or from surface buoyancy fluxes, arising from heat and fresh water (precipitation–evaporation) exchange with the atmosphere. These buoyancy fluxes change the temperature and salinity in the surface layer of the ocean. However, the horizontal and vertical flow carry these properties far into the interior of the ocean, where they tend to mix with other water masses.

This process of transport and mixing is described by the temperature and salinity equations, *Figures 3.17(b)* and *3.17(c)*, at the centre of each ocean box (*Figure 3.18*). From these two equations the sea water density and thence the pressure can be obtained for each box. The horizontal pressure gradient is then determined for the momentum equation, while the vertical velocity is calculated from the horizontal divergence of the flow. This set of time-dependent equations can then be used to describe all the dynamical components of the flow field, provided that suitable initial and boundary conditions are specified.

3.18

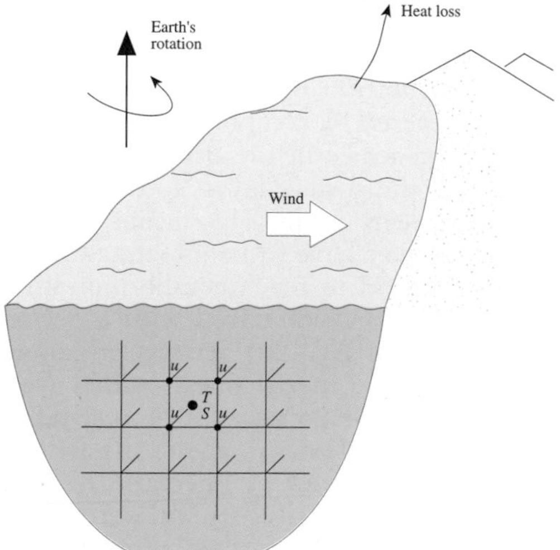

Figure 3.18 A schematic of the model boxes in an ocean general circulation model. The equations for momentum are solved at the corners of the boxes (*u*), while the temperature (*T*), and salinity (*S*) equations are solved at the centres of the boxes. The model is forced by climatological wind stress, surface heat, and fresh water fluxes.

Box 3.2 The Fine Resolution Antarctic Model (FRAM)

FRAM was developed[26] to investigate the role of eddy processes in the circulation of the Southern Ocean, in particular the Antarctic Circumpolar Current. The Southern Ocean comprises 30% of the global ocean and is an important region for the transfer of heat and fresh water between the Antarctic Ice Sheet and the northern land masses. The dynamics of the Antarctic Circumpolar Current are not well understood, although there is strong evidence to suggest that energetic ocean eddies, on horizontal scales of the order of 100 km, play an important role in the dynamics of the current.

The FRAM model forms the basis for the development of a global ocean model, which will be used for climate change experiments. This global model will resolve these energetic ocean eddies and the major frontal zones of the ocean.

The FRAM model subdivides the ocean, south of 22°S to the Antarctic continent, into a regular set of boxes. Each box has a horizontal length of 0.25° latitude by 0.5° longitude (approximately 27 km x 27 km at 60°S). Beneath each surface box a string of boxes reaches to the ocean floor. The thickness of each box varies from 20 m in the surface layer to over 200 m in the deepest parts of the ocean. Within each box, equations for the northward and eastward horizontal components of momentum, temperature, and salinity are specified. There are 5 million boxes which represent the southern ocean, and therefore 20 million prognostic variables to calculate. These variables are calculated by integration in time of the equations from an initial cold, saline motionless ocean. In the first 6 years of the integration the model was forced by the annual mean wind stress and by the observed temperature and salinity. After this period the model was free to run for a further 6 years, subject to seasonal wind forcing, annual mean temperature and salinity at the ocean surface, and an open northern boundary. An example of the model 'output' is shown in Figure 3.19 and a comparison with satellite data is shown in Figure 3.20. Estimates of the meridional heat flux in the model are shown in Figure 3.21.

3.19

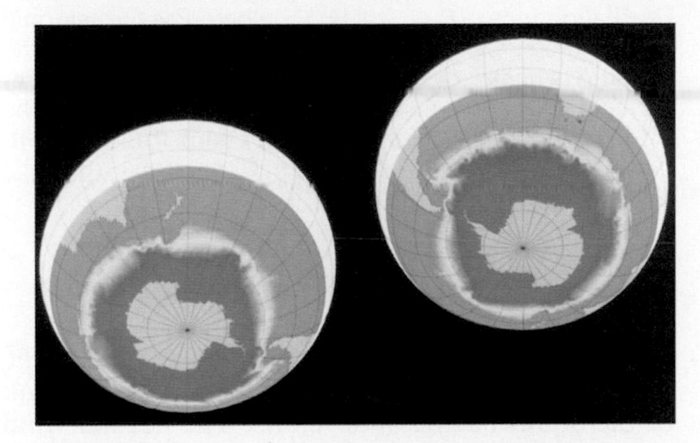

Figure 3.19 Contours of the instantaneous stream function in FRAM. The stream function shows the depth-averaged flow circulating clockwise around the Antarctic continent. The flow is most intense in the Antarctic Circumpolar Current (in the yellow and neighbouring green regions). The flow is unsteady due to the presence of 'eddies', mainly in the Antarctic Current and Agulhas Current south of South Africa.

3.20

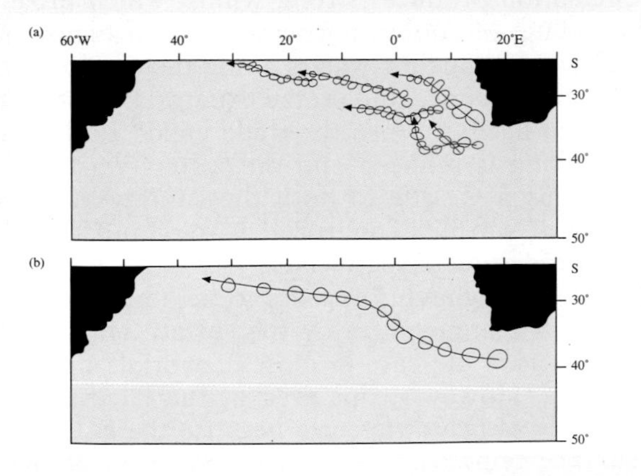

Figure 3.20 (a) Tracks of eddies detected by the Geosat satellite in the South Atlantic. (b) Tracks of eddies in the FRAM. Note that the model produces a more regular eddy track than the observations show.

3.21

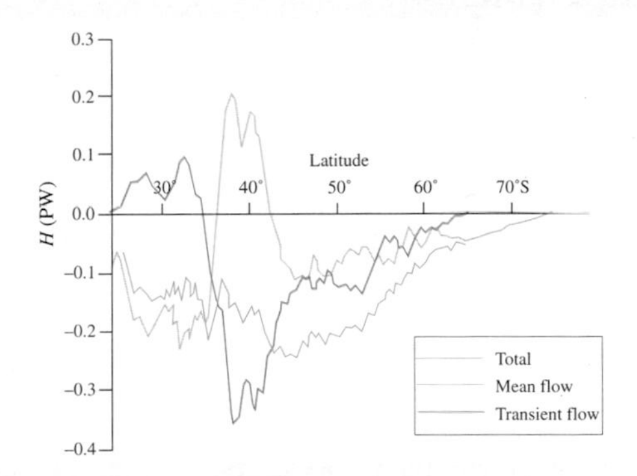

Figure 3.21 The latitudinal variation in meridional-heat transfer (H) by the FRAM model. Northward heat fluxes are positive. The total heat transport is directed southward (negative) toward the pole, although the time-mean circulation drives a heat flow toward the north, between 37°S and 43°S, in the region of the Antarctic Circumpolar Current. The 'eddies' in the flow, however, drive a stronger heat flux toward the south, and thus result in a total heat transport toward the pole. The eddies play an important climatological role in the model.

the present machines (which are capable of 10^9 floating point instructions per second) will be required to manage this task. The recent development of parallel computer systems is expected to deliver this power by end of the twentieth century. Indeed, one of the important tasks of ocean modellers is to develop methods for the analysis and interpretation of the huge quantity of data that these models will produce. Because of their complexity, there will be also be a need to develop simpler models to investigate interactions more completely. To understand and predict the behaviour of sea ice, for example, the sea-ice models will have to be coupled to models of the upper ocean and overlying atmosphere.

A third problem is the importance of ocean chemistry and biology to climate change. The ocean is a depository for the greater part of the Earth's exchangeable fraction of carbon. It is not known how carbon is regulated by the ocean, though it is clear that phytoplankton blooms produce a lowering of the partial pressure of carbon dioxide at the ocean surface and thereby have the ability to alter the flux between ocean and atmosphere (see Chapters 6 and 12). Modelling of the ocean basin ecology and chemistry has commenced in recent years and it is expected that these processes, as they become understood, will be incorporated into the general climate models.

3.22

(a)

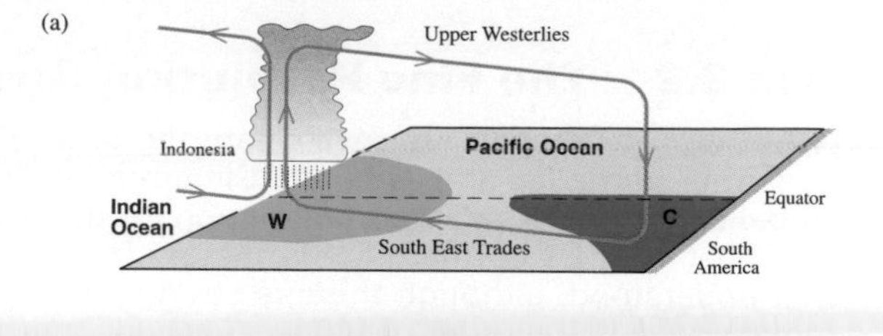

(b)

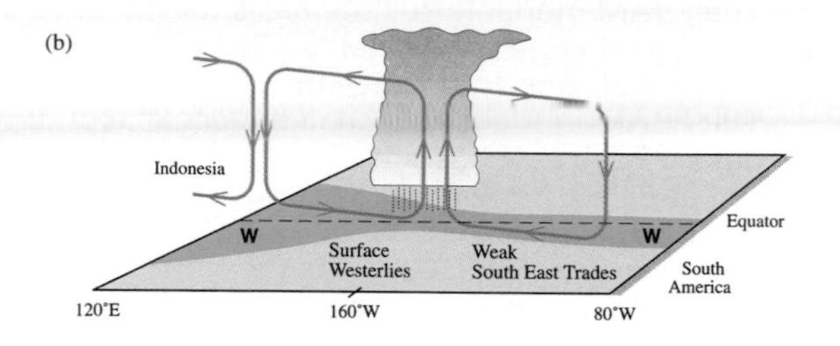

Figure 3.22 The tropical atmospheric circulation over the Pacific Ocean during (a) normal conditions and (b) El Niño conditions. During normal conditions the surface pressure is low over Australia and Indonesia (high rainfall) and high over the southeast Pacific, so the surface trade wind circulation is strong and the Southern Oscillation index ($P_{DARWIN} - P_{TAHITI}$, where P is the seasonal surface pressure) is high. During El Niño conditions, the pressure is higher over Australia and Indonesia (low rainfall) and lower in the southeast Pacific; consequently the trade wind circulation is weaker and the Southern Oscillation index is low (W, warm; C, cold).

The El Niño–Southern Oscillation (ENSO) Phenomenon – An Example of Ocean Prediction?

The ENSO phenomenon is now recognised as the largest contributor to the perturbation of the climate on a global scale over a period of a few years, and is known to be a natural oscillation of the atmosphere–ocean system. It is a coupled interaction between the atmosphere and the upper layers of the tropical Pacific Ocean, which can result in changes of global surface temperature of a few tenths of a degree Celsius on a time-scale of one year. This is a similar change in temperature to that attributed to the 25% increase in atmospheric carbon dioxide in the past 100 years.

The El Niño is the ocean component of the interaction; it is a general warming of the upper layer of the eastern and central Equatorial Pacific Ocean. [The name El Niño comes from the fact that the impact is felt on the coast of South America around Christmas time and hence it is referred to in Spanish as the (Christ) child.] A major consequence of the oceanic warming is the decline of biological productivity and hence of fish stocks, which are a major source of livelihood for the local population.

It is associated with a reduction in the strength of the tropical trade wind system, in particular the southeast trades which occur on the eastern flank of the South Pacific anticyclone. The normal wind circulation produces strong winds, which drive an upwelling of cooler, nutrient-rich, and highly productive thermocline waters along the tropical coast of South America and on the equatorial band of the Eastern Pacific. When the trade winds weaken, the upwelling is reduced and the waters warm by as much as 5°C, due to both the southward movement of warmer equatorial waters and the high solar radiation at the surface.

At first sight this appears to be a one-way forcing of the atmosphere by the surface wind on the ocean, as is the case in most of world's upwelling regions. However, this area of the Pacific Ocean behaves rather differently because the rise in sea-surface temperature over a vast area of ocean changes the distribution of atmospheric heat sources and sinks, which in turn drives the trade wind circulations. The trade winds carry water vapour, evaporated from the ocean, into areas of tropical atmospheric convergence where high rainfall occurs. These convergence zones tend to be

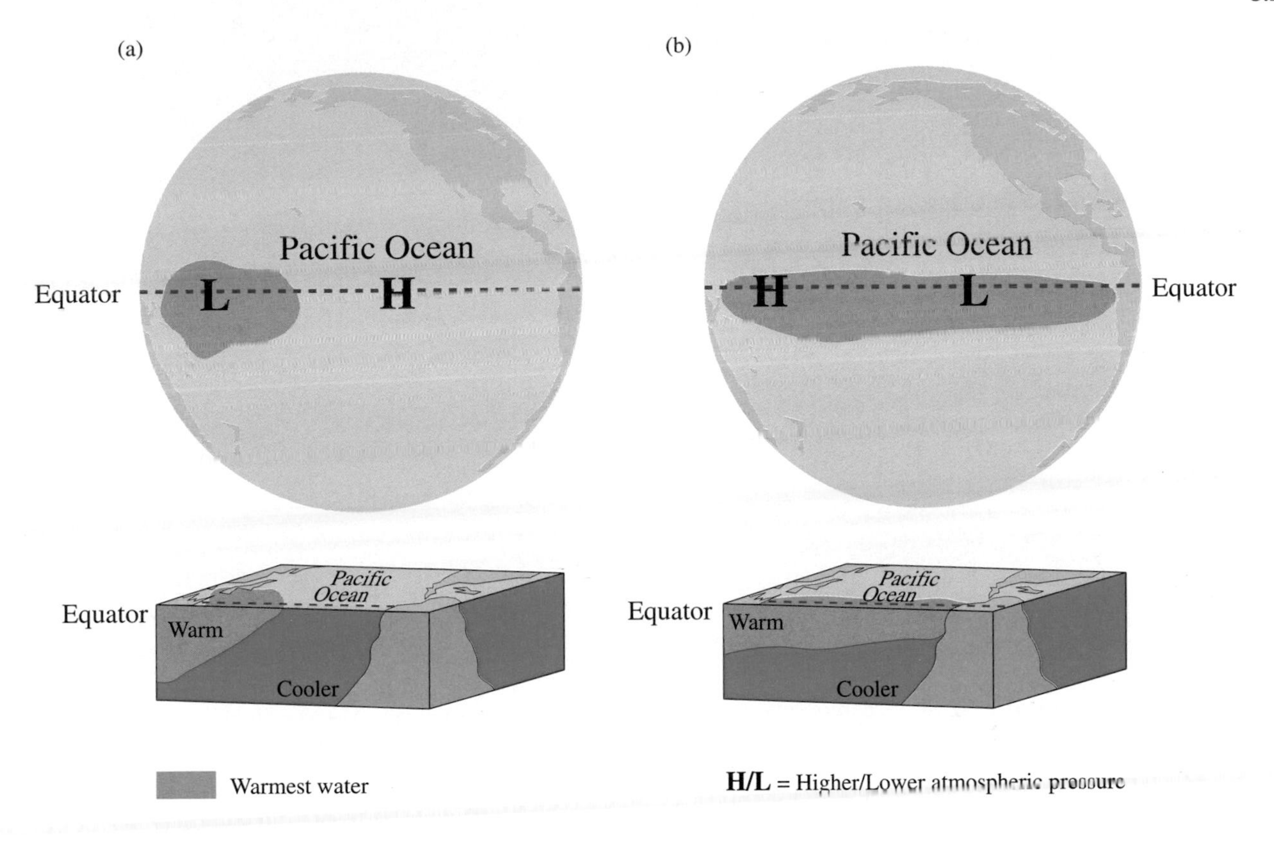

Figure 3.23 The tropical Pacific Ocean during (a) normal and (b) El Niño years. During normal years the strong trade winds drive the warm water westward and intensify upwelling of cooler subsurface waters in the east Pacific. During El Niño years the weaker atmospheric circulation allows the warmer lighter water to flow eastward, replacing the cooler upwelling waters.

located in regions of maximum surface temperature, that is to the north of the equator and in the tropical southwest Pacific. During El Niño events these convergence zones tend to move southward across the equator and eastward (see *Figures 3.22* and *3.23*). The South Pacific anticyclone becomes weaker and the trade winds weaken. Simultaneously, the surface atmospheric pressure over Indonesia and Australia tends to rise and rainfall decreases.

This see-saw of surface pressure between the southeast Pacific and Indonesia is known as the *Southern Oscillation*. Although its major influence is in the tropical Pacific, its effect is felt throughout the world. For example, in Zimbabwe both rainfall and maize yields are highest during El Niño years[6]. The ENSO phenomenon is a coupled interaction between the surface layers of the ocean and the world wind systems, which occurs two or three times a decade.

The monitoring of ENSO in recent decades, in particular during the exceptional episode of 1982–1983[22], has provided a stimulus to atmospheric and ocean modellers. These efforts have provided good simulations of both the individual components of the system (the ocean response to observed winds and the atmospheric response to observed sea surface temperature).

Coupled models of the tropical ocean and atmosphere are showing promise for the simulation of the ENSO cycle and a number of Climate Centres in the world are now producing experimental forecasts for ENSO with some degree of success.

These models will be improved when the results of the Tropical Ocean Global Atmosphere experiment (an experiment to measure and understand some of the complex processes in the tropical atmosphere and ocean) in the western equatorial Pacific Ocean are analysed. The proposed mooring

3.24

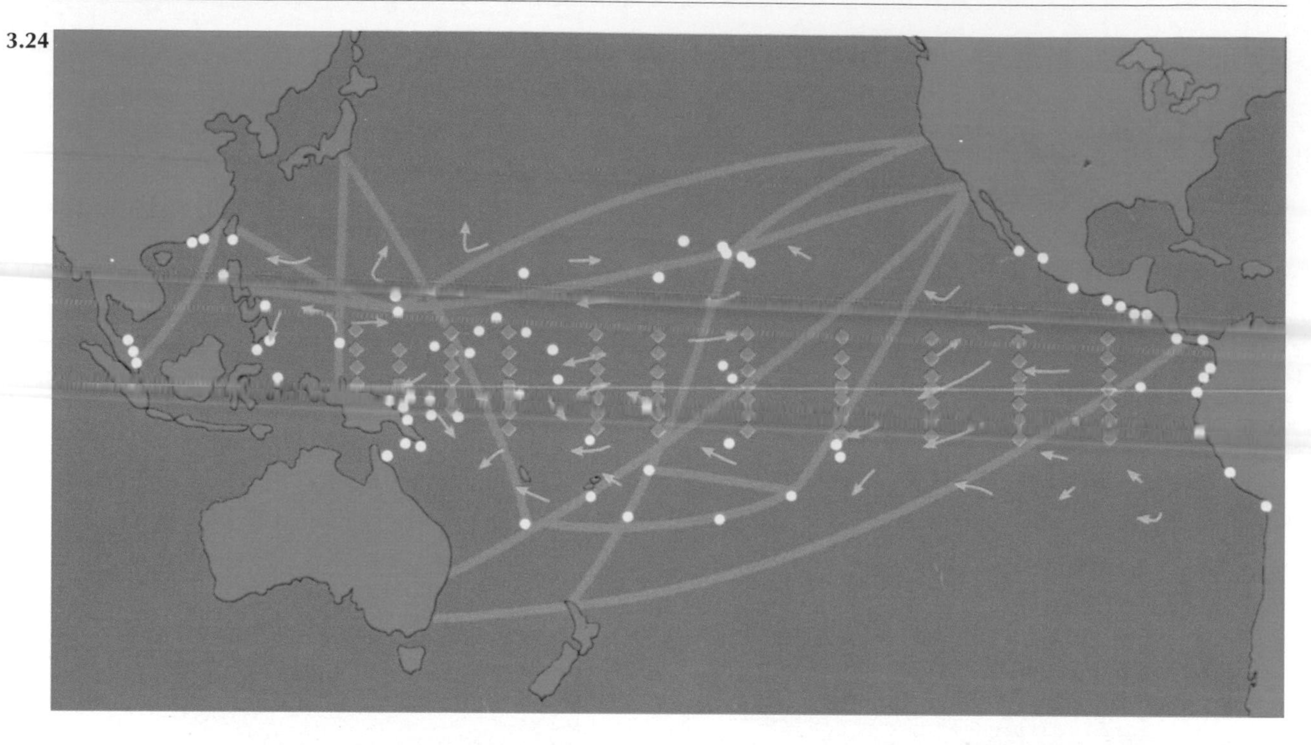

Figure 3.24 The tropical Pacific Ocean observing system. The orange diamonds are ATLAS surface buoys from which continuous temperature measurements are obtained from the surface to a depth of 500 m. These observations together with surface meteorological measurements (e.g., wind, temperature and humidity) are transmitted by satellite to a receiving station. Currents (orange squares) are measured routinely along the equator. The Tide gauge network (yellow circles) provides sea-level height observations, which can be used to calculate horizontal pressure gradients in the oceans. These observing stations are supplemented by routine measurements from satellite-tracked drifting buoys (arrows) and voluntary observing ships (light blue lines).

array (*Figure 3.24*) will provide routine measurements of the upper 500 m of the tropical Pacific Ocean and of surface winds to be used in the ENSO forecast models.

There is also evidence that the intensity of ENSO varies on the longer decadal to centennial time-scales. Longer-term variations in the intensity of El Niño events affecting primary production off California have been related to solar cycles (variations in the intensity of 11–22 year sunspot cycles that cause small changes in incident radiation). Within Californian continental margin sediments, decadal to millennial cycles of preservation of laminated sediments, driven by variation in the intensity of the oxygen minimum zone, have been ascribed to solar cycles affecting the longer-term alternation of El Niño and anti-El Niño[2].

Data on sea surface temperature variations from oxygen isotope studies of Galapagos corals showing 11 and 22 year periods[11] lend support to suggestions of solar cycle modulation of ENSO activity.

The Importance of Climate Change

The most comprehensive study of the scientific evidence for climate change, its potential impact, and the strategies needed to ameliorate its impact has been made by the Intergovernmental Panel on Climate Change (IPCC). While climate change is, in general, regarded as an atmospheric phenomenon, the IPCC reports make it clear that there are direct impacts on and by the ocean and, furthermore, that knowledge of the behaviour of the oceans is central to any climate prediction capability. The potential socioeconomic costs of climate change far outweigh the resources needed to make the measurements and run the models required to improve our ability to predict climate change. The key areas identified in the IPCC report, together with recommendations on improvements in observations and modelling, are given in *Box 3.3*.

Clearly, in all these areas the oceans are important and understanding them represents one of the greatest challenges in the area of climate change prediction.

Box 3.3 The Scientific Uncertainties of Climate Change

IPCC identifies the key areas of scientific uncertainty as:

- *Clouds:* primarily cloud formulation, dissipation, and radiative properties, which influence the response of the atmosphere to greenhouse forcing.

- *Oceans:* the exchange of energy between the oceans and the atmosphere, between the upper layers of the ocean and the deep ocean, and transport within the ocean, all of which control the rate of global climate change and the patterns of regional change.

- *Greenhouse gases:* quantification of the uptake and release of the greenhouse gases, their chemical reactions in the atmosphere, and how these may be influenced by climate change.

- *Polar ice sheets:* affect predictions of sea level rise.

The main observational requirements are:

- The maintenance and improvement of observations (such as those from satellites) provided by the World Weather Watch.

- The maintenance and enhancement of a programme of monitoring, both from satellite-based and surface-based instruments, of key climate elements for which accurate measurements on a continuous basis are required. These include the distribution of important atmospheric constituents, clouds, the Earth's radiation budget, precipitation, winds, sea surface temperatures, and the terrestrial ecosystem extent, type, and productivity.

- The establishment of a Global Ocean Observing System to measure changes in such variables as ocean surface topography, circulation, transport of heat and chemicals, and sea ice extent and thickness.

- The development of new systems to obtain data on the oceans, atmosphere, and terrestrial ecosystem using both satellite-based instruments and instruments based on the surface, on automated vehicles in the ocean, on floating and deep sea buoys, and on aircraft and balloons.

- The use of palaeoclimatological and historical instrumental records to document natural variability and changes in the climate system, and subsequent environmental response.

In the area of modelling the report concludes that any reduction in the uncertainties of climate prediction will be dictated by progress in the areas of:

- Use of the fastest possible computers to take into account coupling of the atmosphere and the oceans in models, and to provide sufficient resolution for regional predictions.

- Development of improved representation of small-scale processes within climate models, as a result of the analysis of data from observational programmes to be conducted on a continuing basis well into the twenty-first century.

General Reference

Wuethrich, B. (1995), El Niño goes critical, *New Scientist*, **145**, 32–35.

References

1. Aagaard, K. and Carmack, E.C. (1989), The role of sea ice and other fresh water in the Arctic Circulation, *J. Geophys. Res.*, **94**(C5), 14485–14498.
2. Anderson, R.Y., Linsley, B.K., and Gardner, J.V. (1990), Expression of seasonal and ENSO forcing in climatic variability at lower than ENSO frequencies: evidence from Pleistocene marine varves off California, *Palaeogeogr., Palaeoclimatol., Palaeoecol.*, 78, 287–300.
3. Birchfield, G.E. and Broecker, W.S. (1990), A salt oscillator in the glacial Atlantic? A 'scale analysis' model, *Paleoceanogr.*, 5, 835–843.
4. Broecker, W.S, Kennett, J.P., Flower, B.P., Teller, J.T., Trunbone, S., Bonani, G., and Wolfi, W. (1989), Routing of meltwater from the Laurentide Ice Sheet during the Younger Dryas cold episode, *Nature*, **341**, 318–321.
5. Bryan, K. (1969), A numerical model for the study of the world ocean, *J. Computat. Phys.*, **4**, 347–376.
6. Cane, M.A., Eshel, G., and Buckland, R.W. (1994) Forecasting Zimbabwean maize yield using east equatorial Pacific sea surface temperatures, *Nature*, **370**, 204–205.
7. Carrissimo, B.C., Oort, A.H., and Van de Harr, T.H.V. (1985), Estimating the meridional energy transports in the atmosphere and ocean, *J. Phys. Oceanogr.*, **15**, 52–91.
8. Cox, M.D. (1984), *A Primitive Equation: 3-Dimensional Model of the Ocean*, GFDL Ocean Group Technical Report No.1, GFDL/NOAA, Princeton University.

4.5

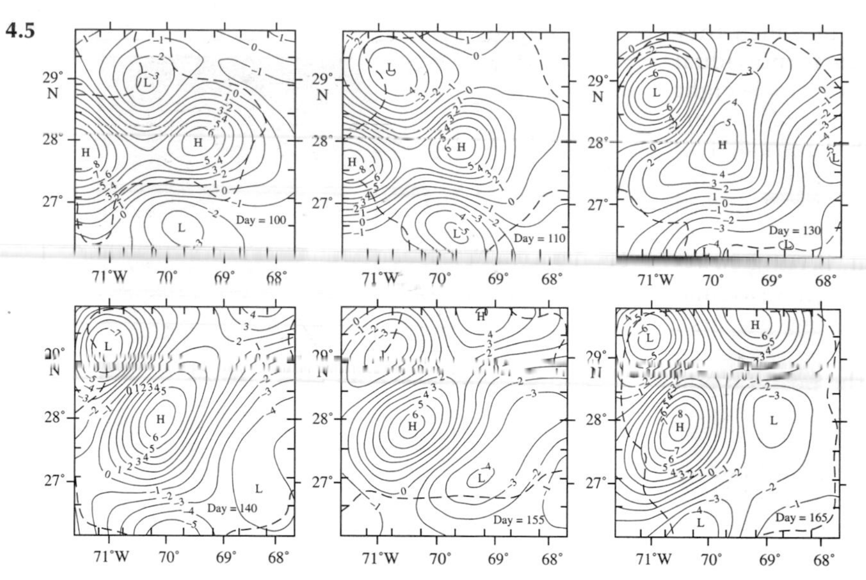

Figure 4.5 The measurements from MODE current meters and neutrally buoyant SOFAR floats were mapped to produce a coherent picture of the flow field, using a technique called objective analysis in which a flow field was devised that had the same statistical characteristics as the data themselves and fitted the observations with minimum error. There is little confidence in the predicted flow field at points outside the observational array marked by the dashed lines[14].

(*Figure 4.5*) and with a complex vertical structure. Features were seen to propagate westward at speeds of about 5 km/day. The complexity of the current field was amply illustrated by the compilation of the SOFAR float tracks into so-called spaghetti plots (*Figure 4.6*), which show the how the eddy-like meso-scale field acted to disperse the floats[10].

Even though the areas of ocean covered by both of these experiments were 200 km across, this is still a relatively small area compared to the size of ocean basins. The experiments, therefore, shed little light on the geographical variability of the characteristics of the meso-scale currents. In order to explore this a joint US–USSR experiment, carried out between August 1974 and April 1975 and called Polymode (a coming together of names as well as scientists), explored the energetics and scales of meso-scale motions over a much wider range of latitudes in the region of anticyclonic circulating water southeast of the Gulf Stream, called the western subtropical Atlantic gyre. In parallel, in the east Atlantic a more limited exploration of eddy variability using long-term moorings (the North East Atlantic Dynamics Study, NEADS) was started by European scientists (one of these current meter sites between the Azores and Madeira is still being maintained by German scientists – the longest direct current meter measurement series[20]).

The Polygon, MODE, and Polymode experiments studied meso-scale features in detail, but from the early 1970s onward evidence accumulated (from the newly developed observational techniques of SOFAR floats, reliable current meters, satellite infra-red images of the ocean surface, and satellite altimetry data; see Chapter 5, *Box 5.4*) for the ubiquitous nature of the meso-scale eddy field (see, for example, *Figure 4.7*). All these observations confirmed the concentration of high kinetic energy of the time-varying currents (the eddy kinetic energy, EKE) in regions near the major current systems.

So a view was formed, which still holds good

4.6

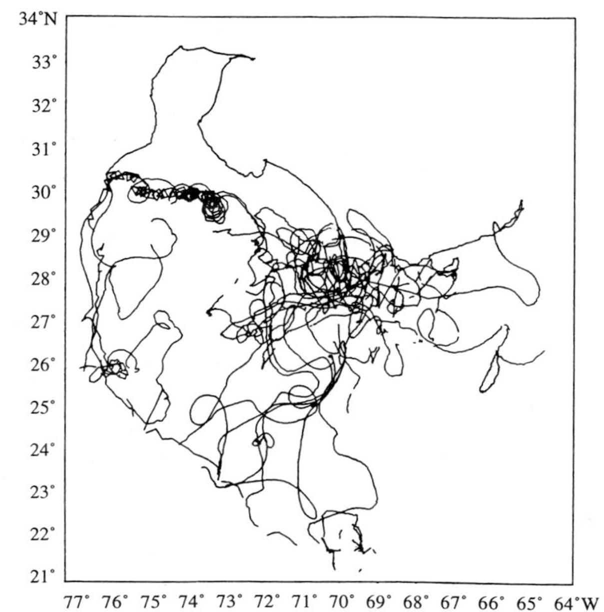

Figure 4.6 A compilation of the tracks (9/1972–6/1976) of all the SOFAR floats launched in the MODE array area near 28°N 70°W (1° of latitude is approximately 111 km). The tracks last several years and show the dispersive nature of currents from a small area to eventually fill much of the western subtropical gyre. In some areas, for instance off the Bahamas and near the Gulf Stream, eddy activity is high. The site chosen for the MODE experiment turned out to one of very low eddy energy[15].

4.11a

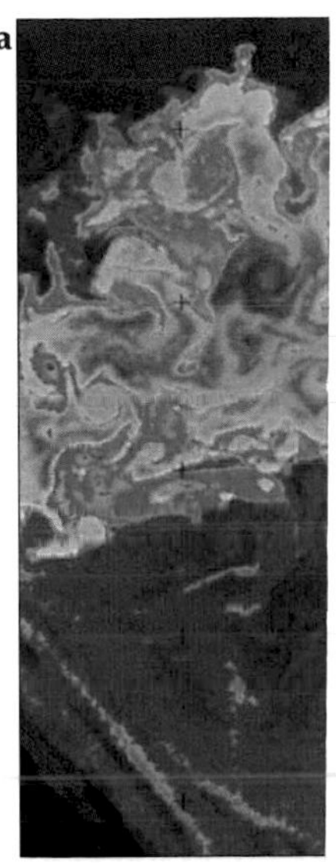

Figure 4.11 The ι
visible reflectanc
reflectance, while
of the phytoplankt
ships report the se
in more detail. He
(the infra-red banc
motion. An anima
centration to prod
properties within t
al days, and even
must also be reme
is not simply a pa
research issue. Th
the production oi
Plymouth Marine

and *Box 5.1, F*
examples), givir
impressive pict
their statistical p
than can be gaii
does not end tl
sea surface field
a subject of on-
mathematical m
of the eddies is
and density var
experiments su
between the ed
the flow condi
quiescent plum
ments are limit
early stage in c
the open ocean.

4.7

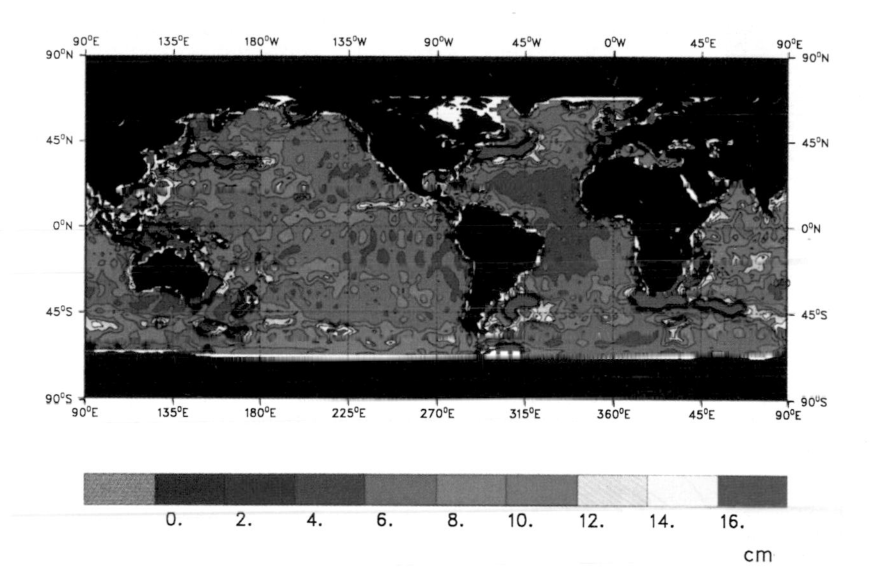

Figure 4.7 The statistics of eddy energy derived from the variability of sea surface slope obtained from the Topex/ Poseidon satellite. High eddy variability is seen in the regions of strong meandering currents, such as the Gulf Stream in the Atlantic, the Kuroshio in the North Pacific, the Agulhas Retroflection off South Africa, and the Antarctic Circumpolar Current in the Southern Ocean. (Courtesy of Prof C. Wunsch, Massachusetts Institute of Technology, Boston, USA.)

today, of the ocean populated with eddies, just as the atmosphere is full of cyclones, anticyclones, and frontal systems.

Beasts in the Eddy Zoo

Meso-scale features in the ocean take on a number of guises. Indeed, in the mid-1970s an article was published entitled *New Animals for the Eddy Zoo*, since at that time almost every eddy studied appeared to have different characteristics. For some, there was clear evidence of their presence at the sea surface, while others were confined within the water column; their diameters ranged from over 200 km to about 10 km, and they appeared to have different formation mechanisms. In a review such as this we can only touch on the characteristics of some of the more abundant animals in the eddy zoo. (The reader is referred to Robinson[13] for a collection of papers giving an overview of our knowledge at that time.)

Gulf Stream rings

Some of the best-documented eddies are those that are formed by 'pinching off' meanders from energetic current systems. The area that has been most intensively studied is the Gulf Stream – its variability downstream of Cape Hatteras is well-known. Even as long ago as 1793 evidence for an isolated body of warm water to the north of the stream was noticed, and in the 1930s analysis of ships' thermograph records started to provide evidence of the population of eddies associated with the Gulf Stream, known as Gulf Stream rings.

Undoubtedly, the detailed study of these features and their formation was strongly influenced by the advent of infra-red sensors flown on satellites that could easily identify both warm and cold core rings north and south of the Gulf Stream on cloud-free images (*Figure 4.8*). The spatial distribution of Gulf Stream rings has been delineated by several censuses (see, for example, *Figure 4.9*), but none is truly comprehensive.

By virtue of being water masses enclosed within water with very different properties the rings can be regarded as isolated ecosystems and the evolution of their physical, chemical, and biological properties over lifetimes of several seasons can be

4.8

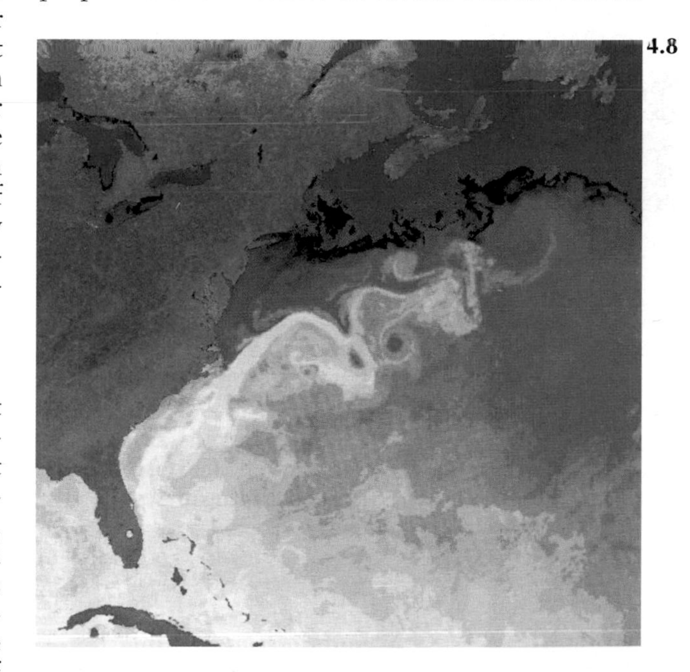

Figure 4.8 The meandering of a current and the location of eddies detached from the current can be clearly seen in satellite infra-red images of sea surface temperature. Warmer hues denote warmer temperatures. (Courtesy of O. Brown, R. Evans, and M. Carle, University of Miami Rosenstiel School of Marine and Atmosphere Science, Miami, USA.)

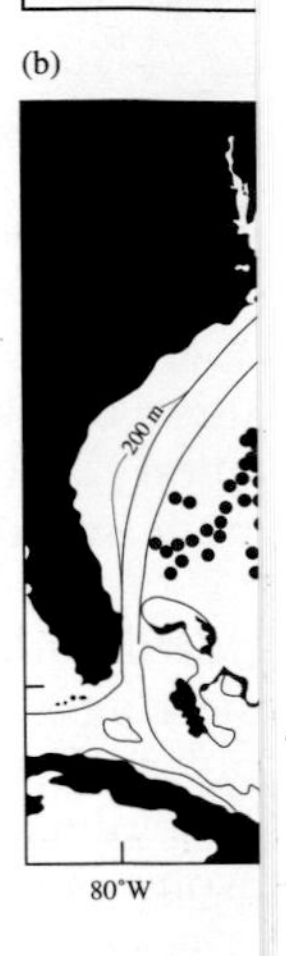

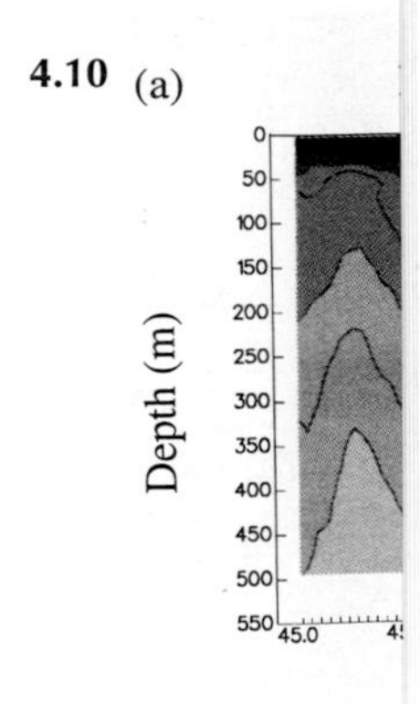

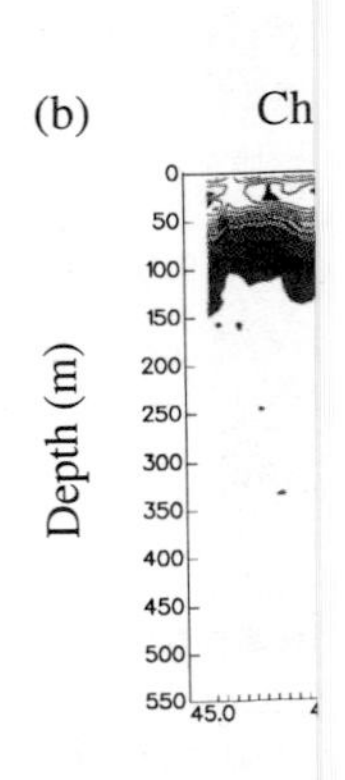

Box 5.3 Synthetic Aperture Radar Imaging of Small-Scale Ocean Processes

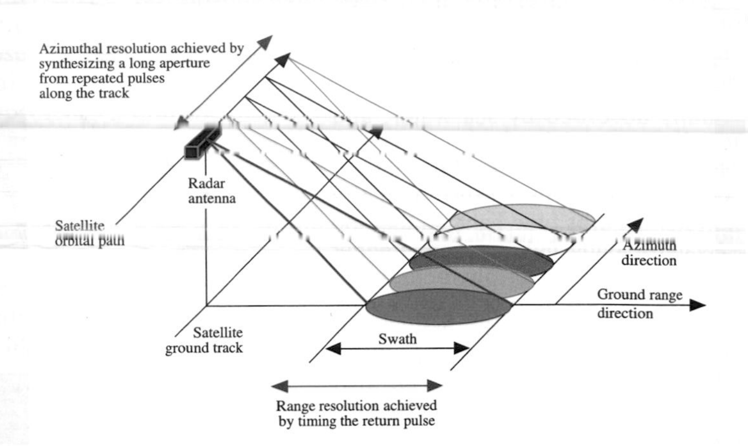

5.17

Figure 5.17 It is not possible with a satellite sensor to focus microwaves in the same way as visible or infra-red radiation. To achieve a high-resolution image in the range direction (the radar pointing direction), the radar uses the timing of the return pulse to determine the precise distance to the patch of sea surface being viewed. To achieve comparably high resolution in the azimuth (along track) direction would require a very large aperture antenna. This cannot be constructed physically, but its effect is synthesized digitally using the recorded pulse returns from many positions of the satellite along its orbit, hence the name Synthetic Aperture Radar. SARs on satellites are capable of resolving down to 20 m. A typical SAR image consists of up to 4000 x 4000 picture elements (pixels). A SAR produces an image corresponding to the magnitude of the radar energy returned from the sea surface. Since the radar views the sea at an oblique angle, the signal relies on detecting back-scattered radiation caused by the interaction between the incident radar waves and the roughness of the sea surface. The microwaves do not penetrate the surface at all, and so the patterns on the radar image are due to variations of the sea surface roughness.

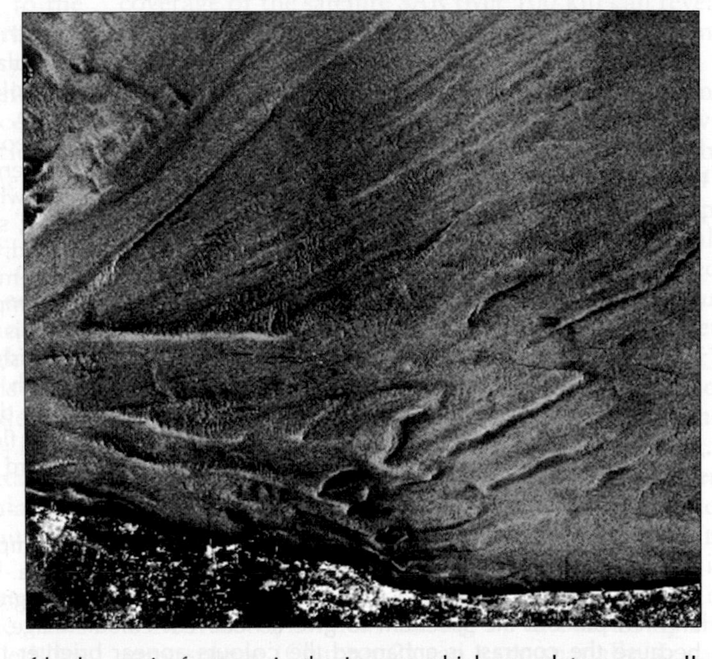

5.18

Figure 5.18 An image of the southern North Sea adjacent to the Dover Straits acquired by SAR on Seasat, 19th August, 1978. The original data have been averaged to a resolution cell of 168 m, eliminating the speckle effect and resulting in a very smooth image. The width of the image represents about 90 km and the direction of North is 26.1° clockwise from the top of the image. The radar backscatter patterns appear to reveal the bathymetry of the sea bed – the image looks rather like a photograph of the sea bed laid bare and illuminated from an oblique angle. Images like this took many oceanographers by surprise when they were first obtained. It must be emphasised that the radar signal does not penetrate through the sea to the sea bed. The imaging mechanism is this: the tidal currents fluctuate in magnitude as they flow over shallow or deep regions, causing horizontal convergence and divergence which concentrates or reduces the surface wave energy, and hence the roughness controlling the radar return. Although such a mechanism seems rather complex, the clarity of bathymetric features in the image, which correlate very well with bathymetric charts, demonstrates that this is an effective method for mapping sandbanks and sandwaves. Current research is determining whether quantitative as well as qualitative information can be recovered, and examining the conditions of tide and wind which are necessary to produce such clear images of the sea bed.

5.19

Figure 5.19 This ERS-1 SAR image of the English Channel off the Isle of Wight was recorded on 2 July, 1993, and the width of the image represents 80 km. Slicks, regions in which the sea surface is smoother than normal, appear as dark regions on the image and may be caused either by the presence at the surface of material such as organic films, or by dynamical features which produce local divergence of the surface current. Either process reduces the amplitude of the short surface waves which influence the radar back scatter. In this image the slicks provide a way of detecting other dynamical processes. In some of the coastal embayments narrow slicks appear to be aligned with the local tidal circulation. In other areas the slicks relate to old ship wakes which have left a trail of smoother water behind. Further offshore the larger slicks are probably patches of surface film discharged from ships. The corrugated shape of these patches is due to the shear associated with the strong tidal streams which flow parallel to the coast. (The data from which this image was derived were supplied by the European Space Agency.)

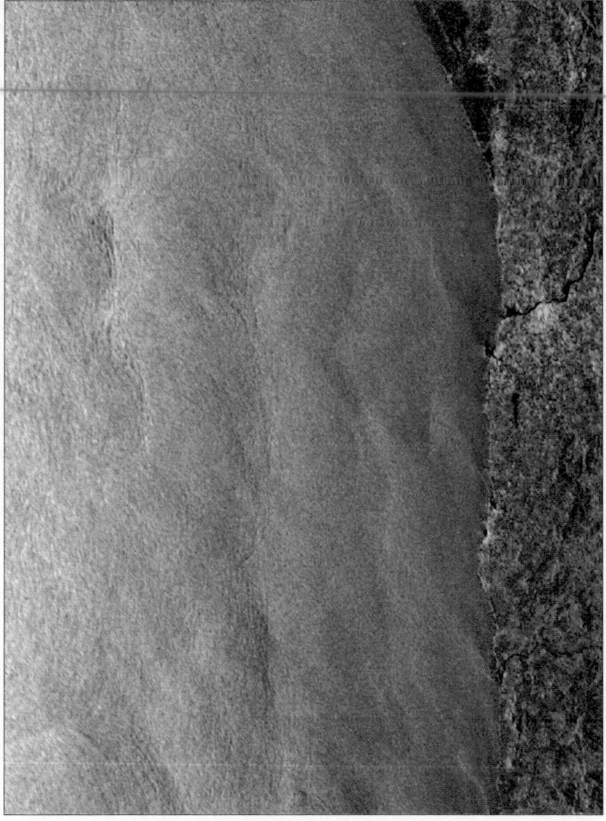

5.20

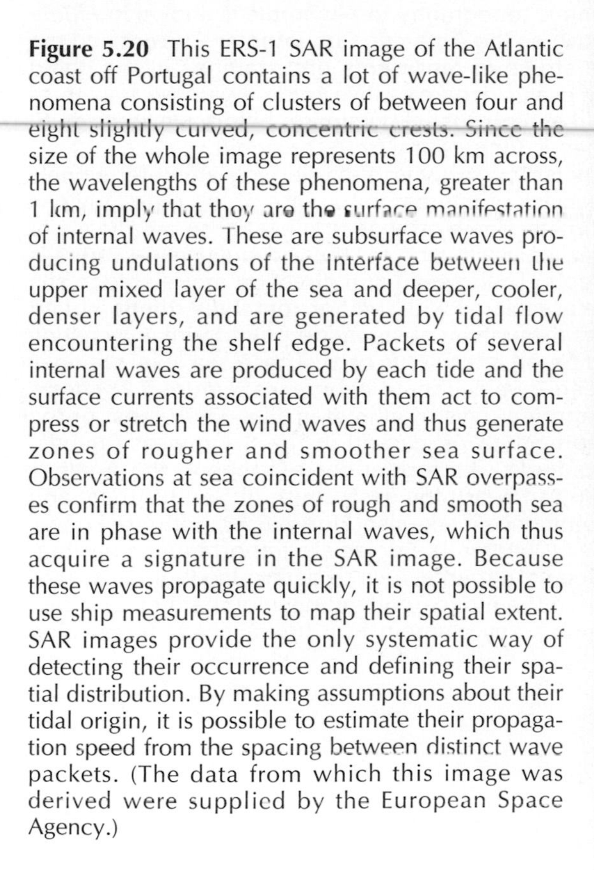

Figure 5.20 This ERS-1 SAR image of the Atlantic coast off Portugal contains a lot of wave-like phenomena consisting of clusters of between four and eight slightly curved, concentric crests. Since the size of the whole image represents 100 km across, the wavelengths of these phenomena, greater than 1 km, imply that they are the surface manifestation of internal waves. These are subsurface waves producing undulations of the interface between the upper mixed layer of the sea and deeper, cooler, denser layers, and are generated by tidal flow encountering the shelf edge. Packets of several internal waves are produced by each tide and the surface currents associated with them act to compress or stretch the wind waves and thus generate zones of rougher and smoother sea surface. Observations at sea coincident with SAR overpasses confirm that the zones of rough and smooth sea are in phase with the internal waves, which thus acquire a signature in the SAR image. Because these waves propagate quickly, it is not possible to use ship measurements to map their spatial extent. SAR images provide the only systematic way of detecting their occurrence and defining their spatial distribution. By making assumptions about their tidal origin, it is possible to estimate their propagation speed from the spacing between distinct wave packets. (The data from which this image was derived were supplied by the European Space Agency.)

Box 5.4 Sea Surface Topography

As implied by its name, the altimeter primarily measures height – in this case, that of sea level. The method is conceptually very simple. A sharp pulse is transmitted vertically toward the Earth's surface from an antenna on the satellite. Some of the signal is reflected back to the same antenna from those parts of the surface within the 7 km footprint which are aligned perpendicular to the beam. The key to the technique is the accurate measurement of the time delay between transmission and reception of the pulse by the antenna. If the signal's propagation speed is known then the distance between the satellite and the surface can be obtained. The height of the satellite orbit can be determined by tracking stations and the difference between the two provides an estimate of sea level. As the satellite circles the Earth, so sea level changes in space and time can be monitored.

In practice, it is very difficult to achieve the accuracy required for oceanography (better than 10 cm). Corrections must be made for the effects of the atmosphere on the propagation speed and, because the reflected pulses are very noisy, sophisticated processing techniques are needed to time the arrival of such pulses with high precision. Even the task of precisely determining the position of the satellite presses orbital dynamics computations to their limit, because large satellites in low altitude orbits are influenced significantly by such processes as air drag and solar radiation pressure. It is, indeed, remarkable that the present generation of altimeters flying at altitudes of 1000 km can measure sea level to 4 cm – a precision of 1 in 25 million!

What can we learn from the resulting sea level data? The largest variations by far, amounting to more than 200 m, are those of the geoid, a surface connecting points of equal gravitational potential. The Earth's gravity field is rather uneven because of the way in which mass is distributed, and this is manifested in sea level. Although of great interest to solid Earth geophysicists – the influence of ocean trenches and seamounts is often readily discernible – it is a nuisance to oceanographers wanting to use altimetry because it prevents the calculation of mean currents. What is required is global mapping of the geoid, independently of altimetry. This surface must be subtracted from the altimeter data to yield the oceanographic signal, referred to as the dynamic topography. An example is shown in *Figure 5.21*. Dominant large-scale oceanographic features, such as the Antarctic Circumpolar Current and the Subtropical Gyre, can be identified, but most of the structure represents uncertainties in the geoid which mask the relatively weak signals produced by ocean dynamics. We must await the launch of satellites dedicated to mapping the Earth's gravity field at high spatial resolution before we can significantly improve our ability to separate the ocean's dynamic topography from the geoid.

Happily, the need to know the geoid can be eliminated by studying time-varying altimeter signals of the ocean, since the geoid is constant on the time-scales of interest to us. One of the most obvious changes in sea level with time is that due to tides. Altimeters now provide the most accurate means of mapping open-ocean tides on a global basis. However, after allowing for these regular rises and falls there remain the irregular changes due to ocean currents. Their speed and direction are related to sea level in the same way as winds in the atmosphere are associated with the horizontal distribution of air pressure. Thus, as an oceanic eddy or meander passes through an observation point there is a change in sea level (depressed for a cyclonic eddy and raised for an anticylonic one). These sea level changes can be expressed as departures from a long-term mean – we call them anomalies. *Figure 5.22* shows sea level anomalies in the South Atlantic obtained from ten days of altimeter data. Over most of the region the signal is less than 5 cm, but two areas (off South Africa and in the Southwest Atlantic) show positive and negative anomalies of 40 cm magnitude. The former corresponds to the Agulhas Retroflection (where the Agulhas Current turns and flows toward the east, south of South Africa) and the latter to the exit region of the Brazil–Falklands Confluence – see also *Figure 5.12*; both are known to be highly energetic and to spawn meso-scale eddies (Chapter 4). Eddies are important in contributing to horizontal heat transports and, through upwelling of nutrients, help to determine biological variability. Altimeter data offers the possibility of monitoring the movement and development of such features and, by assimilation into models, should provide a key element in a future ocean-forecasting system.

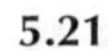

5.21

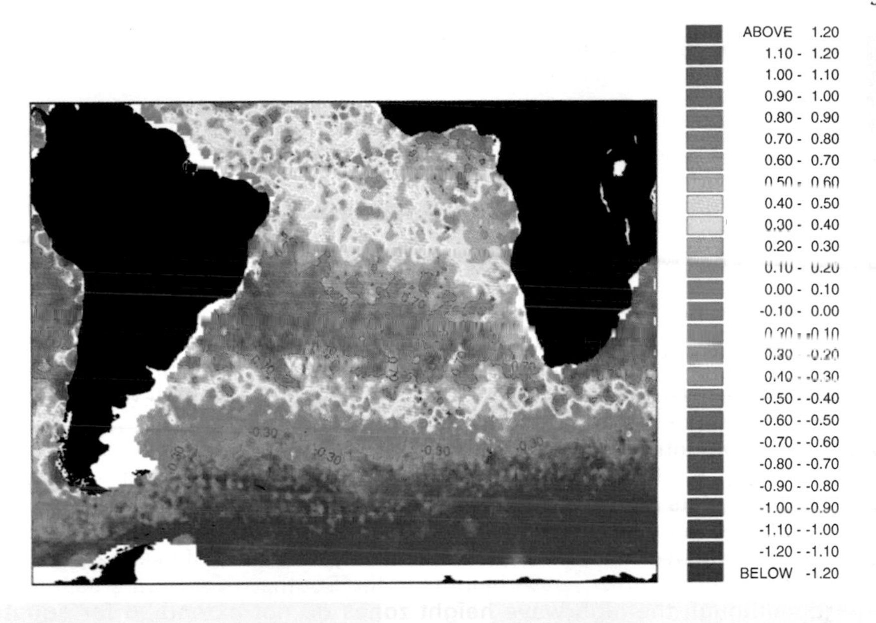

Figure 5.21 Mean sea surface height of the South Atlantic in metres above the geoid. The high in the centre corresponds to the Subtropical Gyre; to the south the Antarctic Circumpolar Current shows up as a region of increased north–south gradient. (Courtesy of Matthew Jones, James Rennell Division for Ocean Circulation, SOC, and Mullard Space Science Laboratory, UCL.)

5.22

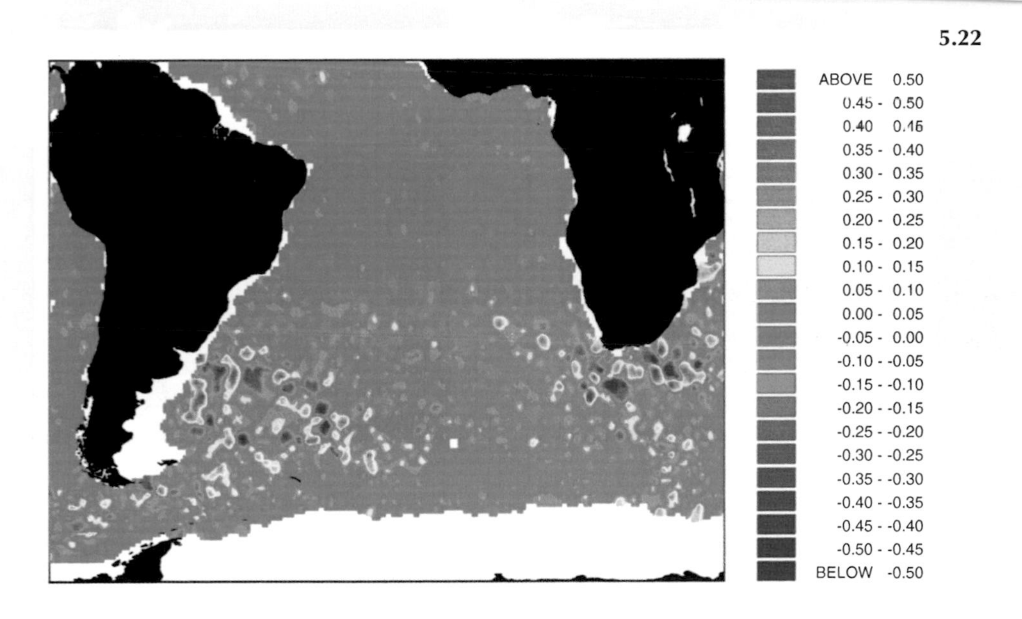

Figure 5.22 Sea surface height anomaly field derived from ten days of TOPEX/POSEIDON data, calculated with respect to a 2-year mean. The colour scale shows heights of the anomalies in metres. (Courtesy of Matthew Jones, James Rennell Division for Ocean Circulation, SOC, and Mullard Space Science Laboratory, UCL.)

Box 5.6 Scatterometer

Like the altimeter, the scatterometer is also a radar sensor, but instead of measuring time delay or the distortion of pulse shape, it relies on using the amount of back-scattered power to infer an oceanographic quantity. From the early days of radar it has been known that when the sea surface is viewed obliquely 'ground clutter' exists, which increases with sea state. In glassy calms there is no detectable return because the incident radiation is reflected away by the mirror-like surface, with no return along the beam. As the wind speed increases small waves are set up which scatter some energy back toward the antenna. The 'cat's paws' often seen on the sea surface when a breeze springs up are an example. Microwave radars operating at wavelengths corresponding to this ripple scale (ca 2 cm) are found to be most sensitive to wind variations. Furthermore, the back-scatter is not the same in all directions, being a maximum in the direction of the wind and a minimum across it. This is due to the effect of longer wind waves which are aligned more or less across the local wind.

Such knowledge is exploited in scatterometry in that broad swaths are illuminated with microwave radiation at incidence angles of 20–55º and with spatial resolution cells of side 50 km. The effect of individual waves on the back-scatter is averaged out on this scale. Several antennae are deployed at various azimuth angles so that the same area of ocean can be viewed from different directions over a period of a few minutes, allowing wind direction to be inferred. There is no accepted model of the wind dependence of the measured back-scatter and so the retrieval algorithm is empirically determined from simultaneous radar back-scatter and *in situ* wind measurements (see *Figure 19.13*). In theory this model function can be achieved by comprehensive airborne measurements prior to the launch of the satellite, but in practice it has proved necessary to carry out considerable modifications using the satellite data themselves.

Despite the lack of fundamental understanding, comparisons of scatterometer wind velocities have shown agreement to 2 m/s and 20º with independent *in situ* data. The accuracy, spatial resolution, and all-weather coverage of scatterometer data make them very useful for studying individual meteorological events and for obtaining the time-averaged wind fields needed to calculate surface momentum and heat and water vapour fluxes for large-scale ocean circulation studies. *Figure 5.25* is an example of the detailed winds in a hurricane, generated from scatterometer data and superimposed

5.25

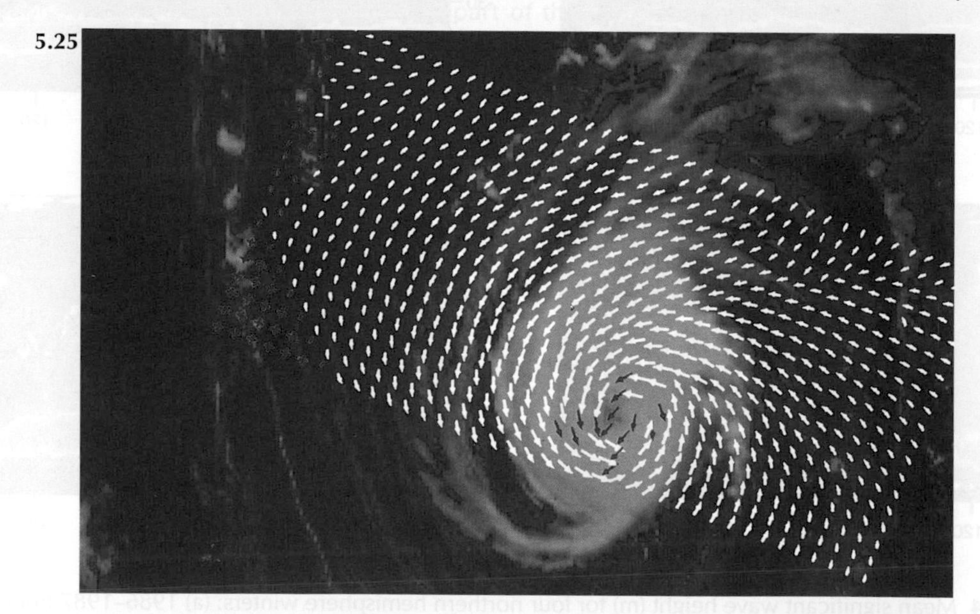

Figure 5.25 ERS-1 scatterometer surface wind vectors across Hurricane Emily on 30 August, 1993, superimposed on a coincident Meteosat cloud image. The length of the vectors is proportional to wind speed and those in red near the eye of the hurricane correspond to winds greater than 15 m/s. (Courtesy of the European Space Agency.)

on an infra-red cloud image obtained by a geostationary weather satellite. The realistic distribution of wind speed relative to the hurricane and the counterclockwise flow centred on the eye cloud add credibility to the scatterometer data. In some cases tropical storms have been detected in ERS-1 scatterometer data before they have appeared on analysis charts, presumably because of the sparseness of the routine weather observations in relation to the relatively small scale of these tropical systems. For some oceanographic purposes, e.g., study of localised coastal upwelling or deepening of the surface mixed layer, the scatterometer wind data should prove extremely valuable, especially in remote ocean areas where other forms of data may not be available. Global wind fields can be produced by combining data from consecutive passes of the satellite, as in *Figure 5.26*. The Trade Winds can be observed converging on the Equator. At higher latitudes the winds are stronger and more variable in direction, being influenced by the passage of depressions. The largest areas of strong westerlies are found in the Southern Ocean.

5.26

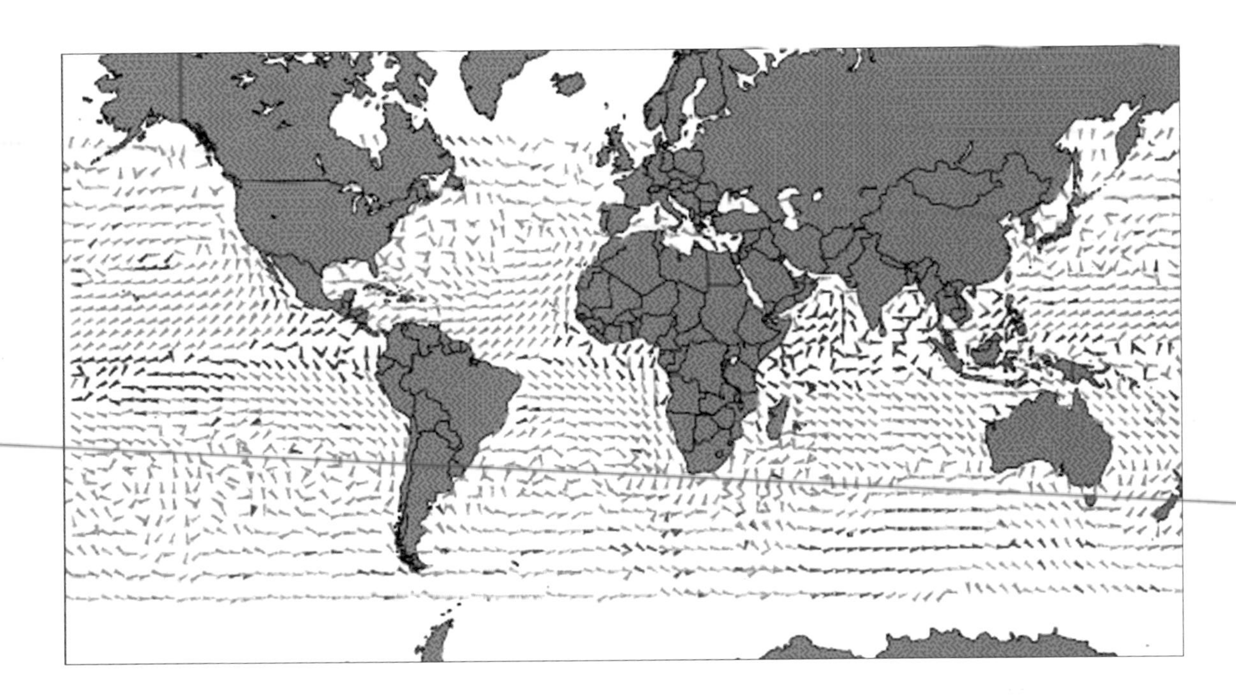

Figure 5.26 Average global wind field for April 1993 derived from ERS-1 scatterometer data. The strength of the wind is indicated by the length of the arrows (lightest winds are in blue, strongest are in red). (Courtesy of the European Space Agency.)

CHAPTER 7:

Snow Falls in the Open Ocean

R.S. Lampitt

When I think of the floor of the deep sea, the single, overwhelming fact that possesses my imagination is the accumulation of sediments. I see always the steady, unremitting, downward drift of materials from above, flake upon flake, layer upon layer – a drift that has continued for hundreds of millions of years, that will go on for as long as there are seas and continents ... For the sediments are the materials of the most stupendous snowfall the earth has ever seen ...
Rachael Carson[5]

Introduction

It is hard to improve on this inspiring prose, describing so graphically as it does a process which continues to excite and tantalise marine scientists. It is widely accepted that the means by which material is transported down to the abyssal depths and the rates at which this occurs present a range of fundamental questions. These are of direct interest to a wide variety of scientists, from those interested in global biogeochemical cycling (see Chapter 11) to, for instance, those keen to understand the population dynamics of the deep sea benthic fauna (see Chapter 13).

In this brief review I acknowledge the technical and philosophical developments which have occurred over the past 120 years, but focus primarily on the progress which has been made during the last 10, progress which has transformed our understanding of the processes involved in material flux. The means by which small particles at the top of the ocean, with almost negligible sinking rates, are transformed into larger, rapidly sinking aggregates which reach the deep-sea floor in only a few weeks continues to provide exciting insights into the complex processes of material cycling in the oceans.

In the language of the travel agent, the open ocean has become a smaller place, and the linkages between the top sunlit zone and the dark, cold, deep sea interior are now thought to be far closer than previously imagined. Rachael Carson's stupendous snowfall[5] is no less awe-inspiring now than it was to her in 1951, and without doubt the questions surrounding its elucidation are as demanding.

Historical Developments of the Concept

We should not think that interest in the way material is transported and modified has only recently been kindled. In the second half of the ninteenth century, there was a vigorous debate among the giants of old (Jeffreys, Thomson, and Lohmann) about the way in which the sediments of the oceans were formed and if, as some believed, there was life in the deep sea, how it was sustained. While Wallich[25] thought that such life was supported by what is currently referred to as chemo-autotrophy, Thomson[22] thought particulate transport of shallow water macrophytes (seaweeds) and terrestrial run-off were the keys to the food supply of the deep-living animals. Jeffreys[10], however, thought that the dead remains of surface dwelling organisms would be an important source of their food. Lohmann[14] made some surprisingly modern calculations about the rate at which the sediments of the ocean were formed by the deposition of planktonic material. He commented on the fact that near-bottom water above the abyssal sea bed sometimes contained a surprising range of thin-shelled phytoplankton species, some still in chains and with their fine spines well preserved. He deduced that they must have been transported there very quickly from their near-surface habitat, and thought that the faecal pellets from some larger members of the plankton (Doliolids, Salps, and Pteropods) were the likely vehicles. His deductions may, in many cases, have been entirely correct and it remains a poor comment on our science that these observations were largely ignored during the next few decades.

The descriptions of Rachael Carson[5] soon found their mark in the minds of a group of Japanese oceanographers using the submersible observation chamber 'Kuroshio', suspended from a fisheries training ship (*Figure 7.1*). The amorphous particles they could see through the portholes were clearly not living and they coined the term 'marine snow' to describe them[21]. The term is still only loosely defined, but is generally recognised to encompass immotile particles of diameter greater than 0.5 mm. In the open ocean these are all biogenic and are thought to be the main vehicles by which material sinks to the sea floor.

The submersible used by the Japanese oceanographers was a cumbersome device and did not permit anything but the simplest of observations to be made. They did, however, manage to collect some of the material and reported that its main components were the remains of diatoms, although with terrestrial material present that provided nuclei for formation.

In spite of these observations and the outstanding questions surrounding material cycles in the oceans, it was, until the late 1970s, a widely held belief that the deep sea environment received material as a fine 'rain' of small particles. These, it was assumed, would take many months or even years to reach their ultimate destination on the sea floor. The separation of a few kilometres between the top and bottom of the ocean was thought sufficient to decouple the two ecosystems in a substantial way, such that any seasonal variation in particle production at the surface would be lost by the time the settling particles reached the sea bed. This now seems to have been a fundamental misconception. Part of the reason for this is the lack of understanding of the role of marine snow aggregates.

We now ask the most basic of questions about this important class of material: What is marine snow, how is it distributed in time and space, and why is it of such significance?

7.1

Figure 7.1 The submersible observation chamber 'Kuroshio' as used to make observations on the distribution and characteristics of marine snow aggregates in the 1950s. The chamber was lowered on a cable from the mother ship, providing a rather cramped view of the sea's interior and sea bottom through the small portholes. Two or three investigators were able to fit into the chamber, descending to depths of up to 200 m for several hours. (Courtesy of Dr Masahiro Kajihara, Hokkaido University.)

A Vertical Profile

Marine snow is found throughout the world's oceans in all parts of the water column. It is not uniformly distributed, either in space or time, but is usually found in higher concentrations in the upper water column and in the more productive regions of the oceans. Although it had been suspected since the early observations of Suzuki and Kato[21] that marine snow concentration decreased with increasing depth, this has been confirmed only recently. The profiles now becoming available do not, however, suggest a simple decrease. There is considerable structure, undoubtedly related to the processes of production, destruction, and sinking. These are all related to the physics, biology, and chemistry of the water column and of the particles themselves.

Figure 7.2 shows some examples of profiles from different parts of the world and using a variety of techniques (*Box 7.1*). Bearing in mind the strong seasonal variation which can occur even well-below the upper mixed layer (see below) and the different techniques employed to obtain these profiles, a common story seems to be emerging. Apart from profiles near the continental slope, where snow concentrations tend to increase near the sea bed due to resuspension, there is generally a rapid fall in concentration over the top 100 m. Peak concentrations are not, however, found throughout the upper mixed layer, but are located at its base, a feature which is directly related to the rates of production and loss of the marine snow particles in this highly dynamic part of the water column.

The upper mixed layer (UML) factory

The mixed layer at the top of the oceanic water column varies in thickness from up to a few hundred metres in winter to a few tens of metres in the spring and summer. It is subject to rapid changes in light, heat, turbulence, nutrient concentration, and depth of mixing on the scale of hours, as well as having distinct seasonal variations. This physical forcing creates changes in the biological processes which depend on them. Furthermore, it is highly variable in the spatial sense, producing a rapidly changing mosaic of physical, chemical, and biological properties not found elsewhere in the water column (Chapters 4 and 5).

It is here in the UML that the primary production of material occurs as a result of phytoplankton growth (see Chapter 6). The cells thus produced (mainly in the range 1–50 μm diameter) are immediately subject to attack from many other elements in the plankton community, but principally from the microplankton (20–200 μm diameter) and the mesozooplankton (0.2–20 mm body length). The

7.2

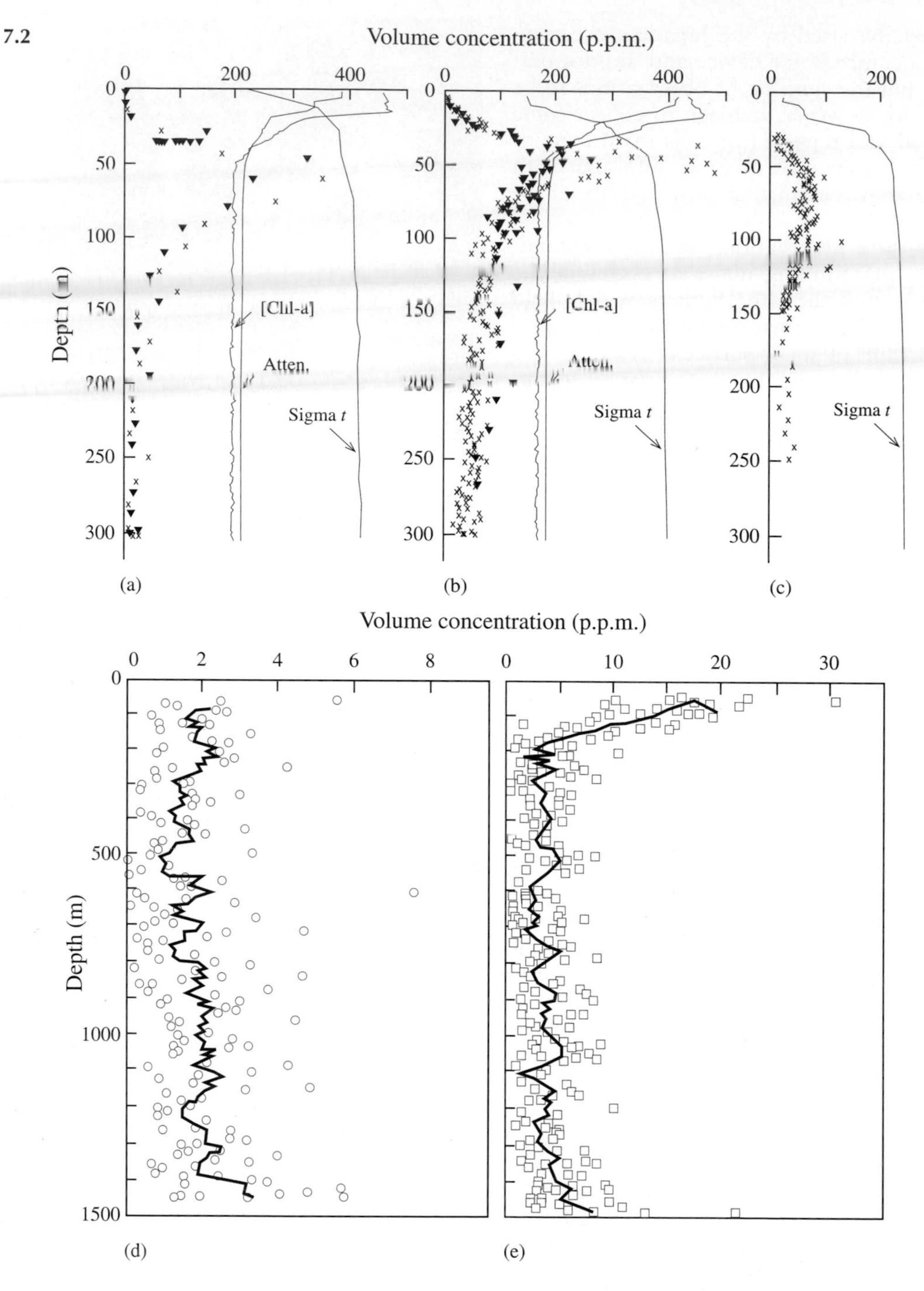

Figure 7.2 Examples of the distribution of marine snow particles greater than 0.6 mm diameter, expressed as volume concentration (p.p.m.). (a) 19 May 1990 (MSP 11), (b) 22 May 1990 (MSP 15), and (c) 26 May 1990 (MSP 19) from the Northeast Atlantic at a water depth of 4800 m (Lampitt *et al.*[12]). Sigma *t* is a measurement of water density, [Chl a] (Chlorophyll a) is the concentration of phytoplankton pigment, and Atten (attenuation) is a measure of the concentration of the smaller particles as determined by their effect on light transmission through the water. In terms of abundance, the maximum concentration of particles at around 50 m depth is about 200/l, decreasing to a deep-water minimum of approximately 30/l. (d) 22 November 1987 and (e) 28 January 1988 from Northwestern Gulf of Mexico at a water depth of 1500 m on two occasions; the lines are 9-point running means[26]. In most instances there is a peak in concentration near the surface, but not necessarily at the surface. Near the sea bed there are also elevated levels. There appears to be a very wide range in concentrations and, although some of this may be related to differences in technique, there can be considerable temporal variability at any one site (*Figure 7.6*); there are also likely to be large regional differences reflecting the structure and dynamics of the biological communities in the different environments.

Box 7.1 How to Examine the Distribution in Time and Space of Marine Snow Aggregates

In situ studies: enumeration, measurement, and collection

Marine snow is a highly variable and ephemeral commodity (*Figure 7.2*). It varies dramatically, not only in its abundance, size, and sinking rate, but also in its origins, composition, and value as a food source. One of the greatest and yet apparently simplest current requirements is to obtain good data on its variability in time and space. Most attempts have used *in situ* photographic techniques. A variety of other sensors are now often attached to the cameras in order to measure other environmental variables, such as water density, fluorescence (a measure of phytoplankton concentration), turbidity (determined primarily by the smaller particles), and oxygen. Subaqua divers have also been used to count particles, but because of the difficulty in seeing the smaller classes of marine snow and the biasing influence of different ambient lighting conditions, divers are mainly reserved for collection and *in situ* experimentation. Manned and unmanned submersibles are also used to observe and capture marine snow. *Figure 7.3* shows examples of some of the photographic and diver-operated techniques.

Both video and emulsion-based photography are used; a light source produces a collimated beam which can then be viewed at right angles by the camera. This produces an image of a known water volume with the particles displayed on a dark background. For long-term observations of temporal changes, such devices have been deployed on moorings[11]. Profiling instruments are generally deployed from research ships[20,20].

7.3a

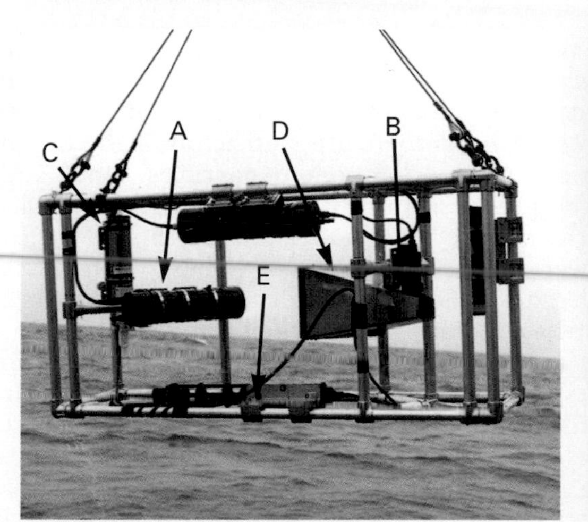

7.3b

7.3c

7.3d

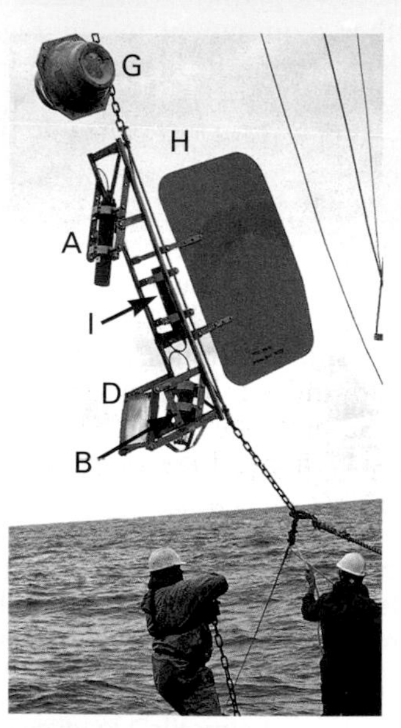

Figure 7.3 A variety of photographic devices used to record the spatial and temporal variation in marine snow aggregates *in situ*. (a) The Large Aggregate Profiling System (LAPS)[8] with collimated light beam: 35 mm camera (A), flashlight (B), CTD (C), fresnel lens (D), and transmissometer (E). A volume of about 20 l is photographed and particles binned into six separate size categories from 0.5 mm to >3 mm. (Courtesy of Drs Gardner and Walsh, Texas A & M University.) (b) The Atlantic Geosciences Centre Floc Camera Assembly (FCA)[20]. This carries three cameras (A) and flash (B), and records particles >0.25 mm in a volume of 0.2 l. (Courtesy of Dr Syvitski, Bedford Institute of Oceanography.) (c) In this case the device is associated with a variety of other instruments, such as a fluorometer to measure the concentration of phytoplankton, conductivity and temperature sensors, an echo-sounder (E), and a transmissometer to measure the concentration of the smaller particles. The grey vertical tubes (F) on top of the device are sampling bottles activated by the conducting wire on which the device is suspended. Particles greater than 0.5 mm are recorded in a volume of 40 l using 35 mm film. These are later analysed on an image analyser (see *Figure 7.6*). (d) In contrast to (a)–(c), this instrument is attached to a mooring to record temporal changes (*Figure 7.6*). It has the same arrangement as in *Figure 7.3(c)* and the frames are similarly analysed. Also shown are a buoy (G), vane (H), and battery pack (I).

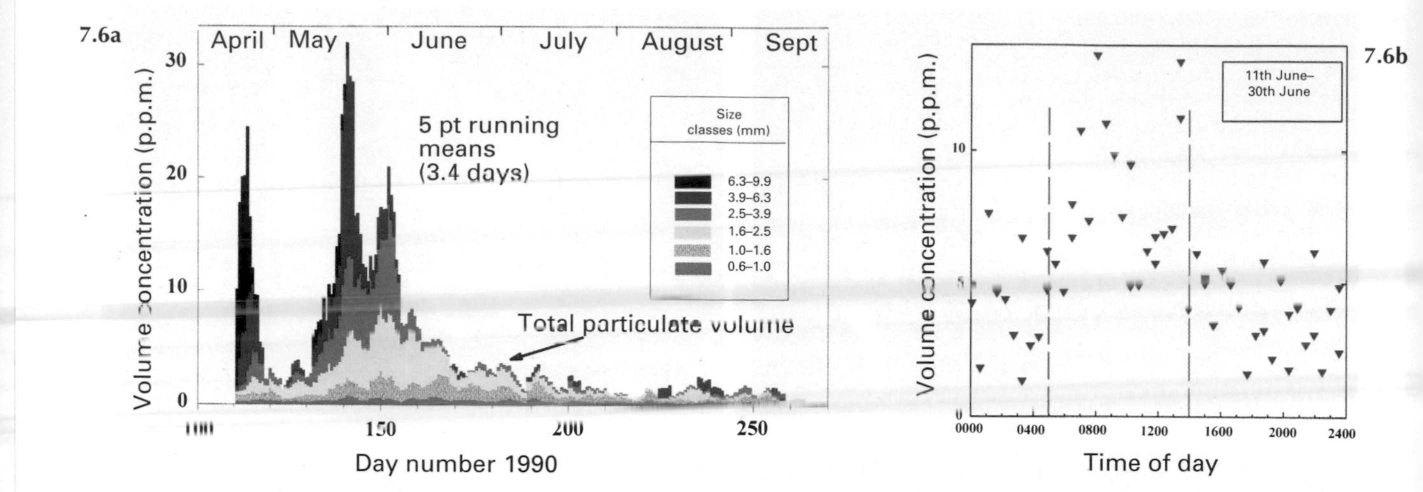

Figure 7.6 (a) Temporal variation in marine snow volume concentration (night values only to remove diel periodicity) at 270 m depth in the northeast Atlantic(48°N 20°W), derived from photographs taken using the marine snow camera system, *Figure 7.3(d)*. The photographs obtained with this instrument are analysed by computer to give a maximum and minimum dimension of each particle in the frame. The volume of each particle is calculated and assigned to one of six size categories. Dramatic peaks in concentration are caused primarily by the larger size classes and occur soon after pulses in productivity of the phytoplankton in the surface water layer above[11]. (b) Accumulated data for volume concentration of marine snow at 270 m depth during the period 11–30 June 1990. During this period concentration levels were significantly higher between the hours of 05.00 and 14.00, identified by the dotted vertical lines.

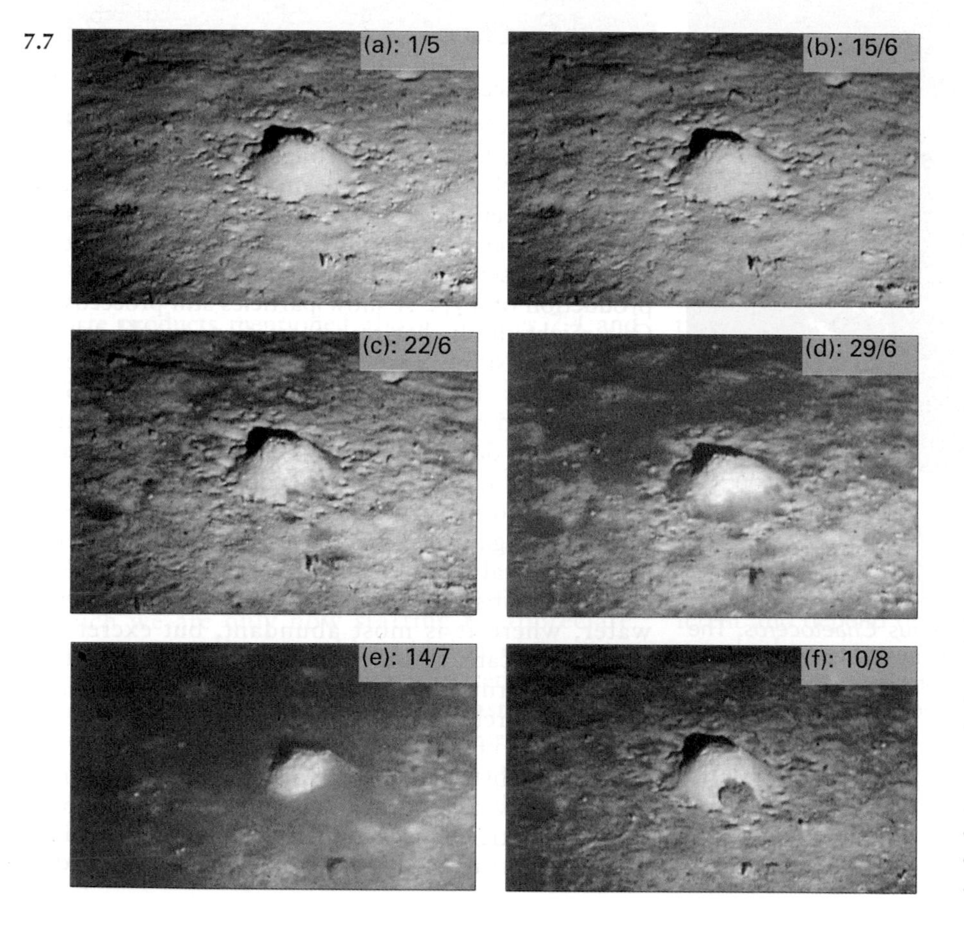

Figure 7.7 Examples of time-lapse photographs of the sea bed at 4025 m depth off the European continental slope, taken using the 'Bathysnap' instrument. Between 1 May and 15 June there is little change in its appearance, but during the rest of the summer there is a progressive increase in the amount of material covering the sea bed, visible as dark patches obscuring the underlying sediment. Between 14 July and 10 August there is a progressive decrease in this covering. The mound in the centre of the frame is 18 cm across.

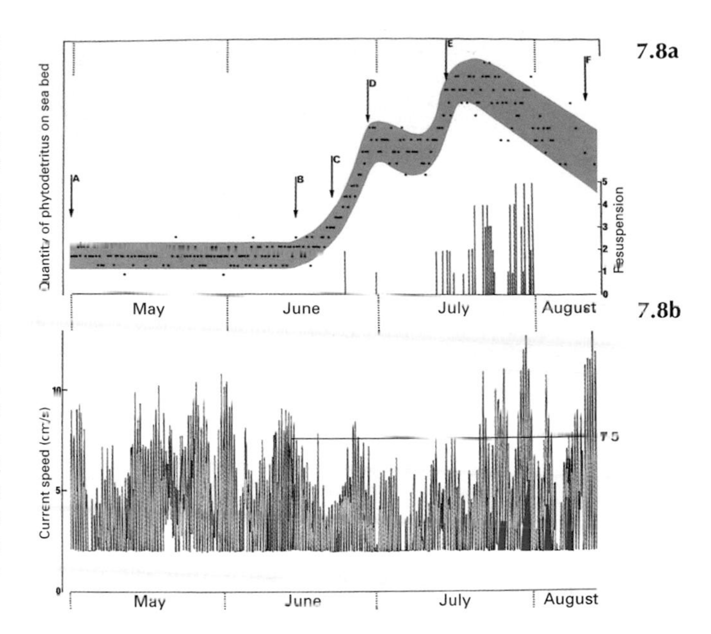

Figure 7.8 The photographs in *Figure 7.7* were used to derive a semi-quantitative measure of the material lying on the sea bed on each frame (a); the green band is to highlight the trend over the year. Also shown in (a) (vertical bars) is a semi-quantitative estimate of the degree of resuspension in each frame, which reduces visibility of the sea bed. It can be seen that resuspension only occurs after the deposition of phytodetritus and is not a constant feature. The letters A–F indicate the times at which the photographs in *Figure 7.7* were taken. (b) The current speed which shows that it is only when currents exceed about 7.5 cm/s that a significant resuspension occurs. It is thought that the material is not resuspended very high in the water column as, from other Bathysnap deployments, resuspension causes an initial loss of material from the sea bed, which is followed within only a few tens of minutes by its redeposition. The current meter rotor stalls at speeds less than 2 cm/s, as indicated. The tidal cycle is clearly evident, but during late July and early August the minimum of the tidal cycle is above the stall speed of the rotor, indicated by red shading. These are particularly energetic periods.

levels of primary productivity in the overlying water and precede peaks in material flux determined by sediment traps 3000 m below. As can be seen in *Figure 7.6(a)*, the values of volume concentration in the spring are more than ten times those later in the year; this is mainly due to the contribution of the larger-size categories. An additional feature of this data set is that there is a distinct diel variability; in *Figure 7.6(b)* data from one period are presented to demonstrate this signal. It is tempting to relate this signal to the diel migration of most of the larger members of the planktonic biosphere, but at present the mechanism is unknown.

Although much of the material sinks at rates of a few tens to hundreds of metres per day, some of it is modified or remineralised in support of the mid-water community of bacteria and zooplankton. Other components dissolve during the descent, especially calcite particles when below the calcium compensation depth (see Chapter 13), but it seems that in spite of this long descent, some of even the delicate structures of phytoplankton spines are still apparent when the deep-sea floor is reached, just as found by Lohmann[14] back in 1908.

The benthic experience

The ultimate repository for much of the material which sinks into the deeper parts of the water column is the deep-sea floor; it is here we now look for evidence of a close link with the surface of the ocean and indications about the role of marine snow.

Rapid changes in material flux in deep midwater and apparent seasonal reproduction by some of the larger benthic fauna posed a serious threat to the established view of weak linkages between the top and bottom of the oceans. Repeated sampling of the deep-sea sediment, particularly the loose surface layer, was technically very difficult and very demanding on ship time. It was therefore fortunate that in the 1980s, time-lapse photographic techniques using an instrument called the 'Bathysnap' became available for use in an area which was subject to very strong seasonal depositions of phytodetrital material. In *Figure 7.7*, recently deposited material can be readily seen on the sea bed as dark patches of loose fluffy material.

The image of the material above the sediment surface is a reflection of supply and demand. It is the difference between the supply of settling particles from above and the demand by benthic community members, which either ingest or bury these. Under less productive parts of the ocean, such as the Madeira Abyssal Plain, no detrital layer is visible on benthic photos; this not only reflects the lower overall level of productivity, but also the reduced variability in particle flux. Here the supply is more constant and the benthic community consumes or buries it at the same rate as it arrives on the sea floor.

Although difficult to quantify on the photographs, an approximation can be made by measuring the density of the film at one particular location on each frame of the time series (*Figure 7.8*). We can now collect the material using a remarkable coring device, the Scottish Marine Biological Association (SMBA) multiple corer. This has the ability to collect sediment samples with virtually no disturbance to the light interfacial layer, and we can state with certainty that the dark patches seen on the photographs are primarily of phytoplankton

Box 7.2 How to Collect and Study Marine Snow Aggregates

Sediment traps can be used to collect settling particles, but it is unlikely that the integrity of the individual particles so-collected is maintained, even when they are immersed in a preservative as is usually the case. Because of the fragility of some types of marine snow, and because of their low abundance and high sinking rates, traditional water-bottle techniques are usually of little value. In spite of the inherent difficulties of subaqua diving (limitations of depth, sea state, and personnel), it remains one of the best methods to obtain undamaged snow particles for experimentation (*Figure 7.12*).

Large bottles have been developed by some groups with the specific aim of collecting marine snow, *Figure 7.13(a)*. This 100-litre water bottle is closed at depth by sliding a weight, the 'messenger', down the supporting wire. On recovery it is left on deck for several hours to allow the snow aggregates to settle to the bottom of the vessel. The top 95 litres are drained off and the bottom chamber removed for collection of the particles in the laboratory, *Figure 7.13(b)*. This has proved to be a useful and successful method where diving is not possible.

7.12

Figure 7.12 Subaqua diver behind a very large marine snow aggregate in the northern Adriatic sea. (Courtesy of Dr Stachowitsch, University of Vienna.)

7.13a

7.13b

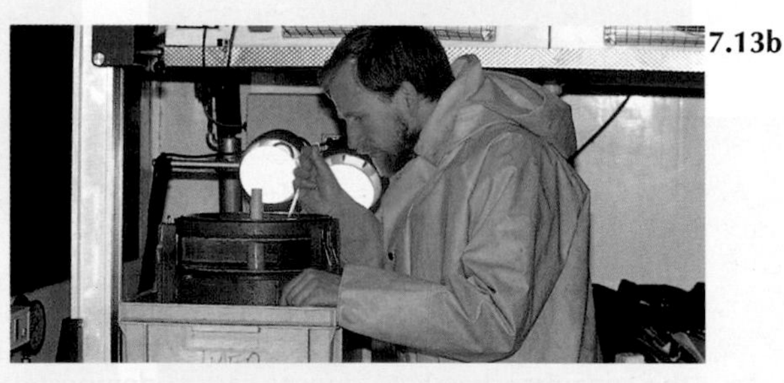

Figure 7.13 The large-volume marine snow catcher, the 'snatcher', for collecting undamaged samples of the aggregates. (a) The entire device just prior to deployment from RRS Discovery. (b) The lower chamber of the 'snatcher' in the cold laboratory on board ship. Aggregates are being collected from the base of the chamber using a wide-bore pipette to avoid damage. The lights are directed through the transparent sides of the chamber for ease of collection.

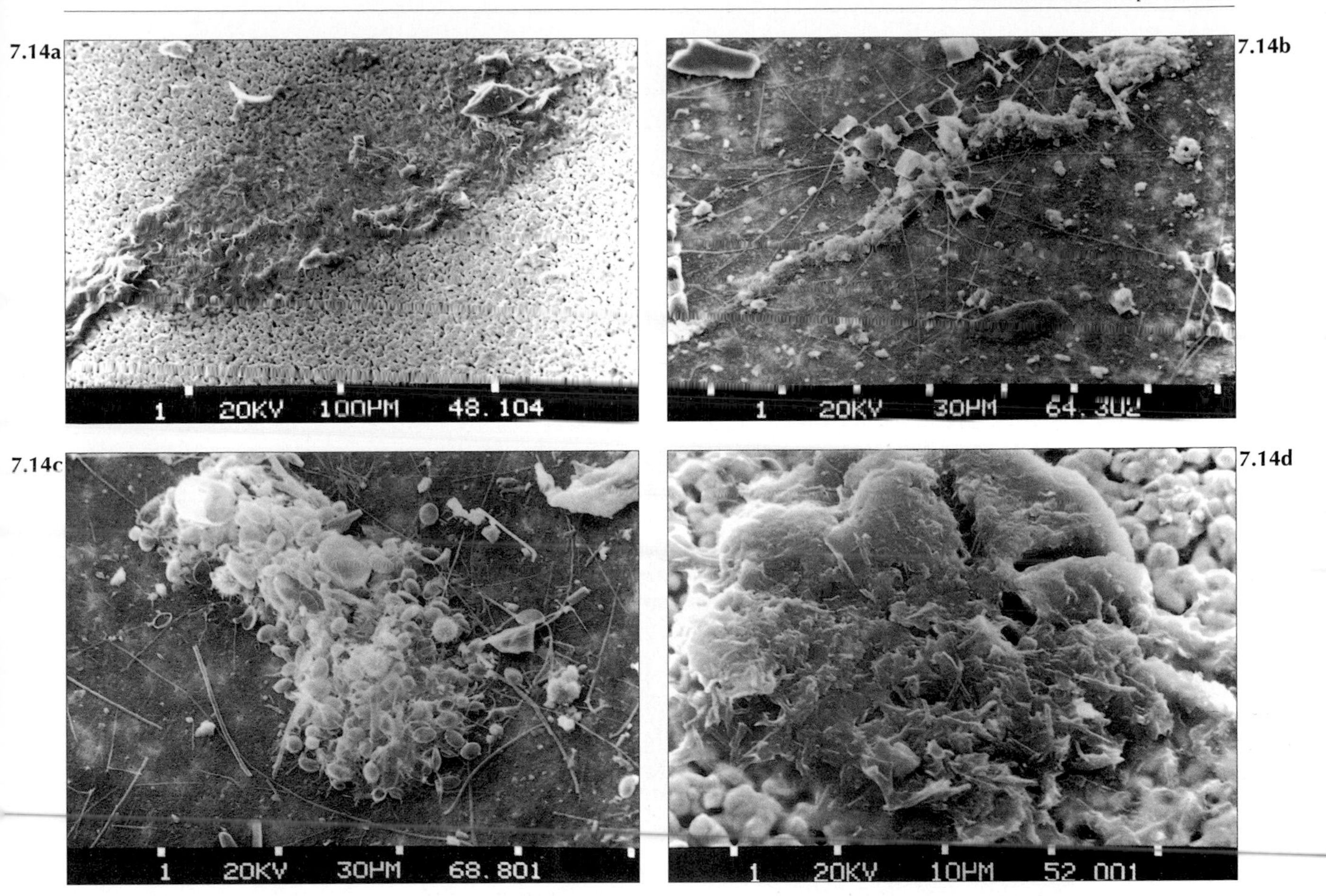

Figure 7.14 Scanning electron micrographs of marine snow particles collected off Baffin Island, showing a wide range of morphologies and composition. (a) Large mucoid aggregate collected at 30 m depth. (b) 'Stringer' collected at 1 m depth. (c) Mixed agglomerate dominated by biogenic material collected at 100 m depth. (d) Aggregate dominated by mineral matter collected at 5 m depth. (Courtesy of Dr Azetzu-Scott, Bedford Institute of Oceanography.)

excess density over that of the surrounding water controls the speed with which they descend through the water column. The proportion of free water in an aggregate, its porosity, determines how fast its internal environment changes in response to varying external conditions; for example, during sinking, porosity controls the rate of exchange of water within the aggregate with that outside. Porosity also influences the rate at which small particles, such as clays, accumulate on the aggregate and the rate at which surface-active elements, such as thorium, are adsorbed onto the snow. Rates of adsorbtion and desorbtion are of considerable relevance to particle-cycling models, which use adsorbtive radioisotopes, such as ^{234}Th, as proxies of solid material (see later).

The physical properties of an aggregate are clearly closely tied to its chemical and biological components. However, as a general pattern, Type B aggregates develop as bloom conditions are reached and nutrients become limiting (see Chapter 6). It is during this stage that the concentration of TEP increases[16] and the particles probably become more sticky. With increasing size of aggregate, excess density tends to decrease[1], so that the increase in sinking rate is not as pronounced as might be expected. Simultaneously, and when still in the euphotic zone, primary production may be enhanced within the aggregate due to an efficient use of the ammonia released within them[9]. In fact, production may proceed so fast that bubbles of free oxygen are created which, in turn, cause the snow to rise in the water[17]. There is also good evidence that the bacteria[19] and protozoa[13] find the micro-environment within the aggregates attractive, sometimes producing anoxic microzones[18]. This may reduce sinking rates if free gases are produced. The activity of the microbiota may change, however, during the descent of the aggregate; it now seems that this activity may be inhibited by increasing pressure[23].

The composition (*Figure 7.14*) of Type B aggre-

7.15

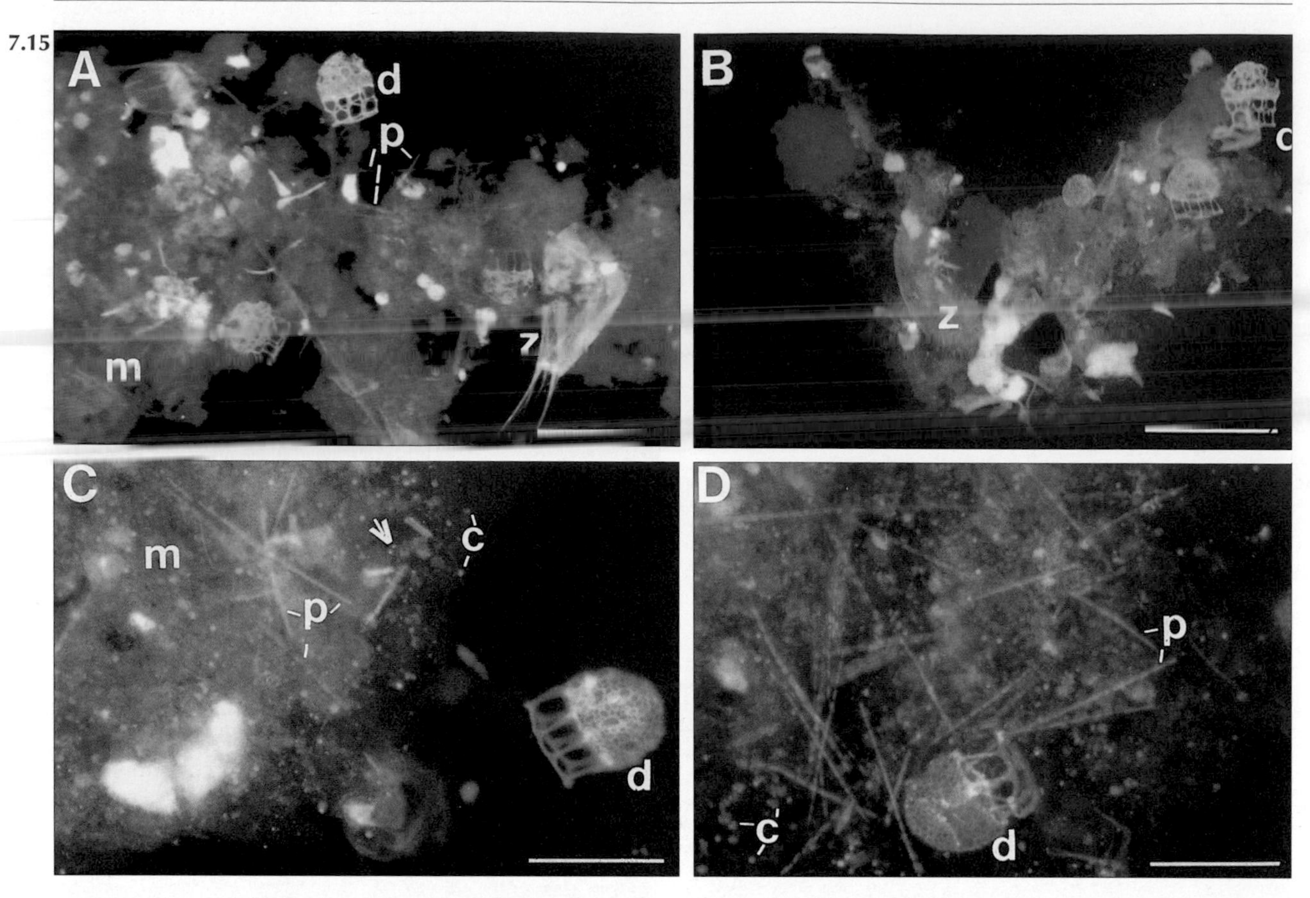

Figure 7.15 Marine snow aggregates collected using the 'Snatcher' (*Figure 7.13*) during May 1990 in the northeast Atlantic over the Porcupine Abyssal Plain from depths of either 45 m (A, C, and D) or 300 m (B). The composition of the aggregates can best be observed under the microscope using ultraviolet excitation, either after staining with acridine orange (A and B) or relying on the autofluorescent properties of the material (C and D). The aggregates contain a wide range of component particles (d, lorica of the tintiniid *Dictyosysta elegans*; p, pennate diatom; z, zooplankton carapace; c, small chlorophyte; m, bacterial matrix). In this instance, the tintiniid lorica tended to be physically damaged when found further down in the water column [scale bars = 100 μm (A and B) and 50 μm (C and D)]. (Courtesy of Dr C Turley, Plymouth Marine Laboratory.)

gates generally reflects that of the suspended particles in the euphotic zone. As such, the aggregates present at any one time at a particular location are of a similar composition[2], whether they be dominated, for instance, by diatoms, faecal material, coccolithophores, or flagellates (*Figure 7.15*).

Whatever the origin of the particles, and however porous the aggregates, it is safe to assume that the environment of a free-living organism or particle changes dramatically when it becomes incorporated into an aggregate. Not only will its chemical environment change, but so too will its sinking rate and, in the case of an organism, its ability to control its depth by, for instance, buoyancy modification. An organism associated with an aggregate will also be subject to quite different predator pressures from its free-living counterparts, as discussed below.

Quantitative Study of Particulate Flux

In order to understand the cycling of biogeochemical components of the oceans, data must be obtained about the time-varying fluxes of the principal compounds, elements, and particle types at different depths in the sea. The concentrations and size distributions of particles can only give a general indication of fluxes, and then only after some major assumptions on sinking rates. There are few ways in which particle flux can be measured; these can be divided broadly into direct measurements using sediment traps and numerical modelling approaches, frequently based on the distributions of various chemicals and particles.

The particle interceptor trap, or sediment trap, is a device akin to the rain gauge, having a funnel into which falling particles are collected. In the case of modern time-series traps (*Figure 7.16*), a rotat-

7.16a

7.16b

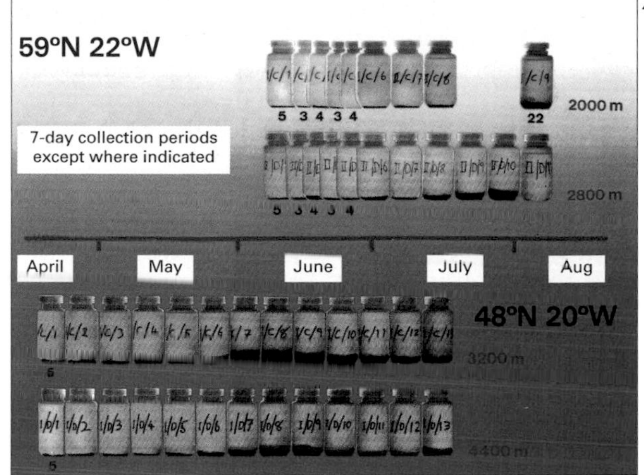

ing carousel moves a fresh collecting jar under the bottom of the trap at predetermined times, such that data may be obtained over several months' duration with a resolution of a week or so. There are several problems related to the accuracy with which sediment traps measure particle flux, but these devices provide the only means of determining directly the flux of material at a particular depth; furthermore, they are the only way in which the material responsible for the downward flux can be collected, described, and analysed.

In 1978, soon after the sediment trap had become a reliable oceanographic technique, a long time-series of measurements was initiated off Bermuda at a depth of 3200 m (*Figure 7.17*). This remarkable series, still continuing, demonstrates a strong seasonal signal of deposition with a peak in the early part of the year. This is certainly not a universal pattern, due to the large physical and biological differences between oceanographic regimes. In this case, the depositional peak lags behind a

Figure 7.16 (a) Time-series sediment trap photographed just after recovery. Settling material enters the yellow cone and from there into the white collecting cups below. The trap had been at a depth of 3200 m for the previous 6 months, in a water depth of 4800 m, and the collected material can be clearly seen in the cups (previously filled with a preservative, formaldehyde). Such devices are suspended on a supporting wire, with a ballast weight on the sea bed attached to the wire by an acoustically operated release mechanism. Buoyancy spheres at the top of the wire carry the entire mooring to the surface after the release has been activated. (b) Sample cups from deep-water sediment traps deployed at two locations in the northeast Atlantic (7 day collection periods except where indicated). The increased flux from the end of May resulted from elevated productivity in the surface waters about 4 weeks previously. (Courtesy of Dr Williamson, University of East Anglia.)

7.17

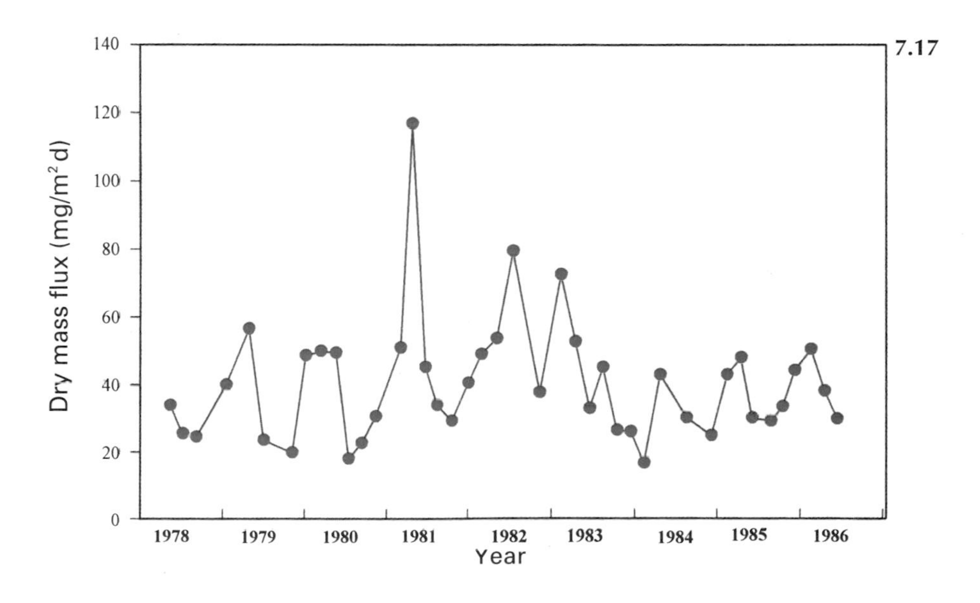

Figure 7.17 Variation in particle flux over an 8-year period at 3200 m in the Sargasso Sea. This is the longest such record from any region and shows the seasonal cycle in particle flux, but with some distinct irregularities during the period 1981–1983. At this location, elevated productivity in the surface is caused by a deepening of the mixed layer, which introduces new nutrients. About 3 weeks later particle flux at 3200 m became enhanced (from Deuser[6]).

8.2

Figure 8.2 Global earthquake activity superimposed on a topographic map of the world (based on a cylindrical equidistant map projection). The epicentres of earthquakes with body-wave magnitudes greater than five are shown as coloured dots, colour-coded according to the focal depth of each earthquake (<50 km, red; 50–100 km, yellow; 100–300 km, green; >300 km, blue). Note how most epicentres are concentrated in bands which delineate the plate boundaries (courtesy of the National Oceanographic and Atmospheric Administration/National Geophysical Data Center).

8.3

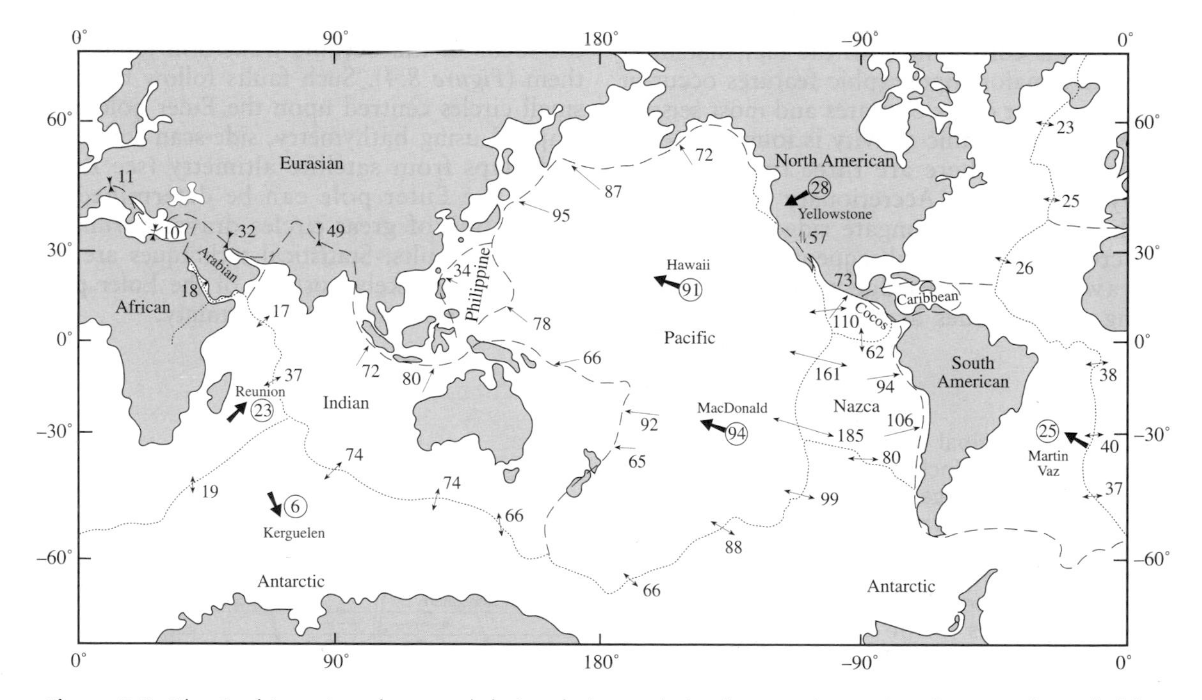

Figure 8.3 The Earth's major plates and their relative and absolute motions. The plates are bounded by oceanic ridges (dotted lines), oceanic trenches, mountain ranges, and transform faults (all as dashed lines). Thin arrows show the directions and rates (in mm/yr) of relative motion at selected points on the plate boundaries [after Bott[6]; reproduced by permission of Edward Arnold (Publishers) Ltd., and based on data in Chase[11]; © Martin H.P. Bott, 1982].

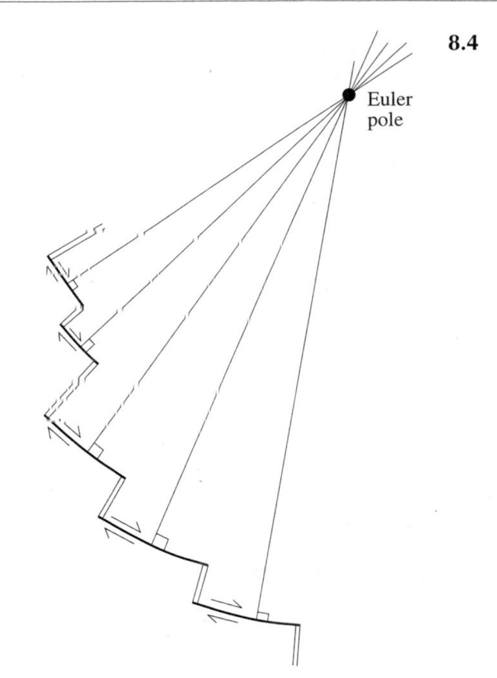

Figure 8.4 Method to determine the Euler pole for a spreading ridge system. Transform faults (thick lines with double arrows), which offset the ridge segments (double lines) describe small circles about the pole. Hence lines drawn normal to the transform faults intersect at the Euler pole.

Rates of contemporary sea-floor spreading at accretionary plate boundaries are also used to constrain the angular rotation rate about the Euler pole, which is required to describe current plate movements. However, active convergent plate boundaries are more problematic and relative velocities have to be determined indirectly. For example, if the relative divergent motions between plates *A* and *B* and between plates *A* and *C* are known, the relative convergent motion between plates *B* and *C* can be determined by simple vector algebra. The above approach can be extended to include all known plate boundaries, with the additional constraint that, globally, all the vectors must form a single self-consistent set (*Figure 8.3*). DeMets *et al.*[14] have completed the most recent

Box 8.1 Satellite Gravity Fields

Short-wavelength (<400 km) gravity anomalies are highly correlated with small-scale topography; because of this it is possible to use high-resolution gravity fields computed from satellite data to map the bathymetry and tectonic features of the sea floor. This is particularly useful in parts of the Southern Ocean where ship-collected data are sparse (Figure 8.5).

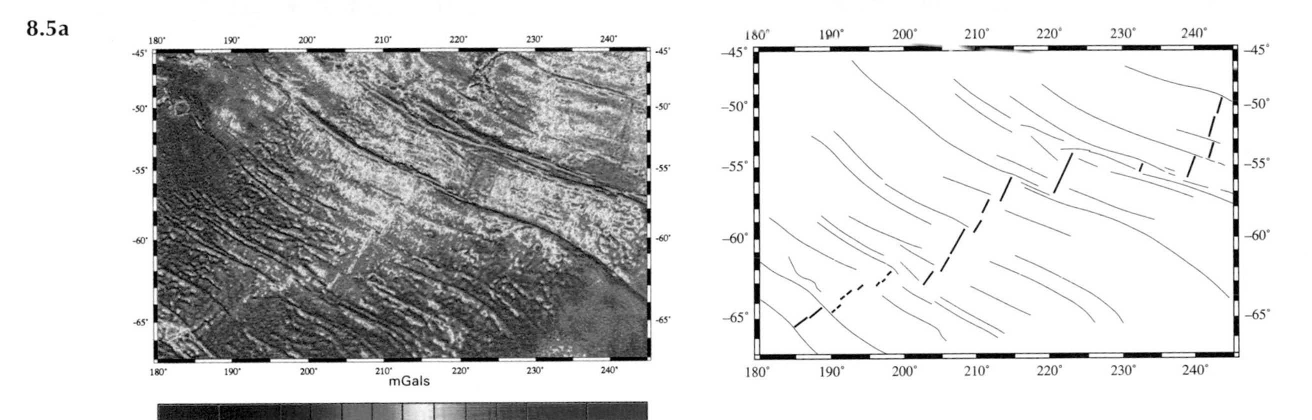

Figure 8.5 (a) An 'illuminated' image of high-resolution gravity anomalies over the Pacific–Antarctic Ridge. This accretionary ridge, outlined in orange and yellow, runs ENE–WSW across the figure, and is offset by numerous transform faults (dark blue and purple; courtesy of W.H.F. Smith, first published in Sandwell and Smith[32]). (b) An interpretation of the gravity image, which reveals features as small as a few tens of kilometres and the complex relationships between spreading segments (thick lines) and small and large offset transform faults and fracture zones (thin lines).

8.17

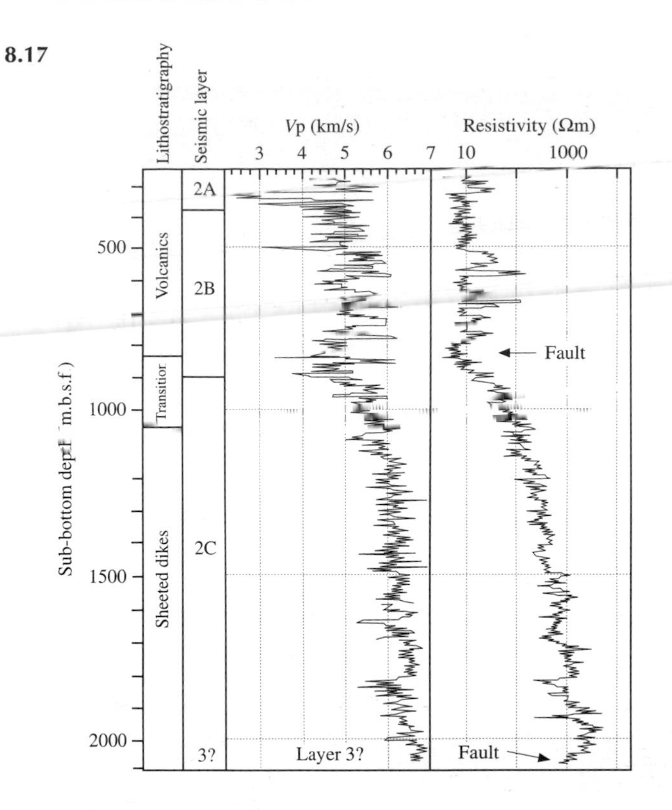

Figure 8.17 Down-hole logs from Ocean Drilling Program Site 504B (as at the end of Leg 148, February, 1993). Note the broad changes in velocity (V_p) and resistivity between the volcanic and sheeted-dike layers, and the short wavelength (ca. 1–10 m) variations in these properties (profiles begin at 250 mbsf; mbsf = metres below sea floor; *Figure 2* in ODP Leg 148 Shipboard Scientific Party[28]; © American Geophysical Union).

sheets that intrude parallel to the ridge axis during sea-floor spreading. Below the dikes there is a layer of 6.7–7.2 km/s material which forms the greater part of the crust. This oceanic layer, or Layer 3, has rarely been drilled *in situ*. From samples, and by analogy with ophiolites (sequences of mainly igneous rocks which contain the same rock types, and in the same order, as are found within the oceanic crust) exposed on land, most geologists think it largely consists of gabbro, a coarser grained rock representing the frozen melt which was the source of the basaltic flows and dikes. At the base of the crust, velocity increases, often abruptly, to around 8.0 km/s, comparable to the Mohoroviçic discontinuity at the base of the continental crust; this marks the top of the Earth's mantle. The velocity–depth structure in *Table 8.1* is an average for normal crust; in fracture zones the crust is often thinner and may be underlain by velocities thought to represent serpentinized peridotite; and near hot spots, such as Iceland, it is thicker. The density of rocks is strongly correlated with their seismic velocity (*Figure 8.18*); this useful property means that we can infer density from velocity and check our seismic models by computing their gravitational effect and comparing this with independent gravity observations (e.g., *Figure 8.14*).

Another important property of the igneous crust is its magnetisation. When magma, containing a few percent of certain iron oxides, cools it acquires a remanent magnetisation, in the direction of the contemporary Earth's field, which is stable over millions of years (see above). This magnetisation provides the 'memory' in the rocks, whereby sea-floor spreading records reversals of the Earth's magnetic field. The remanent magnetisation slowly decreases with time as some iron oxides undergo further oxidation, a process accelerated by hydrothermal circulation. Older basalts acquire a stronger, secondary, possibly chemical, magnetisation (*Figure 8.19*). Hence a minimum magnetisation occurs in rocks which are 8–20 Myr old. The magnetic susceptibility of the crustal rocks is usually relatively insignificant, so that it does not have to be included when

8.18

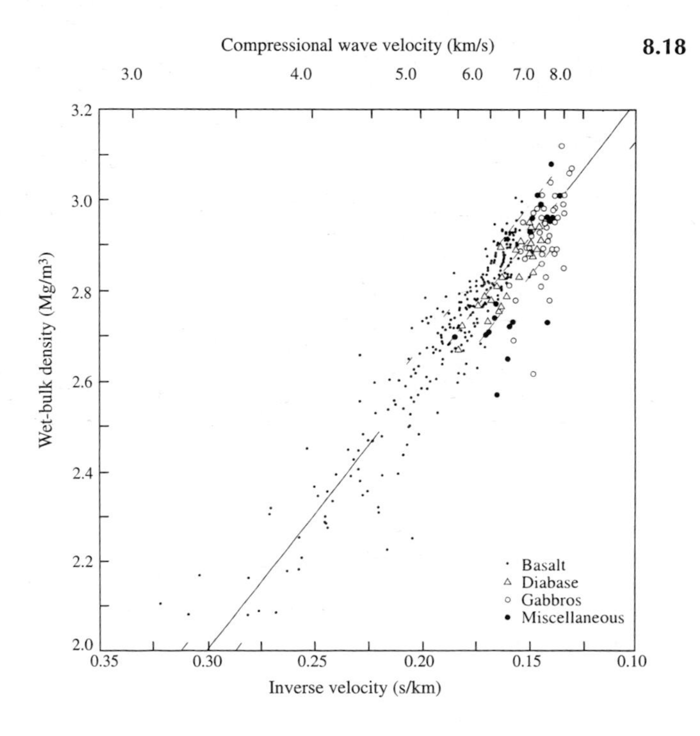

Figure 8.18 Wet-bulk density of samples of oceanic crust and ophiolites plotted against the inverse of compressional-wave velocity. The solid line represents a least squares fit, with the standard error indicated by the dashed lines (*Figure 1* in Carlson and Raskin[10]; reprinted with permission from *Nature*, © 1984 Macmillan Magazines Limited).

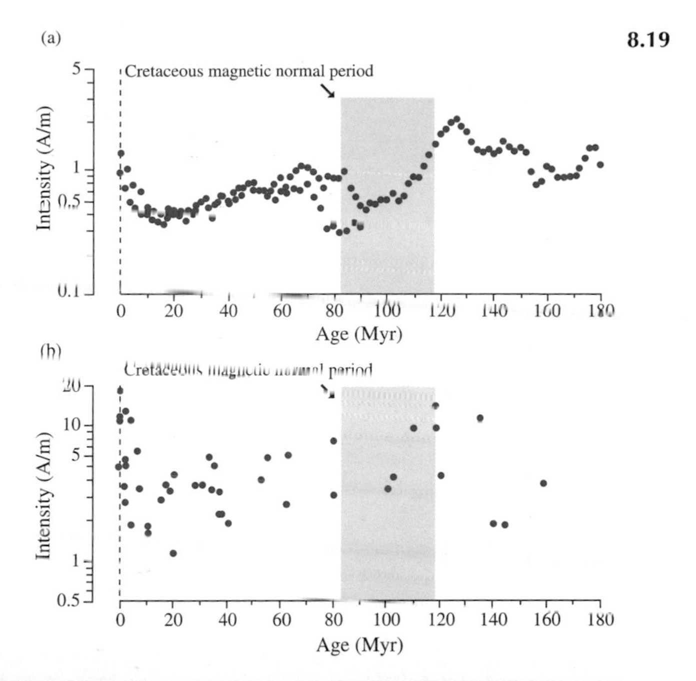

Figure 8.19 Summaries of crustal magnetisation against crustal age. (a) Obtained from the inversion of marine magnetic anomalies. (b) Derived from measurements of normal remanent magnetisation made on drilled basalt samples. Both curves show the decrease in magnetisation caused by the oxidation of magnetic minerals in the first 20 Myr and the subsequent increase indicating the acquisition of a remanent (chemical?) magnetisation (*Figure 4* in Sayanagi and Tamaki[33]; © the American Geophysical Union).

computing the magnetic effect of these rocks.

Other important physical properties of crustal rocks include electrical resistivity, shear wave velocity, permeability, and thermal conductivity; many are anisotropic.

Recently, our perception of the igneous crust has been improved by seismic-reflection profiling[26]. Using the same equipment and ships as in the search for oil and gas, but with specially designed configurations, intriguing reflecting surfaces have been detected below the Atlantic Ocean, deep within the crust (*Figure 8.20*). On profiles acquired along isochrons, these surfaces tend to have low

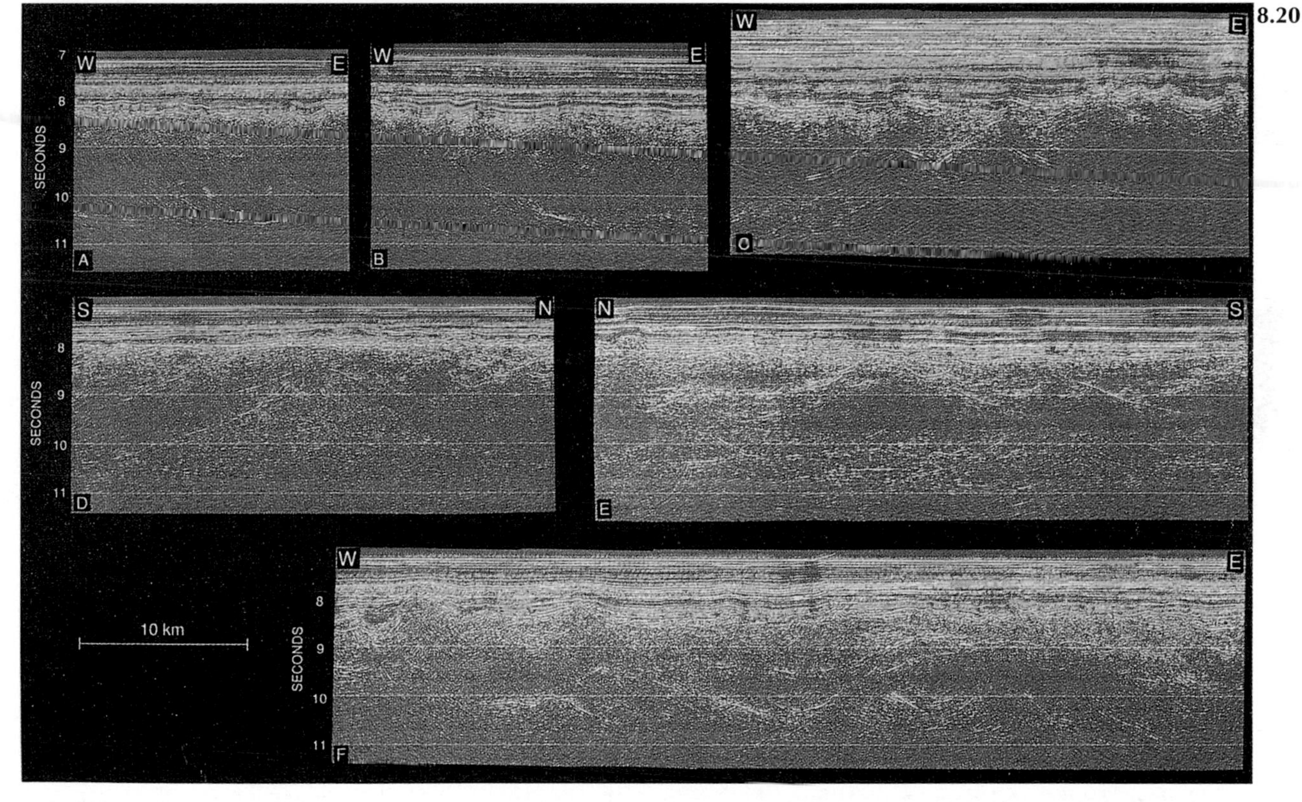

Figure 8.20 Multichannel seismic-reflection profiles of oceanic crust produced at slow-spreading rates in the North Atlantic Ocean, illustrating a variety of forms of reflectivity (the vertical scale in seconds is the time required for sound to be reflected back to the sea surface). A, B, D, E, and F are from the western North Atlantic; C is from the eastern North Atlantic. A and B are from flow-line profiles, C is oblique to the spreading direction, D and E are isochron lines, and F is along the trough of the small-offset Blake Spur fracture zone. Typically, the shallow crust contains distinct sub-horizontal reflections and the middle crust is almost reflection-free. The lower crust exhibits the strongest and most diverse reflectivity, including banded patterns of dipping reflectors. The dipping reflectors may have a tectonic or igneous origin. A distinct Moho reflection is seldom seen; the reflective lower crust typically merges downward to a reflection-free upper mantle (*Figure 4* in Mutter and Karson[26]; © 1992 by the AAAS).

8.21

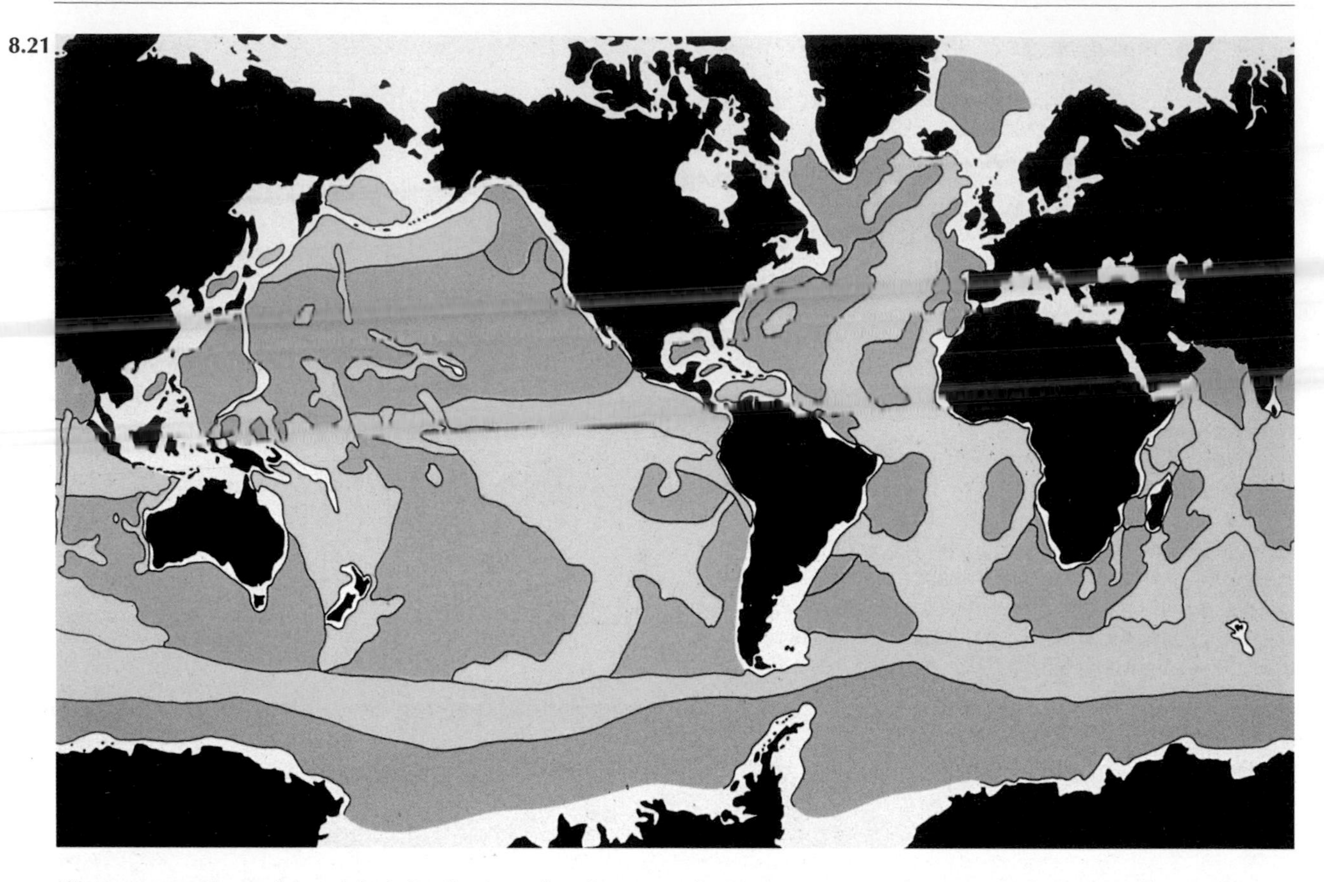

Figure 8.21 Chart of the global distribution of sediment in the ocean (green, calcareous sediments; yellow, siliceous sediments; brown, terrigenous sediments; blue, glaciogenic sediments; pink, deep-sea clay; white, margin sediments; drawn by R.G. Rothwell).

dips, be highly reflective throughout the igneous crust, and dip bi-directionally; on profiles parallel to the spreading direction they are steeper, usually dip toward the spreading centre, and offset the basement surface. The former may represent either contrasts developed during the igneous creation of the crust or faults; the latter are probably ridge-parallel normal faults, mostly active in the early development of the crust.

Sediment Provenance and Transport Processes

Sediments cover most of the ocean floor. Our knowledge of these sediments, and the Earth history they record, has increased markedly in the past three decades, through gravity and piston coring (see *Figure 19.4*), and deep-sea drilling. The sediments comprise, in varying amounts, detrital material derived from the weathering of the continents, biogenic debris derived from planktonic organisms, and clay-size material. Sediment types are distinguished by particular constituents; different sediments show well-defined global distributions (*Figure 8.21*; see also *Figure 11.8*). Sedimentary material is transported to the ocean floor by a number of mechanisms and processes (*Figure 8.22*; also see Chapter 9).

Rivers form the main pathway of terrigenous sediment to the oceans, although wind transport is particularly important for fine-grained detrital material. Glacier input is important at high latitudes. The main factors controlling the flux of sediment derived from continental erosion are climate, precipitation, type of weathering, character of the coarse-grained material, topography, and land area in the source regions. When sea-level was low, such as during glacial periods, deep-sea terrigenous sedimentation was especially dynamic. At such times, the mechanical erosion of continents and the sediment loads of rivers were much greater. Much of the continental shelves were exposed as coastal plains, resulting in rivers that transported their loads to the outer edge of the continental shelf for more rapid deposition into the deep-sea basins.

Pelagic sediments are typically dominated by biogenic material, but vary considerably with latitude and water depth[20]. They include the deep-sea calcareous and siliceous oozes, composed largely

8.22

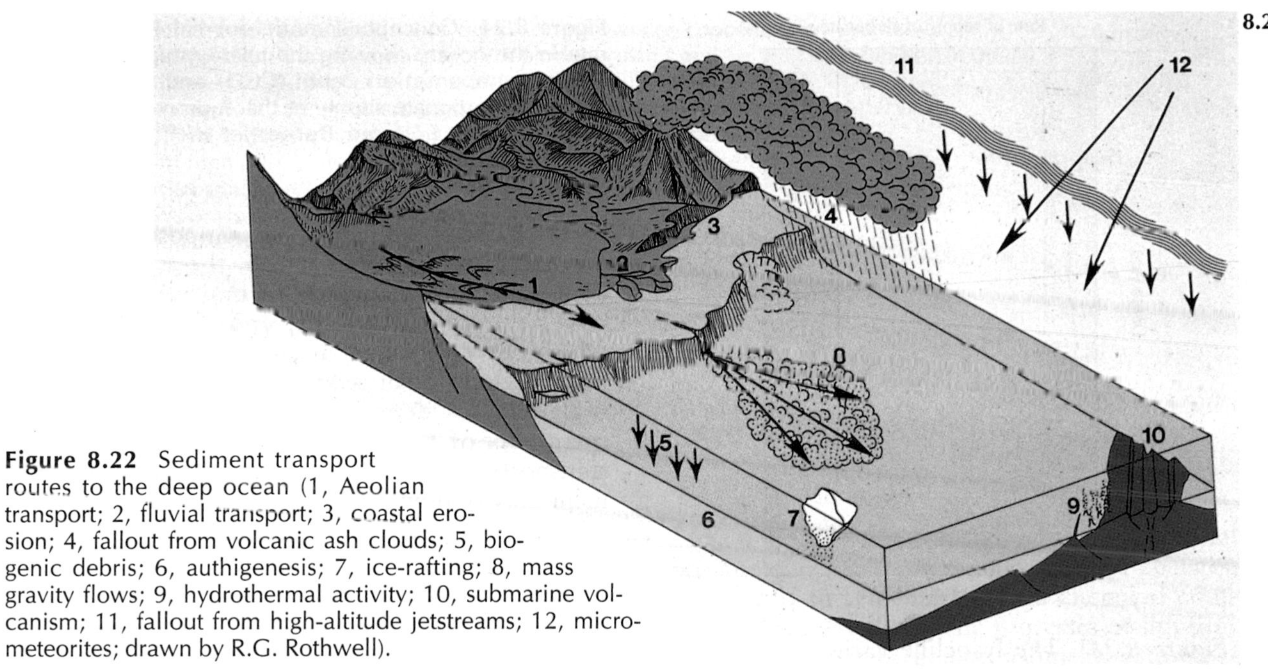

Figure 8.22 Sediment transport routes to the deep ocean (1, Aeolian transport; 2, fluvial transport; 3, coastal erosion; 4, fallout from volcanic ash clouds; 5, biogenic debris; 6, authigenesis; 7, ice-rafting; 8, mass gravity flows; 9, hydrothermal activity; 10, submarine volcanism; 11, fallout from high-altitude jetstreams; 12, micrometeorites; drawn by R.G. Rothwell).

of the remains of planktonic organisms, and deep-water clays (*Figure 8.23*). Four main processes control the character of biogenic oozes: the supply of biogenic material, its dissolution in the water column, its dilution by nonbiogenic material, and subsequent diagenetic alteration (see later). Pelagic sedimentation can be viewed as a form of interaction between the near-surface ocean and the deep ocean. Locally, these two distinct environments interact through wind-driven upwelling of deep water, and the consequent downwelling of near-surface water, and through the constant 'rain' of skeletons from dead planktonic organisms (*Figure 8.23*). This 'rain' of biogenic particulate matter forms the primary sink in the ocean basins (Chapter 7).

The spatial distribution of calcareous oozes is controlled by depth due to dissolution (Chapter 11). Calcite, which forms the main skeletal material of many planktonic organisms (such as foraminifera and coccolithophores), shows increasing solubility with water depth; this is related to increased hydrostatic pressure, increasing CO_2 content within the water, and decreasing temperature. Therefore, there is a depth, called the lysocline, that separates well-preserved from poorly preserved, solution-etched foraminifera and coccolithophores

8.23

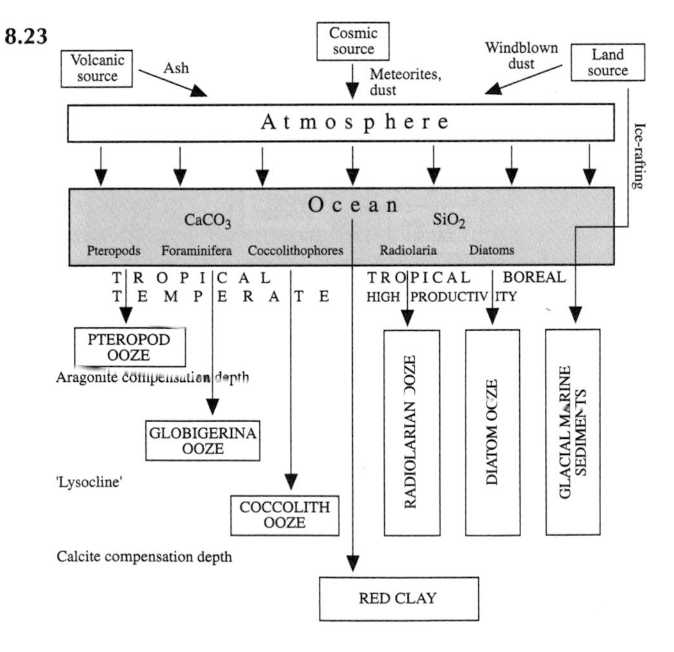

Figure 8.23 Sources and pathways of pelagic sedimentation in the oceans (after Hay[17]).

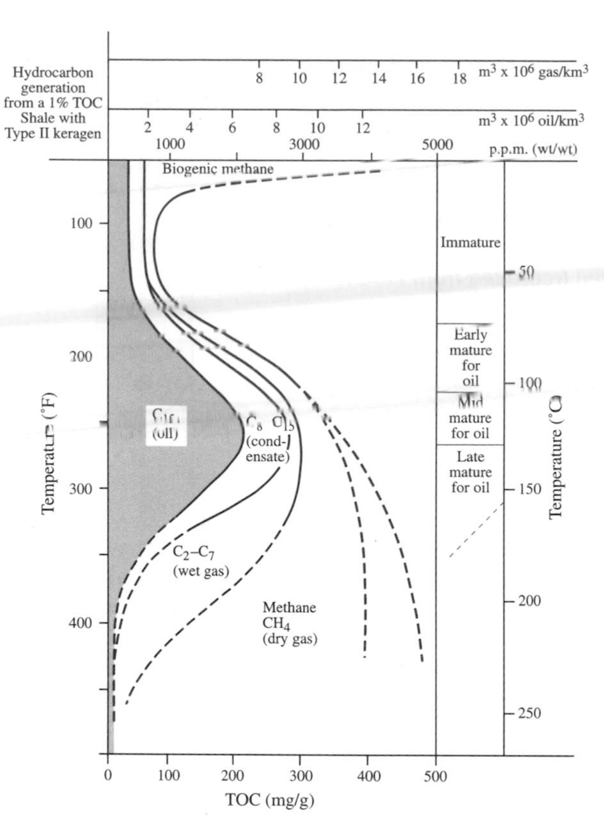

Figure 8.27 Calculating the volume of hydrocarbons generated from a given source-rock unit. The yield of hydrocarbons generated per 1% total organic carbon (TOC) in the source rock is indicated on the horizontal axes in p.p.m. (wt/wt) and other units. As shown, C_8–C_{15} and C_{15+} hydrocarbons (the major components of a typical North Sea oil) are generated in large quantities from 80–130°C, while over this temperature range the light hydrocarbons (C_2–C_7 and methane) are present in relatively small quantities. Once heavy hydrocarbon generation has ceased, at about 130°C, a presumed cracking reaction takes over, increasing the yield of the C_2–C_7 fraction and CH_4 at the expense of heavier hydrocarbons (from Brooks *et al.*[8], reproduced with permission).

the Deep Sea Drilling Project and the Ocean Drilling Program from the Atlantic and some parts of the Pacific, commonly contain organic-rich black shales, testifying to periods of probably quite brief, but widespread, anoxia. The black shales may have been caused by the lack of a regular supply of cold, dense, well-oxygenated water to the deep oceans, due to the absence of ice-caps, and by abundant biological production encouraged by the warm conditions and extensive continental shelves of the time. Recently, it has been suggested that widespread volcanism during the Cretaceous may have played a role in causing contemporary deep-

Box 8.3 Oxygen Isotope Stratigraphy

Oxygen has three stable isotopes (^{16}O, ^{17}O, and ^{18}O) with atomic mass numbers of 16, 17, and 18:

- ^{16}O makes up 99.763% of natural oxygen.
- ^{17}O makes up 0.033% of natural oxygen.
- ^{18}O makes up 0.204% of natural oxygen.

Oxygen makes up 90% of water by weight; the ^{16}O isotope is lighter than the ^{18}O isotope. Therefore, ^{16}O is preferentially evaporated relative to ^{18}O.

During glacial periods, ^{16}O-enriched water vapour is precipitated as snow which builds up to form glacier ice and ice caps. This ice is relatively depleted in ^{18}O. The oceans, however, become relatively enriched in ^{18}O, because of evaporation of ^{16}O-enriched water vapour. The larger the ice caps, the larger the proportion of ^{16}O removed from seawater, and the more the ^{18}O:^{16}O ratio of the sea water increases.

Marine organisms, such as foraminifera, which form skeletons or tests of calcium carbonate, incorporate different proportions of ^{16}O and ^{18}O from the water, according to the temperature; but, more importantly, according to the background ratio of ^{18}O:^{16}O in the sea water, which reflects global ice volumes. Measurements of the small differences in the ^{18}O:^{16}O ratio in different samples using a mass spectrometer allow the sequence and age of warm and cold conditions to be determined.

The ^{18}O:^{16}O ratio of foraminifera, especially benthic species which live in low-temperature bottom water (and hence are not affected by temperature changes), can therefore be taken as a measure of the amount of water held in ice sheets at any given time, and hence also as an indicator of global sea level.

Figure 8.28 Combined plot of global production of oceanic crust, high latitude sea-surface palaeotemperatures, long-term eustatic sea-level, black-shale deposition, and rate of production of the world's oil resources against geological time (from Late Jurassic to Pleistocene; *Figure 1* from Larson[21], by permission of the author).

8.28

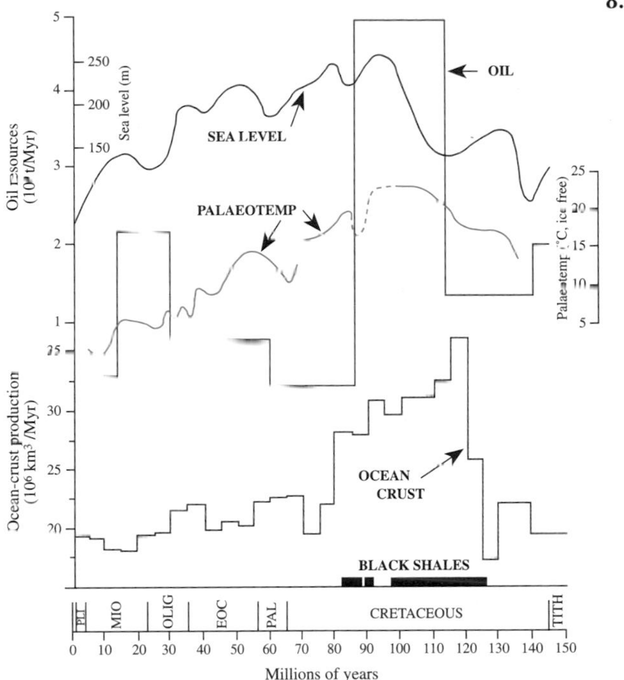

sea anoxia. The resulting increase in outgassing of mantle volatiles and CO_2 and increased ocean crust production (*Figure 8.28*) at this time may have resulted in an enlarged supply of nutrients and carbon to the ocean. The 'greenhouse' effect of increased CO_2 in the atmosphere led to relatively high sea-surface temperatures. The resultant explosion in productivity led to vast increases of organic carbon in the marine system.

During the Cenozoic, the CCD fluctuated widely (*Figure 8.29*), which may have been related partly to changes in sea level. Continental shelves, being shallow, are favourable places for carbonate accumulation in times of high sea-level and may, therefore, act as carbonate traps, thereby removing $CaCO_3$ from the oceanic chemical cycle. However, changes in productivity also lead to changes in the CCD, since biogenic production of $CaCO_3$ skeletons lowers saturation. Therefore, fluctuations in the CCD may possibly reflect productivity fluctuations too.

Oxygen isotope studies (*Box 8.3*) of the skeletons of benthic foraminifera have shown a general cooling trend in the oceans since the Cretaceous. Palaeoceanographic changes, caused by plate motions over the same period, led to greater partitioning of the world ocean system with time and played a major role in causing this trend. Palaeoceanographic studies show that polar cooling was particularly pronounced toward the end of the Eocene (*Figure 8.28*), presumably due to the thermal isolation of the Arctic and Antarctic oceans from the rest of the world ocean, but also perhaps due to albedo changes resulting from changing vegetation and snow cover. This led to the equatorward shifting of climatic belts and, on high latitude shelves, to the cooling of water, which became cold and dense enough to sink and fill the deep ocean basins. The late Eocene cooling therefore resulted in a new type of world ocean – one characterised by the development of marked contrasts between high and low latitudes, between different oceans, and between the deep sea and the ocean margins. Large amounts of ice-rafted debris in sediments around Antarctica since the middle Miocene testify to the build up of the Antarctic ice cap. The microfossil record, particularly of siliceous types, indicates that fundamental changes in deep-water circulation were occurring in the Miocene, concurrently with the build up of Antarctic ice and the world-wide cooling of abyssal waters. Subsequent changes in palaeogeography, particularly the northward drift of land masses and possibly the closing of the Panama seaway (about the middle Pliocene), reinforced by other mechanisms, such as mountain building, resulted in the onset of northern hemisphere glaciation 2.5–3.5 Myr ago and in the oceans we know today.

8.29

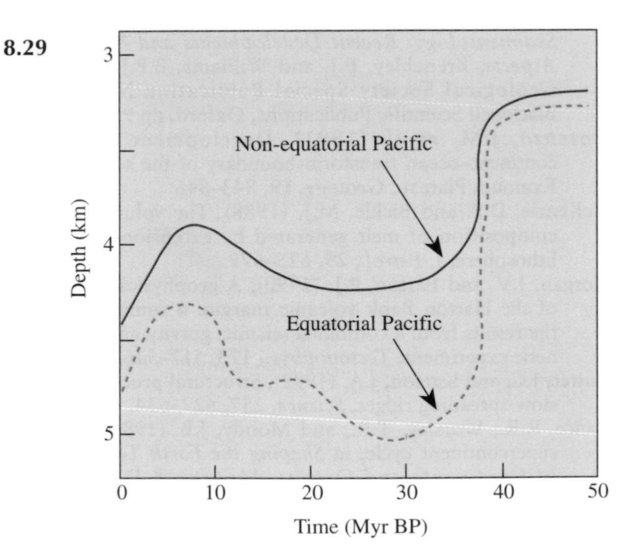

Figure 8.29 Reconstructions of past fluctuations in calcite compensation depth in the Pacific Ocean (reconstruction from Van Andel[37]; after Seibold and Berger[34]).

CHAPTER 9:

Slides, Debris Flows, and Turbidity Currents

D.G. Masson, N.H. Kenyon, and P.P.E. Weaver

Introduction

Gravity-driven flows, in a variety of forms ranging from turbulent suspensions to coherent sliding masses, are the major agents of downslope sediment transport in the deep sea. They sculpt the continental slopes into complex shapes, carry land-derived sediment into the deep ocean basins, and redistribute biogenic sediment on a vast scale. Slope failures and resultant flows are often near-instantaneous events, capable of the destruction of marine installations and submarine telecommunications cables and, in some extreme cases, of generating deadly tsunamis. In ancient rocks, sand bodies once deposited by gravity flows, such as the sands found in submarine sediment fans, are a major reservoir facies for oil and gas, and have considerable economic importance.

The study of gravity flows in the deep ocean has progressed rapidly since the early 1950s. Perhaps the best-known study is the analysis of cable breaks caused by the turbidity current (a sediment-laden flow) associated with the 1929 Grand Banks earthquake, from which the first velocity estimate for a turbidity current was produced[5] (*Figure 9.1*); this is discussed later. Understanding of downslope sediment transport processes has developed, not only through studies in the modern ocean, but also through studies of ancient marine sequences now exposed on land, and through experimental work in the laboratory. Important contributions include the concept of sequential deposition of fining-upward sediment in individual turbidites (the sedimentary layers or deposits laid down by turbidity currents[1]), and the comprehensive theoretical analysis of turbidity current flow and turbidite deposition[12–14]. The development of seismic profiling and side-scan sonar equipment over the past 40 years (see Chapter 20) has revolutionised the way in which we analyse the sea floor, leading to a new appreciation of the extent and importance of gravity flows and their deposits. The discovery of huge sediment slides on the flanks of the Hawaiian

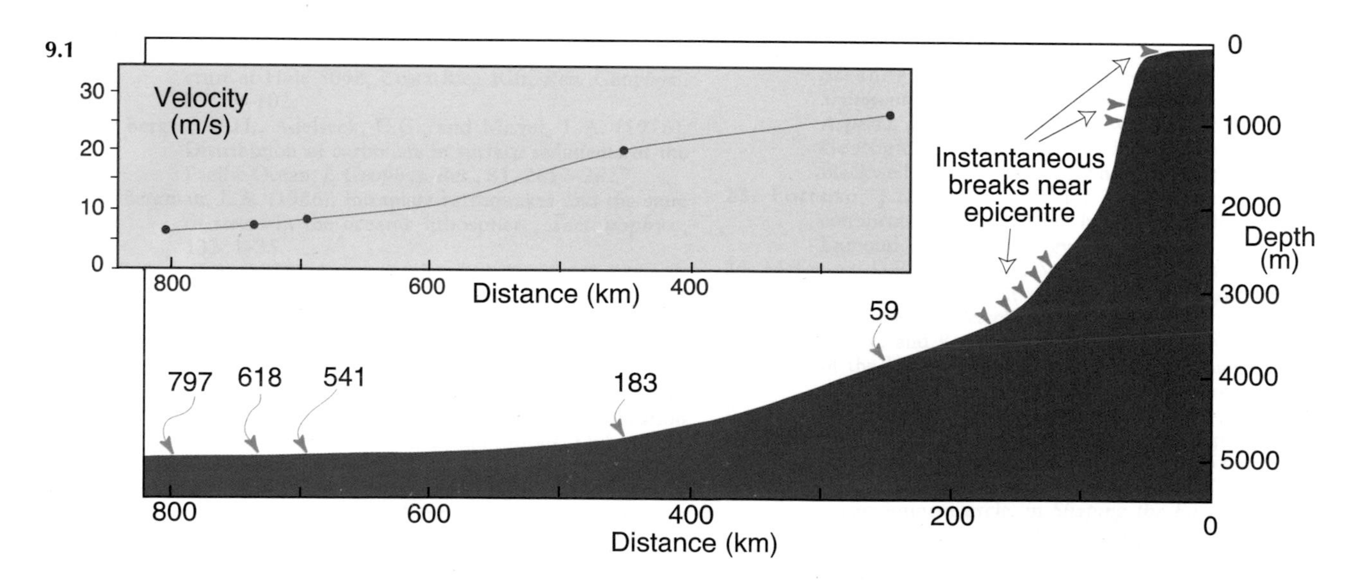

Figure 9.1 Cross-section through the continental slope and rise south of Newfoundland in the area affected by the 1929 Grand Banks earthquake and turbidity current. Green arrows mark the positions of cable breaks, with the time after the earthquake in minutes. Inset shows the turbidity current velocity as deduced from the timing of cable breaks (redrawn from Heezen and Ewing[5]).

Table 9.1. Statistics of some major slides, debris flows, and turbidity currents.[a]

Name/Location	*Waterdepth (m)*	*Area (km²)*	*Length (km)*	*Thickness (m)*	*Volume (km³)*	*Slope*
Nuuanu Slide (Hawaii)[b]	0–4600	23,000	230	up to 2000	5000	?5→–0.1°
Storegga Slide/Debris Flow[c]	150–3000	112,500	850	up to 430	5580	?1.5→0.05°
Saharan Slide/Debris Flow	1700–4800	48,000	700	5–40	600	1.5→0.1°
Canary Debris Flow	4000–5400	40,000	600	up to 20	400	1→0°
f turbidite (Madeira Abyssal Plain)	?–5400	>60,000	1000+	up to 5	190	?0.2° average
1929 Grand Banks Turbidite	600–6000	160,000	1100	? up to 3	185	?→0.01°

[a] Turbidite areas are those covered by deposit only, slide and debris flow areas include scar and deposit.
[b] Nuuanu Slide flowed uphill for final 140 km.
[c] total of three slide events.

Islands using the GLORIA long-range side-scan sonar (see Chapter 19) is one example of the application of these technical advances[15].

Classification of Gravity-Driven Sediment Flows

Gravity-driven sediment transport includes a wide variety of processes, such as slumping, sliding, debris flow, grain flow, and turbidity currents. However, sediment slides, debris flows, and turbidity currents are the three major gravity-driven processes which transport significant volumes of sediment over large distances in the deep ocean.

A slide is defined as the movement of an upper layer on a basal failure surface. It can result in the downslope transport of large coherent blocks of material, with internal deformation ranging from negligible to severe.

Debris flow has been described as the movement of granular solids, sometimes mixed with minor amounts of entrained water (or, on land, air) on a low slope. A common and effective analogue is with the movement of wet concrete.

A turbidity current is a type of gravity or density current driven by gravitational buoyancy forces resulting from the difference in density between two fluids. To the geologist, it is the downslope flow, under the influence of gravity, of a suspension of sediment in water. The sediment particles, kept in suspension primarily by turbulence, provide the excess density which drives the flow.

Size and Scale

The largest slope failures on earth occur around the margins of and in the ocean basins. This is a consequence of the relief and shape of ocean basins (see Chapter 8), as well as the huge quantities of unconsolidated or partially consolidated sediment which occur on smoothly sloping continental margins, often under geotechnical conditions only marginally in favour of slope stability. Should failure occur, the ocean floor offers unimpeded slopes and flat-floored basins hundreds of kilometres in length, allowing flow over enormous distances.

Individual sediment slides and debris flows can involve many thousands of cubic kilometres of material (*Table 9.1*). The largest of the huge slides on the flanks of the Hawaiian Islands is up to 2 km in thickness, with a volume of 5000 km³. A volume as great as 20,000 km³ has been ascribed to the Agulhas Slide, off South Africa, but the available evidence perhaps suggests that this is a complex of failures rather than a single gigantic event. The Storegga Slide[2], off Norway, and the Canary and Saharan Debris Flows[11], off West Africa, all have runout distances of 600–800 km, much of this on slopes less than 0.5°.

The volume of sediment carried by the largest known turbidity currents is an order of magnitude less than that of the largest sediment slides and debris flows. Individual turbidites of 100–200 km³ are known from several abyssal plain basins in the Atlantic. However, transport distances can be

9.10

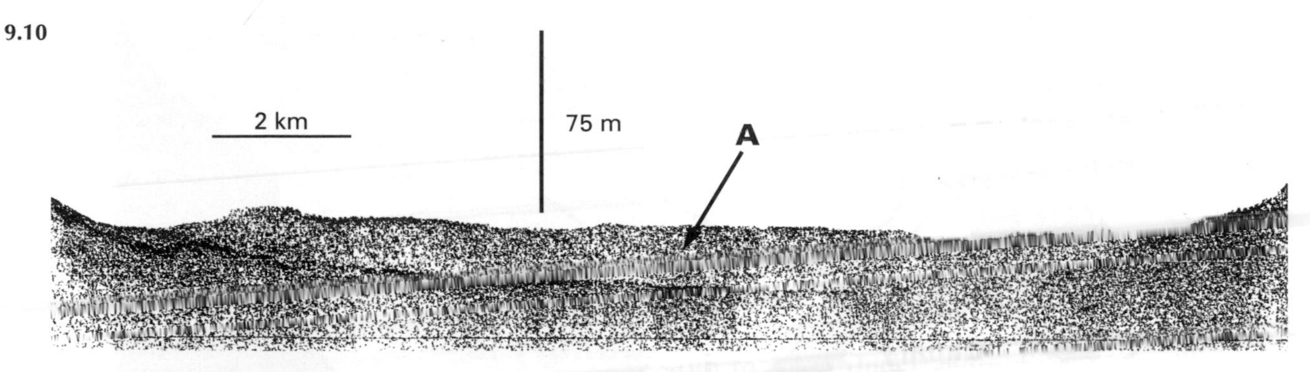

Figure 9.10 3.5 kHz high-resolution profile across the Saharan debris flow deposit (A), immediately southwest of the western Canary Islands (see *Figure 9.9*), showing the typical expression of debris flow deposits (for survey methods see Chapter 20). In this location, the 25 m thick debris flow deposit sits within and partially fills a broad channel across which the profile has been taken.

slope and upper continental rise. The sea-floor gradient decreases downslope from about 1.5° in the source area to as little as 0.1° near the end of the flow. The debris flow incorporated around 600 km^3 of sediment. Its failure scar is bounded by a complex scarp 20–80 m in relief. Southwest of the Canaries, on the continental slope below 4000 m water depth, the debris flow deposit forms a narrow tongue about 25 km wide, ranging in thickness from 5–40 m (*Figure 9.10*). In this area, high-resolution side-scan sonar data show spectacular images of flow banding, longitudinal shears, lateral ridges, and transported blocks (*Figures 9.6* and *9.11*).

The Canary Debris Flow originated on the western slopes of the Canary Islands, at about 4000 m water depth. It produced a relatively broad (60–100 km wide), but thin (usually <20 m thick) debris sheet, which extends for 600 km from the source area to the edge of the Madeira Abyssal Plain. This sheet has an average thickness of 10 m, a volume of about 400 km^3, and covers an area of 40,000 km^2. Gradients decrease from 1° in the source area to effectively 0° at the edge of the abyssal plain. The Canary Debris Flow has a complex outline, which appears to have been strongly influenced by even the gentlest topography, particularly at its distal end, where very subtle topographic lows (e.g., pre-existing shallow channels)

9.11

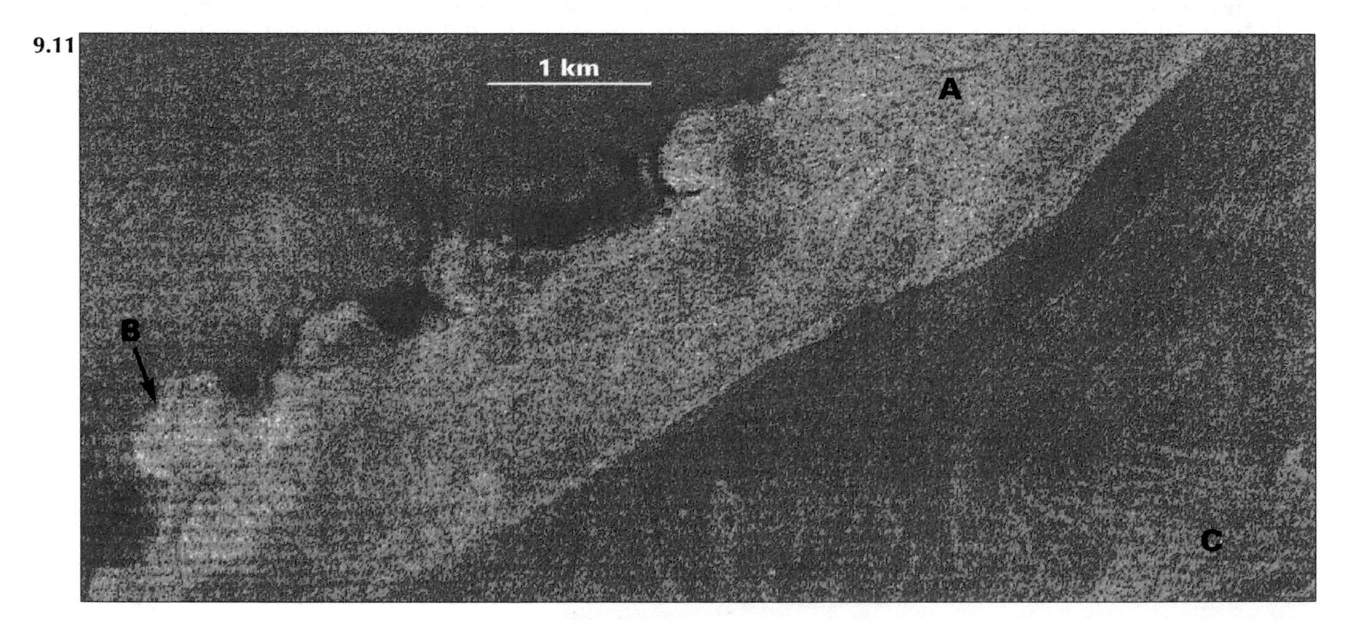

Figure 9.11 False-colour TOBI 30 kHz side-scan sonar image showing the edge of the Saharan debris flow deposit southwest of the western Canary Islands (see *Figure 9.9*). Blue is low back-scatter, and yellow is high back-scatter. The lateral ridge (A), which has a relief of about 5 m, is believed to comprise chaotic rubble deposited along the edge of the flow (B). It is separated from the main part of the flow deposit by a distinct longitudinal shear. The main flow has a characteristic 'woodgrain' fabric (C), which may be evidence for drawing out of the debris into a flow-parallel banding.

clearly control the path of narrow tongues of debris. The head of the Canary Debris Flow is somewhat unusual because no clear headwall scarp is present. Instead, there is a broad zone of apparent shallow rotational faults some 30 km in width. The sediment surface between the faults shows disruption increasing downslope until a featureless, apparently completely homogenised facies is reached. Within this facies, rafted blocks of undisturbed sediments up to 5 km across are seen. Shears within the debris surrounding the blocks, extensional depressions adjacent to their downslope margins, and trails of fragments behind some blocks all suggest that the blocks moved more slowly than the bulk of the flow, presumably because they were in contact with, and dragging on, the underlying sea floor.

Storegga Slides

Some of the world's largest known catastrophic earth movements are found beneath the sea off the heavily populated coasts of northwestern Europe[2]. They are in an area of the mid-Norwegian margin known as the Storegga ('great edge'), because the 290 km wide headwall of a submarine slide forms the top of the continental slope.

The last major slide event occurred about 7000 years ago. It involved erosion of up to 300 m thickness of slope sediments and displacement of about 1700 km^3 of material. The deposits include very blocky slide deposits, containing some huge, largely intact slabs up to 10 km x 30 km in plan view and 200 m thick. The blocky nature is due to the failure having cut down into more consolidated sediments. A very thick (up to 20 m) homogeneous fine-grained turbidite covers the entire deep basin of the Norwegian Sea. It is believed to be related to this latest slide event. Probable tsunami deposits on the coasts of Scotland and Norway are also believed to be related to the latest slide. They confirm the age of about 7000 years.

An earlier slide affected an even wider area and displaced about 4000 km^3 of sediment. It involved shallower, less consolidated sediments and the resulting deposits are less blocky. This event occurred about 30,000 years ago. Two older slide events are now known to have occurred in the region and there were probably others whose scars have been removed by later erosion, but whose deposits may be present in the deep basin.

The slides are believed to have been triggered by earthquakes and the decomposition of gas hydrates (see later). Gas hydrates have been recognised on seismic profiles from the slope just to the north of the slide.

Turbidity Currents

Processes

Turbidity currents may be generated directly from sediment suspensions delivered to the shelf edge by agencies such as tidal currents, rivers, or storms. Many others appear to have their origins in sediment failures on the continental slope, although the actual initiation mechanisms remain poorly understood. One possibility is that they evolve from debris flows, in effect continuing the process of disintegration which first led from slide to debris flow. This evolution requires the dilution of the flow by incorporation of water and a transition from laminar to turbulent flow. One elegant mechanism, which has been demonstrated in flume experiments, is the generation of turbidity currents by erosion of the steep snout of a debris flow as it moves downslope[4]. An alternative mechanism involves mixing of water into the body of the flow, perhaps due to internal flow turbulence. This latter process is attractive because it offers a mechanism for transforming entire debris flows into turbidity currents. However, it has proved difficult to reproduce in experiments, and its occurrence in nature remains hypothetical.

Proving turbidite–debris flow relationships in the modern ocean basins is difficult, the main problem being the collection of appropriate samples from deposits which may be spread over thousands of square kilometres. One set of cores, collected across the snout of the Canary Debris Flow deposit at the edge of the Madeira Abyssal Plain, does, however, conclusively demonstrate such a relationship (*Figures 9.12* and *9.13*). Here, the debris flow actually occurs within the turbidite, interrupting the latter's fining upward depositional sequence. It appears that the faster moving turbidity current began depositing sediment at the edge of the plain, perhaps a few hours before the arrival of the debris flow, which then buried the lower turbidite layers. Deposition of the finer fraction of the turbidite then continued on top of the debris deposit. A wider study of this debris flow–turbidity current pairing indicates that the two phases have markedly different (although overlapping) depositional patterns, and that they clearly had divergent paths as they crossed the lower continental slope and rise[10]. This strongly suggests that the debris flow did not continuously spawn a turbidity current as it moved downslope, but that the two phases evolved in or near the source area and then travelled independently.

The flow characteristics of large turbidity currents in the modern ocean have not been directly observed. Flow models are based on indirect obser-

9.12

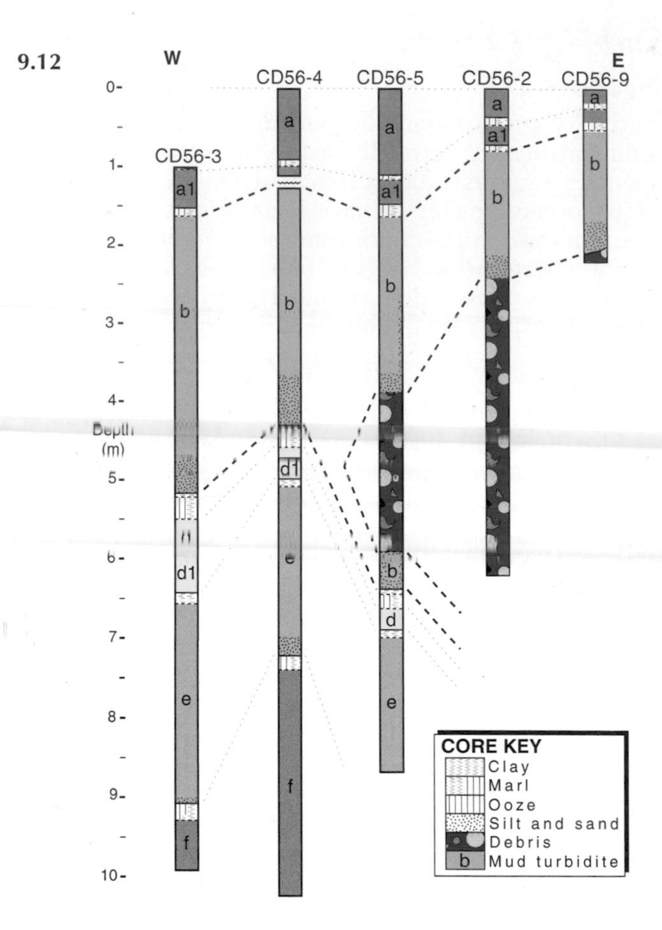

Figure 9.12 Diagrammatic logs of a transect of cores (for a description of coring devices, see Chapter 19) across the snout of the Canary debris flow deposit, showing the relationship between the debris flow and a coeval turbidite, identified as 'b' in the Madeira Abyssal Plain turbidite sequence (see *Figure 9.15*). Lettered units are turbidites which can be correlated across the Madeira Abyssal Plain. The feather edge of the debris flow deposit occurs within 'b', indicating that the two flow phases must be part of the same event (see text for a more detailed explanation).

vations of both the flow (e.g., cable breaks, *Figure 9.1*) and the structures it leaves along its path (e.g., channels, levees; see later), geological evidence from ancient rock sequences (*Figure 9.14*), and flume experiments. The available evidence suggests that flows range widely in concentration, thickness, turbulence, and velocity, although some of these parameters are obviously linked. For example, one model for the emplacement of the thick, ungraded mud turbidites found on abyssal plains is based on thin (<20 m thick), high concentration (50–100 kg/m^3) and low velocity (<1 m/s) flows, which, as they approach their point of deposition, are almost nonturbulent. In contrast, a model for thin, fine grained turbidites, such as those which characterise channel overbank sequences, indicates thick (up to several hundred metres), low density ($\approx$ 2 kg/m^3) flows, but with similar velocities. Overall, velocities are known to vary from <1 m/s to at least 25 m/s.

The classic turbidite, described in minute detail from ancient rocks, is a fining-upward sequence with grain size ranging from sand to mud. Turbidites are usually described in terms of the Bouma sequence[1], a five-fold division based on sedimentary structures and grain size. Using flume experiments, the progression of sedimentary structures has been shown to relate to the progression of bedforms seen under a decelerating flow. In nature, few turbidites exhibit a complete Bouma sequence. For example, the thick fine-grained turbidites which characterise many abyssal plains may consist entirely of only the upper two Bouma divisions.

Case studies

The 1929 Grand Banks turbidity current

In November 1929, a magnitude 7.2 earthquake occurred beneath the upper continental slope just south of the Grand Banks of Newfoundland. Six submarine telephone cables in the immediate vicinity of the epicentre were broken instantaneously and a further six, in an orderly downslope sequence, over the next 13 hours 20 minutes (*Figure 9.1*). In attempts to explain these observations, various theories, based on sea-floor faulting or movement of

9.13

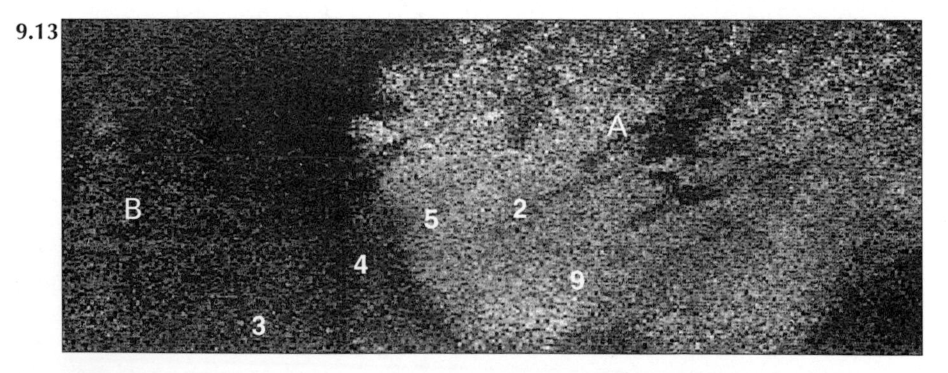

Figure 9.13 Gloria 6.5 kHz side-scan sonar image of the Canary debris flow snout. The chaotic nature of the debris-flow material (A) generates a high level of back-scatter (light tones) and ensures a strong acoustic contrast with the flat (low back-scatter) abyssal plain (B). Numbers show locations of the CD56 cores shown in *Figure 9.12*.

Figure 9.14 Cyclical turbidite sequences believed to be characteristic of sandy fan lobes (reproduced from E. Mutti's *Turbidite Sandstones*[16], by permission of AGIP, Milan, Italy).

sediment leaving sections of cable unsupported, were proposed during the next 20 years. None, however, satisfactorily explained the orderly sequence of cable breaks, the fact that substantial sections of cables were buried (but only on the deeper, less steep, area of the continental slope), or the lack of damage to cables on the continental shelf. The hypothesis that the cables were broken by a turbidity current is the only explanation for this combination of observations, as was realised by Heezen and Ewing[5] in their classic paper. Subsequent sampling in the Sohm Abyssal Plain to the south of the cable break area proved the existence of a basin-wide turbidite, underlain by Holocene sediments, at the sea floor. This turbidite covers an area of some 160,000 km^2, has a maximum thickness in excess of 3 m, and a volume of about 185 km^3.

From the timing of the cable breaks, Heezen and Ewing realised that it was possible to calculate the velocity of the turbidity current. They estimated a maximum velocity of about 25 m/s on the continental slope, decreasing, over a distance of about 500 km, to about 6 m/s at the edge of the abyssal plain. Lack of precise knowledge of the point of origin of the turbidity current, within the general source area, casts some doubt on their maximum value. Most authorities, however, agree that velocities of at least 15–20 m/s were attained. Similar velocities have since been calculated for turbidity currents associated with both the 1954 Orleansville (Algeria) earthquake and the 1979 slope failure off Nice (southern France).

The Madeira Abyssal Plain

The turbidites of the Madeira Abyssal Plain (for location, see *Figure 9.9*), the deepest part of the Canary Basin off northwest Africa, are probably the best-studied in the modern ocean basins[21]. This turbidite sequence has been created by enormous turbidity currents which carry sediment from the African continental margin, often over distances in excess of 1000 km. On the plain, these currents become 'ponded' by the surrounding higher topography and deposit their sediment load. Indeed, the flat plain results from the stacking of numerous turbidites on top of each other, forming a 350 m thick layer which has levelled off the otherwise irregular topography. The upper 35 m of this sequence, corresponding to the past 750,000 years, has been sampled, allowing its depositional history to be determined.

During that time, a turbidity current reached the Madeira Abyssal Plain, on average, once every 30,000 years. Most occurred during periods of rapidly changing sealevel (both rises and falls) associated with Pleistocene glacial cycles (*Figure 9.15*). The turbidity currents derive from the northwest African continental slope, from the flanks of the Canary Islands, and, occasionally, from seamounts to the west of the plain.

The turbidites deposited on the abyssal plain range in volume from a few cubic kilometres to almost 200 km^3, with those over a few tens of cubic kilometres forming layers over the whole plain. They consist predominantly of fine-grained mud; any sand present tends to be deposited at the break of slope at the edge of the plain (*Figure 9.16*). Much of the coarser sediment is transported across the continental rise through deep-sea channels, while, in contrast, the finer material appears to move as an unconfined sheet flow. The incoming turbidity currents cause no significant erosion of the underlying sediments, and successive turbidites

9.15

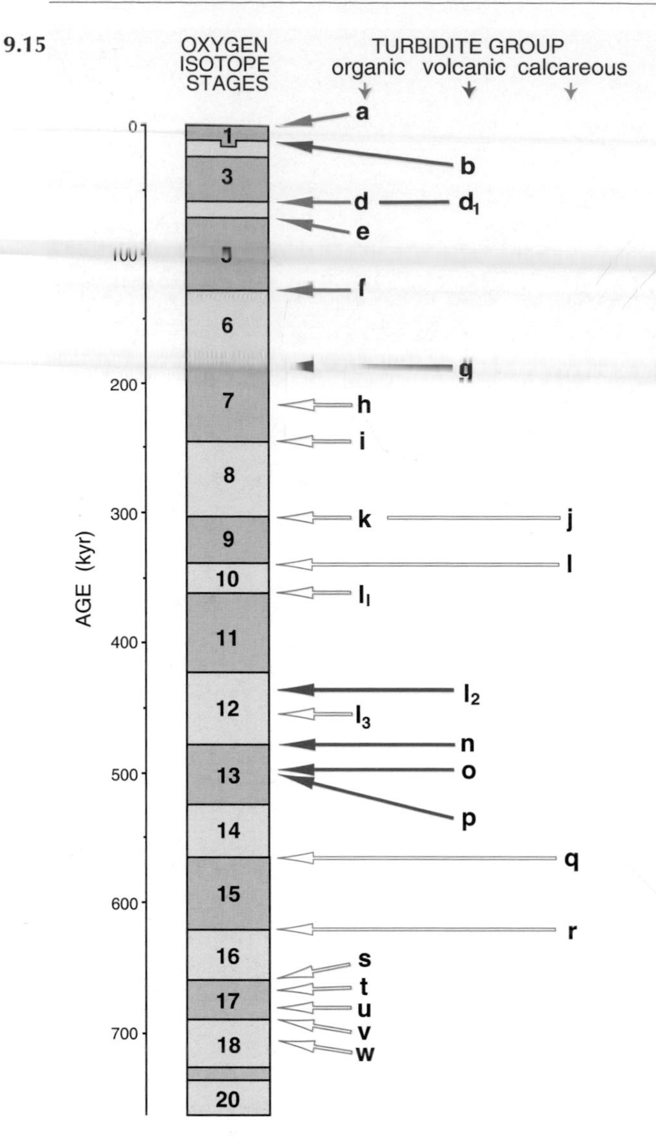

Figure 9.15 Summary of turbidite emplacement on the Madeira Abyssal Plain, showing the ages of individual turbidites (each identified by a letter) and turbidite groups based on source area (organic turbidites – green arrows – from the northwest African continental margin, volcanic – red arrows – from the Canary Islands and Madeira, and calcareous – blue arrows – from seamounts to the west of the plain). These groups can be further subdivided into turbidites from north (filled arrows) and south (open arrows) of the Canaries. Ages of turbidites have been determined relative to the oxygen isotope time-scale[18] (blue and even numbers are glacial periods, brown and odd numbers are interglacials). Note the strong correlation of turbidite emplacement with oxygen isotope stage boundaries, suggesting a relationship between turbidity currents and changing sea level.

are separated by sediment layers built up by a slow pelagic rain of biogenic carbonate and wind-blown dust from the Sahara Desert. Fossils in this pelagic record allow us to date the turbidite sequence, and even to assign approximate ages to individual turbidity currents.

Turbidites such as those sampled on the Madeira Abyssal Plain (*Figures 9.15* and *9.16*), which originate as failures on sedimented slopes, contain a mixture of sediments (and microfossils) with an age range corresponding to that of the failed sediment mass. In theory, if the pelagic fossil record in the source area is well-known, it should be possible to estimate the age range and thus thickness of the original sediment failure. If, in addition, the volume of the resultant turbidite is known, then the area eroded to form the corresponding turbidity current can also be calculated. In the case of the Madeira Abyssal Plain turbidites, this theory has been put into practice, allowing a typical sediment failure on the northwest African margin, 1000 km away from the study area, to be described – this failure is a few

9.16

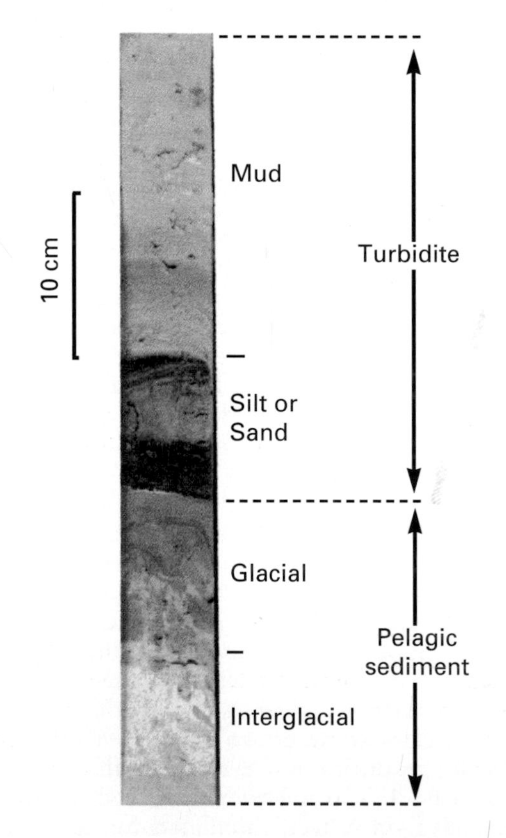

Figure 9.16 Core photograph of some typical Madeira Abyssal Plain sediments. The upper part consists of a turbidite with a black volcanic sand/silt basal unit and a brown mud top. The lower part consists of pelagic sediment, with the brown sediment marking deposition during a cold glacial climate and the white that during a warmer interglacial.

9.17a

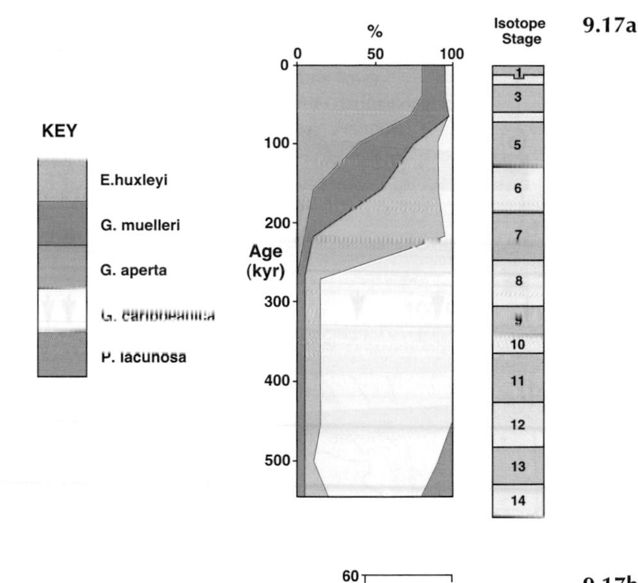

9.17b

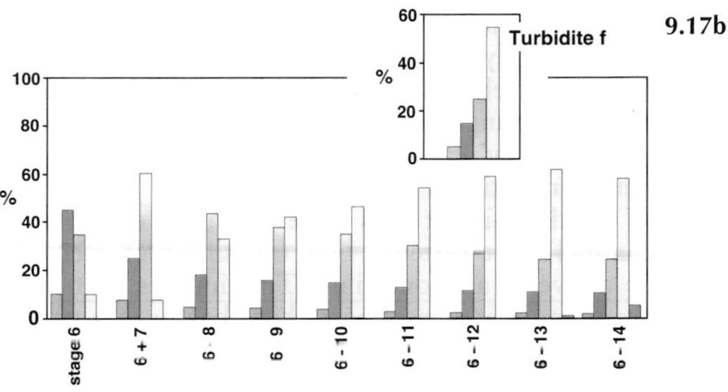

Figure 9.17 (a) The distribution, through time, of five key coccolith species occurring in pelagic sediment in the Madeira Abyssal Plain area. For sediments deposited during the most recent half million years, variation in the ratio of these species gives an age accurate to within one or two oxygen isotope stages, i.e. a few tens of thousands of years. (b) Erosion of sediment representing more than a few tens of thousand years produces coccolith mixtures not seen in the pelagic record, but dependent upon the age range of sediments which were eroded. Using the distribution of coccolith species through time (a), it is relatively simple to calculate what age range any observed coccolith mixture represents. In the example shown, for turbidite 'f' of the Madeira Abyssal Plain sequence (emplaced at the end of oxygen isotope stage 6), synthetic mixtures can be created for the erosion of stage 6 sediments only, 6 + 7, 6–8, and so on. By comparing these with the actual mixture found in 'f' (inset), it can be seen that 'f' contains sediments originally deposited during isotope stages 6–12, i.e. between 130,000 and 480,000 years ago. This corresponds to erosion of about 50 m of sediment from the source region of 'f' on the African margin south of the Canaries.

tens of metres thick, covers an area as great as 6000 km^2, and incorporates sediment with an age range of 50,000–500,000 years (*Figure 9.17*). An important additional observation based on this study is that none of the turbidites examined contains a significant excess of surface sediment, suggesting that, once formed, the turbidity currents which transported them were virtually nonerosional, and that they travelled many hundreds of kilometres in this state.

Turbidites and sediment fans

Turbidites and related deposits, such as debris flows and debris avalanches, have the greatest volume of any types of sediment in basin fills. The processes that form them concentrate the sands into bodies that are potential reservoirs for hydrocarbons. It is necessary for effective hydrocarbon exploration that the size, shape, and relationships of the bodies are known. These deposits usually lie in front of a subaerial and shelf-feeder system, such as a river drainage basin or a glacially carved cross-shelf trough. The largest, high-input, feeder systems, such as the Indus, Amazon, Mississippi, and Nile, can supply enough material to the deep sea to form bodies of sediment (sediment fans) that are 10 km or more thick. Glacial-fed submarine fans, like that in front of the Barents Sea, can be of an equivalent size. However, small fans fed through the almost ubiquitous submarine canyons which dissect the continental slopes are much more common. Submarine canyons can be thousands of metres deep, and are usually fed by submarine tributary systems with a hierarchy of gullies (*Figure 9.18*). The fans at the mouths of these canyons are

9.18

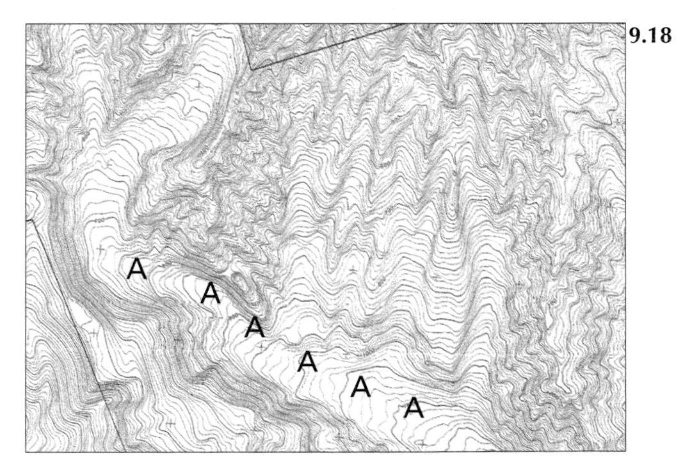

Figure 9.18 A hierarchy of tributary gullies feeding into the main, flat-floored Var Canyon (A), off Nice in the northwest Mediterranean Sea. The main canyon is up to 300 m deep; depth contours are in metres.

ocean ridge system are not yet well-understood[27]. Variations in the composition of volcanic material, speed of plate motion, and rate at which magma is fed to the ridge are certainly contributory. More significantly, we recognise that the localised coalescence of pillow basalt mounds and seamounts into linear ridges several kilometres in length is a fundamental mechanism in the generation of new crust. It is a major thrust of marine geoscientific endeavour in the 1990s to understand the relationship of these volcanic features, and their subsurface plumbing, to the occurrences of hot-fluid circulation systems (hydrothermal activity), and to predict their location along the world's spreading ridges.

Hydrothermal Circulation

One of the most exciting developments since the beginning of the study of oceanography was the discovery, in the late 1970s, that submarine hydrothermal vents were associated with the volcanically active zone at the crest of the mid-ocean ridge system (*Figure 10.1*). The base of the oceanic crust is extremely hot (>1000°C), yet its upper boundary is in contact with sediments and sea water at temperatures that are only a few degrees above 0°C. Since the earliest recognition of mid-ocean ridges and their significance to plate tectonics in the 1960s, geophysicists have known that conductive heat flow measured through young ocean crust could not account for all the heat lost at these tectonic spreading centres. So, they predicted that some alternative *convective* heat-transfer process must also occur near the crests of mid-ocean ridges[33].

It was not until the late 1970s that these predictions were proved correct, with the discovery of, first, low-temperature submarine hydrothermal activity (10–30°C) at the Galapagos Spreading Centre[3] and, second, high-temperature (350°C) hydrothermal activity on the East Pacific Rise[30]. The importance of these discoveries went much further than simply proving the geophysicists' theories correct. To marine geochemists, hydrothermal activity represented a new supply of chemicals to the oceans, comparable in importance to the influence of rivers flowing from the land[4]. For marine biologists, the discoveries were, perhaps, even more outstanding. Here, fed by the chemicals emanating from the vents, living in total darkness, and isolated from almost all other strands of evolution, were extraordinary, undescribed species. These animals were previously unsuspected and certainly not looked for, yet were soon recognised to be living in an entirely novel ecosystem whose origins could be traced back to the earliest life on Earth (see also Chapter 13).

The pattern of hydrothermal circulation is one in which sea water percolates downward through fractured ocean crust toward the base of the oceanic crust and, in some cases, close to molten magma. In these hot rocks, the sea water is first heated before it reacts chemically with the surrounding host basalt. As it is heated, the water expands and its viscosity reduces. If these processes occurred on land, at atmospheric pressure, catastrophic explosions would result as temperatures would rise above 100°C and the water would turn into steam. But, because mid-ocean ridges lie under 2000–4000 m of sea water, at pressures 200–400 times greater than atmospheric pressure, the reacting sea water reaches temperatures up to 350–400°C without boiling. At these temperatures, the altered fluids do become extremely buoyant, however, with densities only about two-thirds that of the downwelling sea water; thus, they rise rapidly back to the surface as hydrothermal fluids. The movement of the fluid through the rock is such that, while the downward flow proceeds by gradual percolation over a wide area, the consequent upflow is often much more rapid and tends to be focussed into natural channels emerging at 'vents' on the sea floor.

Beneath the sea floor, the reactions between sea water and fresh basalt remove the dissolved Mg^{2+} and SO_4^{2-} ions that are typically abundant in sea water, resulting in the precipitation of a number of sulphate and clay minerals. As the water seeps lower into the crust and the temperature rises, metals, silica, and sulphide are all leached from the rock to replace the original Mg^{2+} and SO_4^{2-} ions. The hot and, by now, metal-rich and sulphide-bearing fluids then ascend rapidly through the ocean crust to the sea floor[1]. As soon as they begin to mix with the ambient, cold, alkaline, well-oxygenated deep-ocean waters there is an instantaneous precipitation of a cloud of tiny metal-rich sulphide and oxide mineral grains[6]. These rise within the ascending columns of hot water, giving the impression of smoke. Precipitation around the mouths of the vents over time builds chimneys through which the smoke pours, hence the term 'black smokers' (*Figure 10.8*); hot water gushes out of these tall chimney-like sulphide spires at temperatures of ca 350°C and at velocities of 1–5 m/s. Upon eruption, this hydrothermal fluid continues to rise several hundred metres above the sea bed, mixing with ordinary sea water all the time, in a buoyant turbulent plume (see later).

At slightly lower temperatures (below 330°C), the fluid may cool and mix with sea water sufficiently to deposit some metal-rich precipitates in the walls of the channels up which the fluids rise, before reaching the surface. In such cases, the particulate material formed when the ascending hot water finally emerges at the sea bed is made up predominantly of amorphous silica and various sul-

Figure 10.8 Black smokers on the sea floor of the East Pacific Rise at approximately 2600 m depth, near 21°N. Individual chimneys measure approximately 30–40 cm tall and 10 cm across, and are seen venting hydrothermal fluids at temperatures of 350–400°C and velocities of 1–5 m/s. The fluid within the chimneys is extremely clear, but rich in dissolved metals and hydrogen sulphide. As soon as this fluid mixes with cold oxygen-rich sea water, at the very mouths of the vents, precipitation of a range of sulphide and oxide minerals occurs, giving rise to the clouds of tiny black 'smoke' particles, which are seen billowing upward above the chimneys into the overlying sea water. A previously active vent chimney, which has been cemented solid by mineral precipitation, is at the extreme right. The light, angular area to the top of the structure represents the relatively fresh internal composition of the dark grey, almost cylindrical, chimney structure, where it has been broken open using the robot arm of the submersible (photograph courtesy of Woods Hole Oceanographic Institution and the American Geophysical Union[17]).

10.8

phate and oxide minerals, yielding a white cloud of mineral precipitates; the common name for these slightly cooler vents is thus 'white smokers'[5,31].

Individual high-temperature vents at mid-ocean ridges may only be ca 10 cm in diameter at their mouth, yet over time, growing like stalagmites from the sea floor, they can form chimneys anywhere from 1 m to 30 m tall. A typical vent field might comprise several such chimney structures spread over a circular area ca 100 m in diameter. Throughout this area there may also be a number of lower temperature vents emitting hot, shimmering water from the sea bed. Even these vents are as hot as 10–30°C, which is notably warmer than typical deep-ocean water (2–3°C)[18,25]. It is in the vicinity of these warm and more diffuse emissions that the majority of vent-specific biota are most abundant.

Life at Hydrothermal Vents

Hydrothermal fluids are enriched in dissolved hydrogen sulphide, a substance which is toxic to most forms of life; elsewhere in the oceans it is found only in lifeless, stagnant anoxic basins such as the Black Sea. It is quite remarkable, therefore, that sites of active hydrothermal venting are not barren wastelands where no life can exist. Instead, scientists diving at the very first vent site to be discovered, on the Galapagos rift in the eastern equatorial Pacific, were surprised to find dense concentrations of benthic (sea-bed dwelling) megafauna (large animals) living within the vent-field area. Since that time a number of new vent sites have been discovered around the world – and most have remarkably high concentrations of animals around them[17,32]. The reason why these areas attract such abundant life is even more remarkable. A common observation at all the vent fields that have been studied to date is that the dominant species of animals at any site often appear to be extremely large. This has raised the question: "Where do the animals get food to grow at all, let alone enough to reach such a size?" The answer is that hydrothermal vent animals derive their food from a chain that is driven by geothermal (terrestrial) energy, unlike all the other organisms that rely, directly or indirectly, upon sunlight for their survival. In hydrothermal-vent communities, free-living bacteria, which are anchored on the sea bed or float free in the water column, coexist with symbiotic sulphide-oxidising bacteria, which live within the larger vent-specific organisms; these exploit the free energy of reaction released when hydrogen sulphide present in the vent fluids interacts with dissolved carbon dioxide and oxygen in ordinary sea water to form organic matter (equation 10.1), where $(CH_2O)_n$ is a carbohydrate. Because it is a *chemical*, hydrogen sulphide, which plays the role that sunlight plays in the more familiar process of photosynthesis in the warm surface waters of the oceans and on land, this unique deep-ocean, sunlight-starved process has been given the name *chemo*synthesis[2].

$$n(CO_2 + H_2S + O_2 + H_2O) = (CH_2O)_n + n(H_2SO_4) \qquad (10.1)$$

Approximately 95% of all animals discovered at hydrothermal vent sites are previously unknown species. So far, over 300 new species have been identified and, for many of these, the differences from previously known fauna are so great that new taxonomic families have had to be established in order to classify them satisfactorily[32]. Some of the most exciting examples of hydrothermal vent species discovered include the spectacular tube-worms found along the East Pacific Rise vent sites,

10.9

Figure 10.9 Dense populations of tube worms (*Riftia pachyptila*) and brachyuran crabs inhabiting the Genesis hydrothermal vent site, 13°N, East Pacific Rise (depth, 2636 m). The tube-worms, which can grow up to 2–3 m in length and may be 2–5 cm across, derive their nutrition from *symbiotic* sulphide-oxidising bacteria which live within their gut. The crabs, which may only grow to a modest 5–10 cm across, feed by scavenging on the red haemoglobin-filled 'plumes' of the tube-worms; these can extend up to 10 cm beyond the end of their protective chitin tubes to draw in the hydrogen sulphide rich waters necessary for chemosynthesis, but can be withdrawn rapidly back into their tubes for protection (photograph by R.A. Lutz, Institute of Marine and Coastal Studies, Rutgers University, and the American Geophysical Union[17]).

which can measure 2–3 m or more long, and which typically appear in thick clusters as shown in *Figure 10.9*. Also common along the East Pacific Rise are giant clams and mussels (*Figures 10.10* and *10.11*), which can often reach the size of a large dinner plate. The total biomass at any one hydrothermal site is typically very high. Indeed, hydrothermal vent fields have been likened to submarine oases which punctuate the deserted barren plains of the deep-sea floor (Chapter 13). In contrast, biodiversity at individual vent sites (i.e., the total range of different species present; see Chapter 15) is surprisingly low. Not only that, but the species present at vent sites in the different oceans show remarkably little similarity. For example, no giant tube-worms or giant clams have been found at any of the five known hydrothermal fields discovered so far in the North Atlantic Ocean. Instead, for example, the

10.10

Figure 10.10 Giant white clams (*Calyptogena magnifica*) living in clusters within the crevices between basalt pillows, in an area known as Clam Acres at 21°N on the East Pacific Rise. Other vent fauna include a clump of tube worms (*Riftia pachyptila*), galatheid crabs (*Munidopsis susquamosa*), and limpets. Note the robot arm of the submersible 'Alvin' collecting a clam approximately 25 cm in length (photograph by R.A. Lutz, Institute of Marine and Coastal Studies, Rutgers University, and the American Geophysical Union[17]).

10.11

Figure 10.11 Dense beds of mussels (*Bathymodiolus thermophiolus*) and clams (*Calyptogena magnifica*) at the Rose Garden vent site along the Galapagos Rift. Again, individual clams and mussels are typically 20–30 cm in length. Other vent fauna seen include galatheid crabs and snails (photograph by R.A. Lutz, Institute of Marine and Coastal Studies, Rutgers University, and the American Geophysical Union[17]).

10.12

Figure 10.12 A schematic cross-section of the TAG hydrothermal mound, 26°N Mid-Atlantic Ridge (the co-ordinates give the dive site). Hot vent-fluid (shown in pale blue) flows up an open fissure in the deeply faulted oceanic crust, and then percolates out through the entire sulphide mound along a tortuous network of interconnected channels, giving rise to the highest-temperature (350–365°C) black smoker fluids at the apex (50 m across and 50 m above the sea floor) and to the lower temperature (270–300°C), partly diluted white smokers around the outer section of the mound (200 m across and 20–30 m above the sea floor). Extinct chimneys are also seen across much of the outer mound, where earlier fluid flow has ceased because subsurface mineral precipitation has choked the flow-channels solid. Toward the flanks of the mounds, rubble deposits occur where oxidised and altered material from the hydrothermal mound has been broken up by mass-wasting ('landslide') events and carried out across the sea floor, to be deposited upon more typical volcanic basement and a thin veneer of more typical pelagic sediments (courtesy of Pierre Minon, © National Geographic Society).

southernmost three of these sites are characterised by abundant small shrimp, which cluster in their millions around the black-smoker chimneys.

Such completely isolated biological communities indicate separate paths of evolution over many generations – perhaps stretching back millions of years. However, we know that individual vent fields and chimneys may only remain active for periods of perhaps 100–1000 years at a time, before the chimneys themselves become choked with minerals and the flow of warm fluids becomes blocked. For any one species of hydrothermal organism to have survived down through the generations, therefore, we know that the ability to migrate from one vent site to another must be vitally important. Thus, an important question – particularly given the globe-encircling nature of mid-ocean ridges – is "How do animals move from one hydrothermal field to colonise another on an intra-basinal scale, yet live within communities which, on an inter-basinal scale, are evidently quite isolated?" One key area of research currently underway involves mapping the entire mid-ocean ridge system in sufficient detail to determine the total number of hydrothermal vent sites world-wide, their average spacing one from another, and how that spacing is controlled by the tectonic and volcanic nature of the mid-ocean ridges which host them. The second key issue to explain the biodiversity and separate evolution problems is to understand how animals reproduce and migrate along the lengths of these ridges. It is proposed that the processes of reproduction and migration must be intimately related to each other, because a large majority of vent-specific organisms exhibit quite sessile lifestyles as adults. Therefore, only if these species give rise to planktonic larval stages is there any prospect of them being able to migrate along mid-ocean ridge axes and colonise new vent fields as and when they occur[32].

Hydrothermal Deposits

Hydrothermal deposits on the sea floor range from single chimneys to large, sprawling mounds topped by clusters of chimneys[20]. In general, the size of the hydrothermal deposits appears to be related to the length of time for which active venting persisted at those sites. Pacific vent fields, with very localised chimney deposits, have life-spans of just tens of years to perhaps a hundred years or so before mineral precipitation cements their chimneys solid[11]. By contrast, certain Atlantic sites are characterised by mounds stretching in excess of 200 m across and reaching 30–50 m high, and which dating reveals have remained active for thousands and even tens of thousands of years[13].

One such site that has been studied extensively by the international community is called TAG (named after a North Atlantic basin study in the 1970s – the Trans-Atlantic Geotraverse)[22]. The TAG site comprises one large active mound and several mounds that are now extinct. The deposit is one of the largest active sites known and is approximately 200 m in diameter and 50 m high. There is a vast range in the style of fluid venting and type of mineral deposit at different sites on the mound. Fluid flow through the mound is pervasive, fed by a complex network of channels through which high-temperature fluid flows, rising through the mound from the underlying basement (*Figure 10.12*). Hot

10.13

Figure 10.13 A cluster of inactive chimneys from the outer portion of the TAG hydrothermal mound. The chimneys each measure several metres in height and 10–30 cm in width. They have become clogged with minerals, as the supply of hydrothermal fluid from the underlying network of channels has waned – presumably due to further mineral precipitation subsurface, within the mound. Oxidation of the iron-, copper-, and zinc-sulphide minerals gives rise to the alteration of the chimneys' colour from blue–grey to brown. In time, these chimneys will become sufficiently altered and weakened to collapse, adding to the unconsolidated sulphide rubble which makes up much of the TAG hydrothermal mound. Evidence that this particular site has not long been inactive is also provided, however, by the presence of the luminous pale-blue mineral seen plated to the exterior of many of the chimneys. This mineral is anhydrite (a form of calcium sulphate), which only precipitates in the oceans at high temperatures and will continue to dissolve as the chimneys age and cool (courtesy of Woods Hole Oceanographic Institution).

(360°C) acidic fluids gush from the summit at speeds of several metres per second; the whole mound apex is shrouded in black smoke which is rapidly entrained upward into a buoyant hydrothermal plume. Those chimneys that have been sampled are 1–10 cm in diameter and consist predominantly of copper- and iron-sulphides (chalcopyrite, marcasite, bornite, pyrite) and calcium sulphate (anhydrite)[31]. Lower temperature, 'white smokers' vent at rates of centimetres per second to the southeast of the main black smoker complex. Here the temperatures are up to 300°C, and the chimneys are bulbous and zinc sulphide (sphalerite) rich. The white 'smoke' consists of amorphous silica mixed with zinc- and iron-sulphides. Much of the rest of the mound surface is covered with red and orange iron oxides, through which diffuse, low-temperature fluids percolate. Areas of diffuse flow are delineated by clusters of white anemones and shimmering water[20].

The range in fluid-venting styles is evident from the distribution of fauna over the mound surface. Three different species of shrimp have been discovered at TAG; the most abundant is *Rimicaris exoculata*, which swarms over the black-smoker edifice. The lower-temperature diffuse flow areas host a community of anemones and crabs. The distribution of organisms can give semi-quantitative information as to the fluid flow regime[32].

Hydrothermal activity is intermittent, even at individual sites. Inactive chimneys have been discovered on the outer portions of the main mound and are shown in *Figure 10.13*. The chimneys have been oxidised to orange iron oxides and are beginning to crumble and collapse. Eventually, they will be completely weathered down to the mound surface. The mound itself is unstable and subject to collapse and mass wasting events, in which submarine equivalents of landslides, perhaps triggered by minor earthquakes, sweep altered chimney material down off the slopes of the mound and out onto the surrounding sea-floor sediments. As a result, the sediments immediately adjacent to the flanks of the mound are also full of metal-rich sulphide and oxide minerals, just like the hydrothermal chimneys from which they are derived[7,19]. Warm fluids can continue to percolate up through these flanking sediments, over time altering their mineral assemblage to clays. Despite this mineral alteration, the metal concentrations preserved within these sediments remain extraordinarily high – up to 45% iron and 34% copper, and as much as 10–15% zinc. The potential for mining such deposits in the future is discussed in Chapter 21.

Eventually, hydrothermal sediments are buried beneath the normal background pelagic sediments and transported away from the ridge axis by plate-spreading processes (see earlier). Deep-sea sediments have been drilled by the Ocean Drilling Program at over 700 sites. On those occasions where drilling has reached the basement rock, a metal-rich layer has often been observed at the bottom of the sediment pile, representing ancient metalliferous sediments[14]. Such basal metal enrichments are often the only evidence that hydrothermal activity has been extensive, not only spatially, but also throughout geological time.

Hydrothermal Activity, Ocean Circulation, and Ocean Composition

The effects of hydrothermal activity are not restricted to the immediate vicinity of black smoker vent sites – although this is where their most visually spectacular impact is best observed. As hydrothermal fluids erupt from the sea floor, they remain buoyant as they mix with sea water, and rise, carrying their mix of particles and fluid upward in a conical expanding plume (*Figure 10.14*). As this turbulent, continuously mixing plume rises it eventually reaches a stage where it is no longer more buoyant than the surrounding water column, and so ceases to rise[28]. This particle-rich fluid is then dispersed as an approximately horizontal layer (often referred to as the neutrally buoyant plume), flowing along isopycnal (constant density) surfaces. Dispersion of this material through the oceans is then driven, primarily, by large-scale ocean circulation patterns[16].

It takes the fluid, solutes, and particles in a buoyant hydrothermal plume less than an hour to rise 100–300 m above the sea bed, before being carried away by the prevailing deep-ocean currents. During this time, mixing is so turbulent that the initial vent fluid is typically diluted approximately 10,000 fold by ordinary sea water[24]. Because the initial vent fluid is enriched by factors of up to a million in certain key chemical tracers (e.g., dissolved methane, dissolved manganese, and total suspended particulate matter), strong enrichments can still be detected after emplacement into and dispersion within neutrally buoyant hydrothermal plumes. Recognition of this characteristic of active venting has greatly increased the ability of geochemists to locate new sites of hydrothermal activity[8,9,12,21]. Instead of needing to photograph every square metre of sea bed to look for individual vent fields, it has become sufficient simply to sample the overlying water column, perhaps just once every few kilometres, to detect the presence of any chemical or physical anomalies characteristic of hydrothermal discharge in any particular area.

10.14

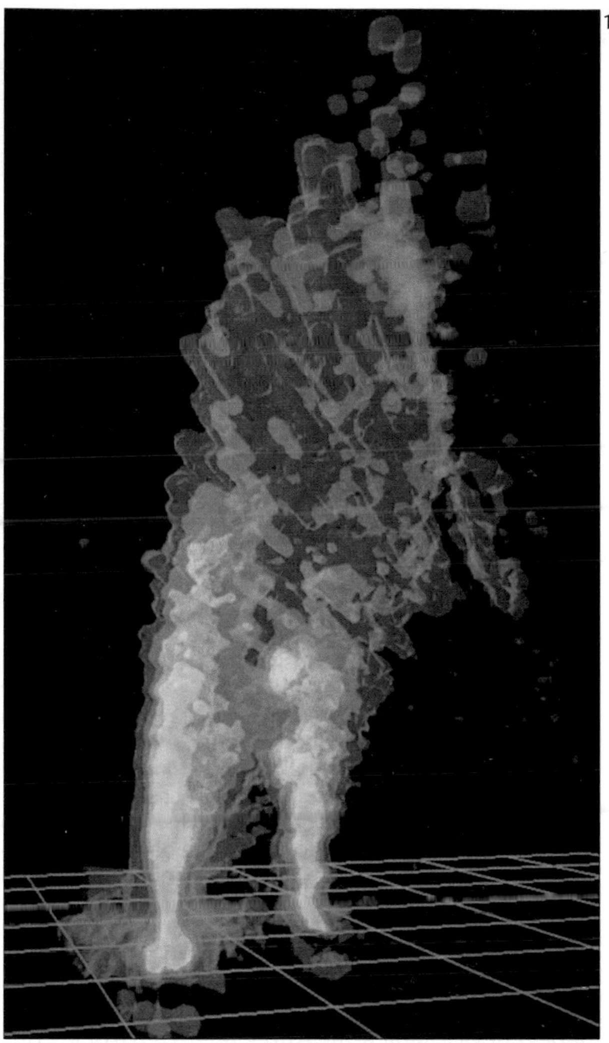

Figure 10.14 A three-dimensional acoustic image of the lower 40 m of two buoyant hydrothermal plumes discharging from adjacent black smoker vents at 2635 m depth on the East Pacific Rise, near 21°N. This image, obtained using a sonar (echo-sounder) system mounted on the sail of the US submersible 'Alvin', shows coalescence of the two plumes as they rise, as well as a 'bending-over' of the uppermost portion of their merged plume in the prevailing ocean-current direction. (courtesy of P.A. Rona, Institute of Marine and Coastal Studies, Rutgers University, and the American Geophysical Union[23]).

5. Edmond, J.M., Campbell, A.C., Palmer, M.R., Klinkhammer, G.P., German, C.R., Edmonds, H.N., Elderfield, H., Thompson, G., and Rona, P. (1995), Time series studies of vent fluids from the TAG and MARK sites (1986, 1990) Mid-Atlantic Ridge: a new solution chemistry model and a mechanism for Cu/Zn zonation in massive sulphide orebodies, in *Hydrothermal Vents and Processes*, Parson, L.M., Walker, C.L., and Dixon, D. (eds), Special Publication, The Geological Society, London, pp 77–96.
6. Feely, R.A., Lewison, M., Massoth, G.J., Robert-Baldo, G., Lavelle, J.W., Byrne, R.H., von Damm, K.L., and Curl, H.C. J. (1987), Composition and dissolution of black smoker particulates from active vents on the Juan de Fuca Ridge, *J. Geophys. Res.*, **92**, 11347–11363.
7. German, C.R., Higgs, N.C., Thomson, J., Mills, R.A., Elderfield, H., Blustajn, J., Fleer, A.P., and Bacon, M.P. (1993), A geochemical study of metalliferous sediment from the TAG hydrothermal mound, 26°08'N, Mar, *J. Geophys. Res.*, **98**, 9683–9692.
8. German, C.R., Briem, J., Chin, C., Danielsen, M., Holland, S., James, R., Jónsodottir, A., Ludford, E., Moser, C., Ólafsson, J., Palmer, M.R., and Rudnicki, M.D. (1994a), Hydrothermal activity on the Reykjanes Ridge: the Steinahóll vent-field at 63°06'N, *Earth Planet. Sci. Lett.*, **121**, 647–654.
9. German, C.R., Parson, L.M., and Scientific Party of RRS *Charles Darwin* cruise CD89'HEAT' (1994b), Hydrothermal exploration at the Azores triple-junction, *EOS, Trans. Am Geophys. Union.*, **75**, 308.
10. Haxby, W.F. (1987), *Gravity Field of the World's Oceans: A Portrayal of Gridded Geophysical Data Derived from SEASAT Radar Altimeter Measurements of the Shape of the Ocean Surface (Scale 1:40,000 at the Equator)*, World Data Centre for Marine Geology and Geophysics.
11. Haymon, R.M. and Kastner, M. (1981), Hot spring deposits on the East Pacific Rise at 21°N: preliminary description of mineralogy and genesis, *Earth Planet. Sci. Lett.*, **53**, 363–381.
12. Klinkhammer, G., Rona, P., Greaves, M.J., and Elderfield, H. (1985), Hydrothermal manganese plumes over the Mid-Atlantic Ridge rift valley, *Nature*, **314**, 727–731.
13. Lalou, C., Thompson, G., Arnold, M., Brichet, E., Druffel, E., and Rona, P.A. (1990), Geochronology of TAG and Snakepit hydrothermal fields, Mid-Atlantic Ridge: witness to a long and complex hydrothermal history, *Earth Planet. Sci. Lett.*, **97**, 113–128.
14. Leinen, M. (1981), Metal-rich basal sediments from northeastern Pacific Deep Sea Drilling Project sites, in *Initial Reports of the Deep Sea Drilling Programme 63*, Yeats, R.S. and Haq, B.U. (eds), US Govt. Printing Office, Washington, pp 667–676.
15. Le Pichon, X. (1968), Sea floor spreading and continental drift, *J. Geohpys. Res.*, **73**, 3661–3697.
16. Lupton, J.E. and Craig, H. (1981), A major ^{3}He source at 15°S on the East Pacific Rise, *Science*, **214**, 13–18.
17. Lutz, R.A. and Kennish, M.J.(1993), Ecology of deep-sea hydrothermal vent communities: a review, *Rev. Geophys.*, **31**(3), 211–242.
18. Mills, R.A., Thomson, J., Elderfield, H., and Rona, P.A. (1993a), Pore-water geochemistry of metalliferous sediments from the Mid-Atlantic Ridge: diagenesis and low-temperature fluxes, *EOS, Trans. Am. Geophys. Union*, **74**, 101.
19. Mills, R.A., Thomson, J., and Elderfield, H. (1993b), A dual origin for the hydrothermal component in a metalliferous sediment core from the Mid-Atlantic Ridge, *J. Geophys. Res.*, **98**, 9671–9681.
20. Mills, R.A. (1995), Hydrothermal deposits and metalliferous sediments from TAG, 26°N Mid-Atlantic Ridge, in *Hydrothermal Vents and Processes*, Parson, L.M., Walker, C.L., and Dixon, D. (eds), Special Publication, The Geological Society, London, pp. 121–132.
21. Murton, B.J., Klinkhammer, G., Becker, K., Briais, A., Edge, D., Hayward, N., Millard, N., Mitchell, J., Rouse, I., Rudnicki, M., Sayanagi, K., Sloan, H., and Parson, L. (1994), Direct evidence for the distribution and occurrence of hydrothermal activity between 27°N 30°N on the Mid-Atlantic Ridge, *Earth Planet. Sci. Lett.*, **125**, 119–128.
22. Rona, P.A., Klinkhammer, G., Nelsen, T.A., Trefry, J.H., and Elderfield, H. (1986), Black smokers, massive sulphides and vent biota at the Mid-Atlantic Ridge, *Nature*, 321, 33–37.
23. Rona, P.A., Palmer, D.R., Jones, C., Chayes, D.A., Czarnecki, M., Carey, E.W., and Geurrero, J.C. (1991), Acoustic imaging of hydrothermal plumes, East Pacific Rise, 21°N, 109°W, *Geophys Res. Lett.*, 18(12), 2233–2236.
24. Rudnicki, M.D. and Elderfield, H. (1993), A chemical model of the buoyant and neutrally buoyant plume above the TAG vent field, 26°N, Mid-Atlantic Ridge, *Geochim. Cosmochim. Acta*, **57**, 2939–2957.
25. Schultz, A., Delaney, J.R., and McDuff, R.E. (1992), On the partitioning of heat flux between diffuse and point-source seafloor venting, *J. Geophys. Res.*, **97**, 12229–12314.
26. Searle, R.C. (1992), The volcano-tectonic setting of oceanic lithosphere generation, in *Ophiolites and their modern oceanic analogues*, Parson, L.M., Murton, B.J., and Browning, P. (eds), Special Publication 60, The Geological Society, London, 65–80.
27. Smith, D.K. and Cann, J.R. (1993), Building the crust at the Mid-Atlantic Ridge, *Nature*, **365**, 707–715.
28. Speer, K.G. and Rona, P.A. (1989), A model of an Atlantic and Pacific hydrothermal plume, *J. Geophys. Res.*, **94**, 6213–6220.
29. Speer, K.G. and Helfrich, K.R. (1995), Hydrothermal plumes: a review of flow and fluxes, in *Hydrothermal Vents and Processes*, Parson, L.M., Walker, C.L., and Dixon, D. (eds), Special Publication, The Geological Society, London, pp 373–385.
30. Spiess, F.N., MacDonald, K.C., Atwater, T., Ballard, R., Carranza, A., Cordoba, D., Cox, C., Diaz Garcia, V.M., Francheteau, J., Guerrero, J., Hawkins, J., Haymon, R., Hessler, R., Juteau, T., Kastner, M., Larson, R., Luyendyke, B., Macdougall, J.D., Miller, S., Normark, W., Orcutt, J., and Rangin, C. (1980), East Pacific Rise; hotsprings and geophysical experiments, *Science*, **207**, 1421–1433.
31. Thomson, G., Humphris, S.E., Schroeder, B., Sulanowska, M., and Rona, P. (1988), Active vents and massive sulfides at 26°N (TAG) and 23°N (Snakepit) on the Mid-Atlantic Ridge, *Canad. Mineralog.*, **26**, 697–711.
32. Tunnicliffe, V. (1991), The biology of hydrothermal vents: ecology and evolution, *Oceanogr. Marine Biol. Ann. Rev.* **29**, 319–407.
33. Wolery, T.J. and Sleep, N.H. (1976), Hydrothermal circulation and geochemical flux at mid-ocean ridges, *J. Geophys. Res.*, **84**, 249–276.

CHAPTER 11:

The Ocean: A Global Geochemical System

J.D. Burton

Introduction

When the first systematic measurements of the chemical composition of sea w[illegible] made as part of the work of the C[illegible] of 1872–1876, the constit[illegible] analysed with any accur[illegible] number. They comprised [illegible] dant dissolved salt com[illegible] certain dissolved gases. [illegible] analytical chemical t[illegible] improved methods of sampling, [illegible] enabled a fuller knowledge of the concentrations of a wider range of constituents to be acquired. These advances accelerated remarkably from about the mid 1970s, so that there is now information not only on the concentrations, but also on the patterns of oceanic distribution for the majority of the elements. Information on the concentrations of some important dissolved elements in sea water is given in *Table 11.1*.

Marine scientists need information on chemical constituents for various purposes. Some constituents are useful as tracers, providing information on the circulation and mixing of water bodies in the ocean. They include the chlorofluorocarbons (freons), added to the atmosphere and thence to the ocean by man's use of them from about 1950 onward, and anthropogenic radioactive materials, some of which entered the ocean from the atmospheric testing of nuclear weapons in the late 1950s and early 1960s, and some of which are due to discharges of low-level waste from nuclear fuel reprocessing. Examples of the kind of information obtained by the use of such tracers are given in *Figure 11.2*. Constituents such as nitrate, phosphate, and dissolved silicon have been studied very intensively because they are important plant nutrients, with an influence on the photosynthetic production of organic matter. Others, including trace metals (such as cadmium, mercury, and lead) and organic micropollutants (such as pesticides and polynuclear aromatic hydrocarbons) are of concern because of their potential impact on marine life in estuaries and coastal waters, which now receive increased inputs of these substances as a result of agricultural uses and the disposal of industrial and domestic wastes.

In a solution with so complicated a composition as sea water, many elements are present in a variety of physicochemical forms. For example, in the presence of the various inorganic, negatively charged ions (anions) in sea water, the dissolved element copper occurs to only a small extent as the positively charged divalent ion, the cupric Cu^{2+} cation, and is largely associated (complexed) with the anions. The distribution of copper between the simple cation and the complexes formed with various

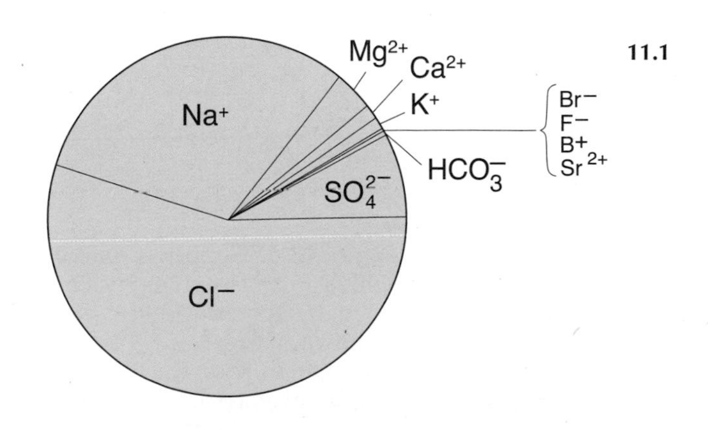

Figure 11.1 Relative concentrations of dissolved major constituents in sea water. The bulk content of dissolved material is expressed by salinity (*S*), a defined dimensionless quantity which approximates to the total concentration (in g/kg) of the substances present in solution. Salinity can be measured at best to 0.001; only the few major constituents in the diagram contribute significantly to it. The actual mean concentrations of the constituents are given in *Table 11.1*. The mean concentration of dissolved silicon exceeds 1 mg/kg (approximately 0.001 in salinity), but its distribution is much more variable than those of the elements shown, and it is in a chemical form largely undetected by the widely employed conductimetric determination of salinity. It is therefore grouped with the minor or trace constituents, many of which occur at concentrations between 1 μg (10^{-6} g) and 1 ng (10^{-9} g)/kg, and others at still lower concentrations.

Table 11.1 Concentrations and most important chemical species of some dissolved elements in sea water.

Element	*Average concentration*	*Range*	*Units (per kg)*	*Most abundant chemical species*[a]
Lithium	174	b	μg	Li^+
Boron	4.5	b	mg	H_3BO_3
Carbon	27.6	24–30	mg	HCO_3^-, CO_3^{2-}
Nitrogen[c]	420	<1–630	μg	NO_3^-
Fluorine	1.3	b	mg	F^-, MgF^+
Sodium	10.77	b	g	Na^+
Magnesium	1.29	b	g	Mg^{2+}
Aluminium	540	<10–1200	ng	$Al(OH)_4^-$, $Al(OH)_3^0$
Silicon	2.8	<0.02–5	mg	$H_4SiO_4^0$
Phosphorus	70	[illegible]	μg	HPO_4^{2-}, $NaHPO_4^-$, $MgHPO_4^0$
Sulphur	0.904	b	g	SO_4^{2-}, $NaSO_4^-$, $MgSO_4^0$
Chlorine	19.354	b	g	Cl^-
Potassium	0.399	b	g	K^+
Calcium	0.412	b	g	Ca^{2+}
Manganese	14	5–200	ng	Mn^{2+}, $MnCl^+$
Iron	55	5–140	ng	$Fe(OH)_3^0$
Nickel	0.50	0.10–0.70	μg	Ni^+, $NiCO_3^0$, $NiCl^+$
Copper	0.25	0.03–0.40	μg	$CuCO_3^0$, $CuOH^+$, Cu^{2+}
Zinc	0.40	<0.01–0.60	μg	Zn^{2+}, $ZnOH^+$, $ZnCO_3^0$, $ZnCl^+$
Arsenic	1.7	1.1–1.9	μg	$HAsO_4^{2-}$
Bromine	67	b	mg	Br^-
Rubidium	120	b	μg	Rb^+
Strontium	7.9	b	mg	Sr^{2+}
Cadmium	80	0.1–120	ng	$CdCl_2^0$
Iodine	50	25–65	μg	IO_3^-
Caesium	0.29	b	μg	Cs^+
Barium	14	4–20	μg	Ba^{2+}
Mercury	1	0.4–2	ng	$HgCl_4^{2-}$
Lead	2	1–35[d]	ng	$PbCO_3^0$, $Pb(CO_3)_2^{2-}$, $PbCl^+$
Uranium	3.3	b	μg	$[UO_2(CO_3)_3]^{4-}$

[a] Refers to the inorganic speciation in oxygenated waters.
[b] Variations are determined entirely or largely by those in salinity, i.e., the element is essentially conservative (see text for definition). For these elements, average concentration given is for sea water of salinity 35.
[c] Concentrations refer to combined nitrogen; element occurs also as dissolved nitrogen (N_2) gas. Species other than NO_3^- are often important in the upper ocean (e.g., NO_2^-, NH_4^+).
[d] Concentrations are affected by inputs to surface ocean of atmospherically transported lead from combustion of leaded petroleum.

Note:
Based mainly upon information in Bruland[4]. Usual ranges for oceanic waters are shown; concentrations of certain elements can be higher in some coastal waters.
Chemical oceanographers often employ molar units instead of the mass units shown here. For sodium (atomic weight 22.99) the concentration given above of 10.77 g/kg can alternatively be expressed as 0.449 mol/kg.

anions is shown in *Figure 11.3*. Copper is one of a number of metals which are also strongly complexed by the organic matter present in sea water, and so the speciation shown in *Figure 11.3* actually applies to only a small fraction of the dissolved copper. The dissolved organic material in sea water comprises a great diversity of molecules dominated by compounds which are described as 'marine humic material', because they show some analogies to the humic and fulvic acids produced in soils by the decomposition of terrestrial vegetation. These marine humic substances have not been fully characterised as regards their molecular structures, but show a range of capacities to form complexes with

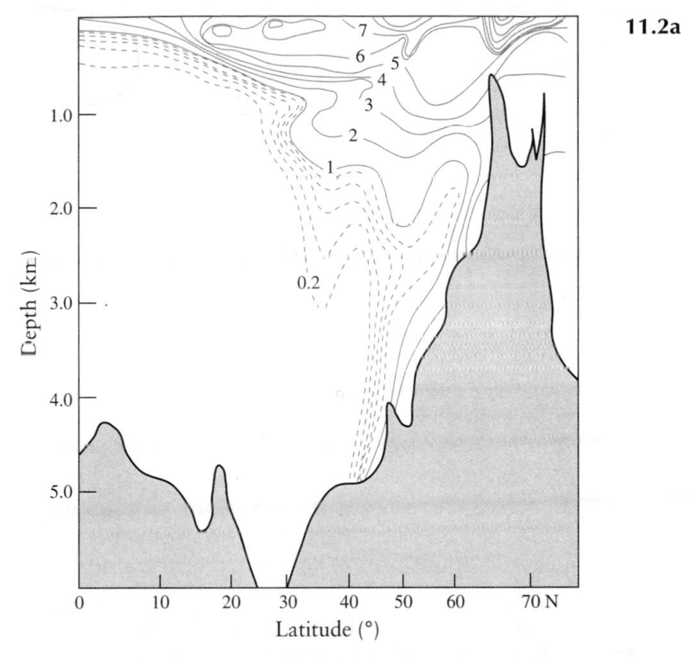

Figure 11.2 (a) Distribution of tritium (^{3}H) in a longitudinal section in the western North Atlantic Ocean in the 1970s (Broecker and Peng[3]; with permission from the authors). Values are given in tritium units (TU; 1 TU = 1 x 10^{-18} atoms of tritium per atom of hydrogen). Surface concentrations are greatly enhanced above natural background values by the input of tritium from the atmosphere, due to thermonuclear weapon testing, particularly in the late 1950s and early 1960s. The penetration of tritium into deeper waters at high latitudes reflects the formation of North Atlantic Deep Water. Because tritium is an isotope of hydrogen and enters the ocean largely as tritium-labelled water molecules, it is an ideal tracer for the advection and diffusion of water from the surface to the deeper ocean. (b) Distribution of a chlorinated fluorocarbon (freon), CFC 11, in a section across the South Atlantic Ocean in the 1990s. The section runs along 45°S from the western margin to the mid-Atlantic Ridge and then in a northeastern direction to the eastern margin at 30°S. The higher concentrations in surface waters reflect exchange with the atmosphere and transfer downward by advection and diffusion. The increase toward the sea bed reflects the inputs to the deepest ocean by transport of Antarctic Bottom Water of more recent surface origin than the overlying North Atlantic Deep Water. This penetrates the Argentine Basin, west of the mid-Atlantic Ridge, more effectively than the Cape Basin to the east (unpublished data provided by D. Smythe-Wright and S. Boswell).

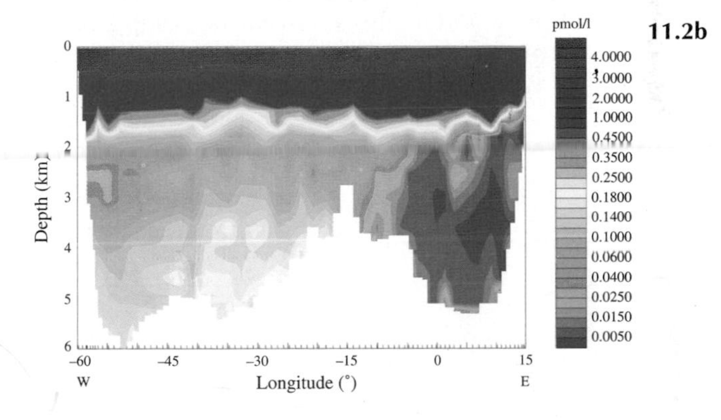

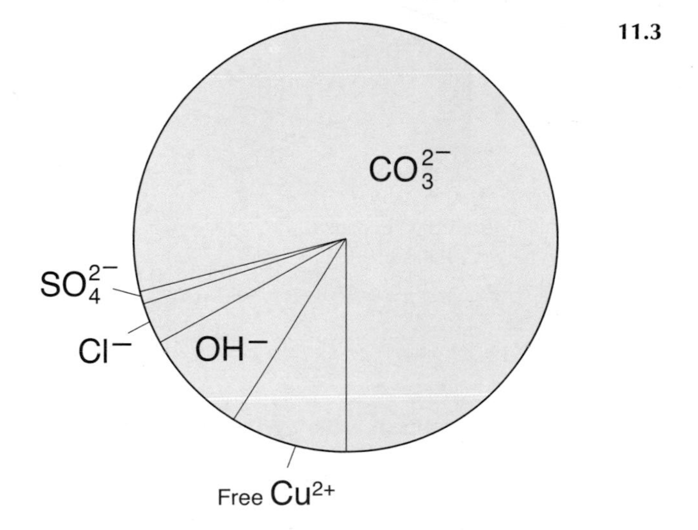

Figure 11.3 The distribution of copper between the free cation and the complexes formed with various inorganic anions in sea water (chloride, Cl^-; sulphate, SO_4^{2-}; carbonate, CO_3^{2-}; and hydroxyl, OH^-). For trace constituents, information of this kind cannot be obtained directly by measurements on sea water, but is derived from models which assume equilibrium between the different species and use data on the stabilities of the various complexes. The nature of the organic complexes that account for a large part of the actual chemical speciation of copper in sea water cannot be similarly specified (see text).

metals. In contrast with copper, the abundant element sodium has a very simple chemistry; it occurs in sea water almost entirely as the Na^+ ion.

The probable main inorganic chemical species for some important elements are shown in *Table 11.1*. Understanding chemical speciation is important because it affects the reactions of the elements in sea water. For example, the toxicity of copper to organisms such as phytoplankton (see Chapter 6), is related to the concentration (or, more strictly, the equivalent thermodynamic quantity, the activity) of the cupric ion rather than that of total copper. In chemical analyses, however, the chemical speciation is rarely resolved, so dissolved concentrations generally refer here to the total element or ion.

Knowledge of the chemistry of the ocean relies heavily on the analytical chemical measurements which reveal the spatial distributions of chemical constituents and their variations in time. Because many constituents of interest are present in sea water at very low concentrations, they can be determined only by sensitive methods; special precautions are necessary to avoid contamination during sampling and analysis. Given the complexity of the mixture and the different levels of concentration at which constituents occur, the potential interference of one in the analysis of another must be avoided. For these reasons, analytical measurements on sea water often involve quite complicated chemical manipulations to concentrate and separate the required chemical forms. Very few chemical constituents can be measured by probing the ocean with a sensor, in the way that salinity can be measured (see Chapter 19). Much of the picture of their oceanic distributions is, therefore, coarsely resolved, but the great progress made in the acquisition of reliable data over the past two decades has, nevertheless, underpinned comparable advances in understanding the chemical processes that determine these distributions.

The Geochemical Context

The oceanographic study of chemical properties and processes is undertaken, as already indicated, from a variety of standpoints. The most central, however, and the one which provides the most valuable unifying insights, is that of geochemistry – the science which addresses questions concerning the occurrence and distribution of the chemical elements within Earth as a whole and within its major components, in space and time. These major components include the crust (lithosphere), the atmos-

11.4

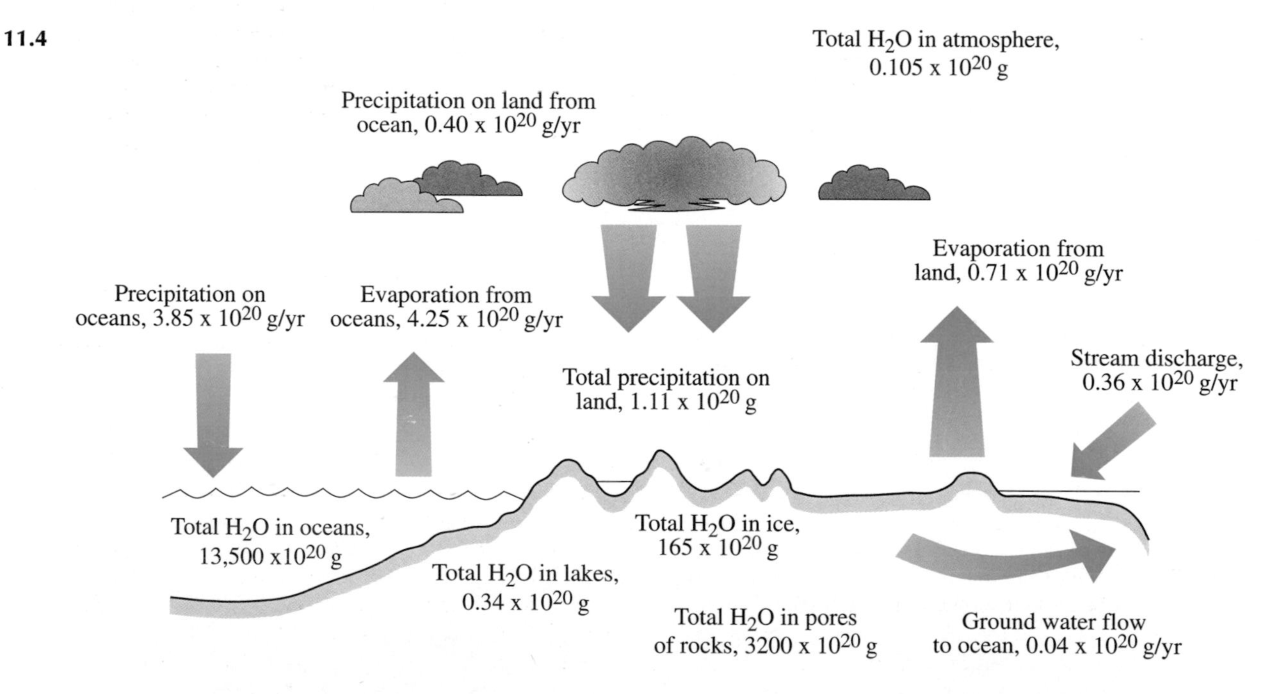

Figure 11.4 The global water cycle – the numbers show the total amounts of water held in the major reservoirs and the fluxes of water between reservoirs. In effect, the cycle acts not only as a transport system for weathered continental material, but also as a liquid extraction system for continental solid phases. Some of the water distilled by evaporation from the ocean, containing only a small proportion of sea-water salts as aerosol (see text), is deposited as precipitation on the continents and returned to the ocean containing dissolved material leached from continental rocks. Repeated cycles create a continuous net transfer of dissolved constituents to the ocean (based on data from Baumgartner and Reichel[2]).

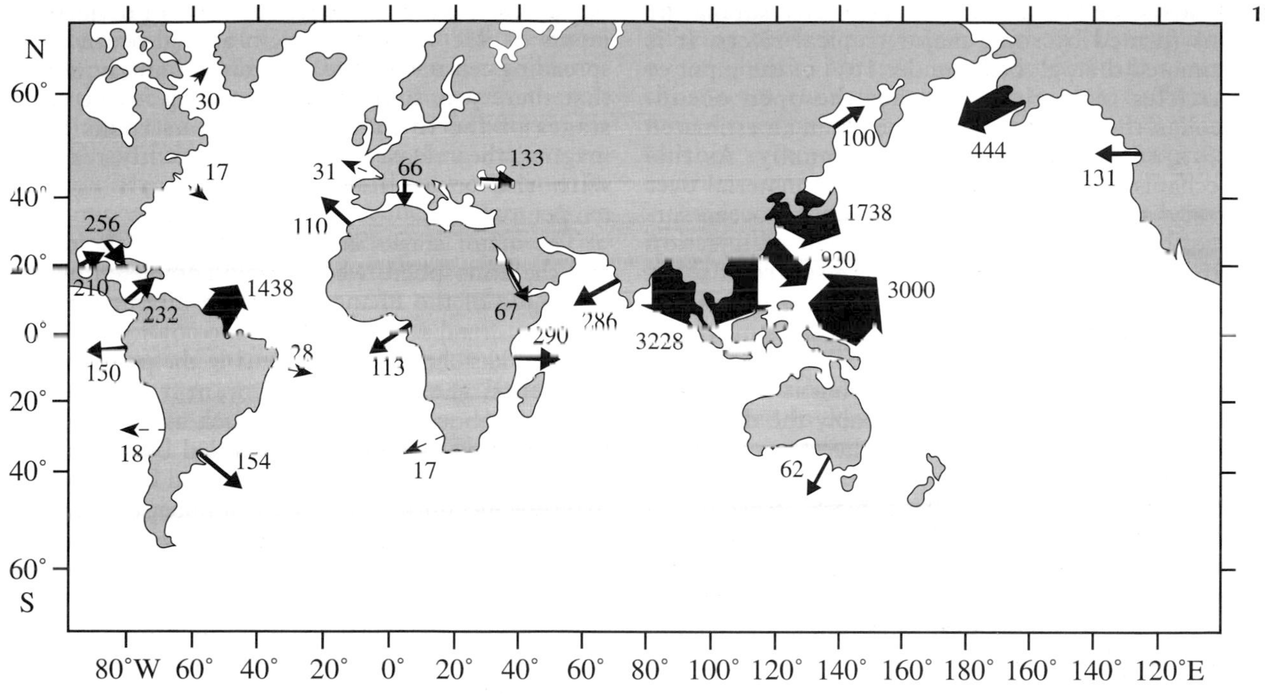

Figure 11.5 Fluxes of river-borne particulate material discharged from the major drainage basins. The widths of the arrows are proportional to the discharge; numerical values are in units of 10^6 tonne/yr. Fluxes to the Arctic Sea, totalling 226 x 10^6 tonnes/yr, are not shown. The directions of the arrows are arbitrary. Continental erosion in South East Asia and adjacent islands accounts for much of the discharge. The Amazon carries about 20% of the total water discharge to the ocean, but is not proportionately important as a source of either particulate or dissolved material. In some catchments, weathered particulate material is efficiently trapped in lakes or artificial reservoirs so that relatively little reaches the coast, e.g., the St. Lawrence (after Milliman and Meade[15]; with permission from the University of Chicago Press).

phere, and the aquatic environments collectively described as the hydrosphere, of which the ocean comprises the major mass, accounting for more than 98% of the free water at Earth's surface. The geochemical perspective emphasises the fact that the ocean is a dynamic system, chemically as well as physically, continuously exchanging material at its boundaries and also distributing it internally among its major water masses.

The major sedimentary cycle

The role of the ocean is central in the major sedimentary cycle. Continental rocks undergo physical weathering, which produces particles (mineral dust) small enough to be lifted from the ground by wind action and transported, particularly from arid regions, through the atmosphere (aeolian transport); particles are also carried by surface water run-off. Water, containing carbon dioxide dissolved from the atmosphere, is a powerful agent of chemical weathering, reacting with rock minerals to produce dissolved constituents and solid-phase products, of which the clay minerals are particularly important.

The global water cycle (*Figure 11.4*; see also *Figure 2.2*) thus drives a continuous flux of dissolved and particulate material to the seas, globally amounting to estimated annual inputs of about 3.7×10^{15} g of dissolved material and about 15×10^{15} g of particles. Estimates of global fluxes are essential for understanding the ocean's material budget, but these values are averages of discharges, which show large differences regionally, as illustrated for particulate material in *Figure 11.5*. About 10% of the riverine flux of dissolved material is recycled from the ocean to the continents, mainly through the agency of bubbles of air: these bubbles are produced largely by breaking waves, are then carried below the sea surface, and, on returning to the surface, burst to create sea-salt aerosols, a fraction of which is transported through the atmosphere and deposited on land.

Much of the particulate material discharged from rivers settles out to form bottom sediments in

11.6

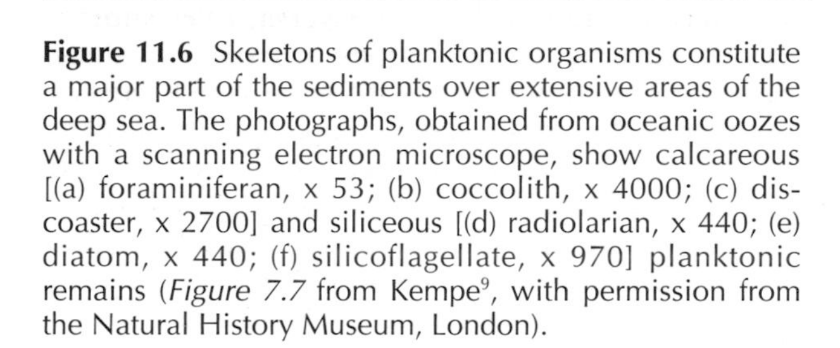

Figure 11.6 Skeletons of planktonic organisms constitute a major part of the sediments over extensive areas of the deep sea. The photographs, obtained from oceanic oozes with a scanning electron microscope, show calcareous [(a) foraminiferan, x 53; (b) coccolith, x 4000; (c) discoaster, x 2700] and siliceous [(d) radiolarian, x 440; (e) diatom, x 440; (f) silicoflagellate, x 970] planktonic remains (*Figure 7.7* from Kempe[9], with permission from the Natural History Museum, London).

11.7a

11.7b

Figure 11.7 Cross sections of ferromanganese nodular concretions ('manganese nodules') from the sea floor. (a) Polished radial section of a nodule from the equatorial North Pacific Ocean, showing asymmetrical growth of layers around a nucleus (dashed white line) of consolidated sediment (note the discontinuity in the dark layer, indicated by white arrows, which shows that the orientation of the nodule on the sea bed has not remained constant; from von Stackelberg[23]; with permission from the author and D. Reidel Publishing Company). (b) Polished radial section of a nodule from the Blake Plateau (Atlantic Ocean). The light-coloured veins consist largely of calcite and clay, which have accumulated in cracks in the growing nodule. The nodule accreted around a piece of phosphorite, which is no longer clearly distinguishable as a nucleus, having been partly replaced by ferromanganese material (photograph by J. Mallinson).

biogenous components, however, are the minerals produced as skeletal material by organisms (*Figure 11.6*). Where these are important sedimentary components, they accumulate typically at rates of about 1 cm every thousand years.

Of less importance in terms of sediment mass, but of great geochemical interest, are the solid phases formed by chemical reactions in sea water; these sediments are sometimes termed hydrogenous (formed in the hydrosphere), but more commonly are described as authigenic. Important examples are the sediments deposited by hydrothermal plumes (see Chapter 10) and the ferromanganese concretions (*Figure 11.7*), which are abundant on the sea bed in some regions and have potential economic value (see Chapter 21). Other notable authigenic phases include phosphorite (carbonate fluorapatite), which occurs particularly in eastern boundary regions of the ocean, where coastal upwelling is frequent, and glauconite (a potassium-rich aluminosilicate). Glassy basalts reacting with sea water are so extensively altered that the minerals formed

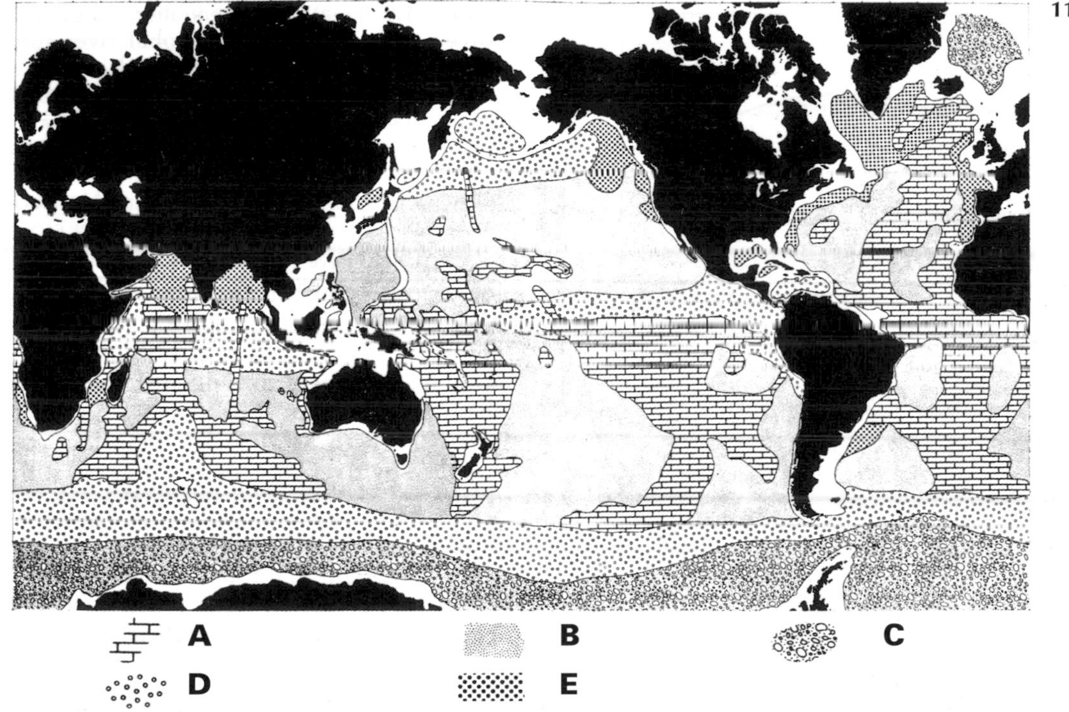

Figure 11.8 Dominant sediment types, classified by origin, in different areas of the deep-sea bed. Low concentrations of calcareous sediments in the North Pacific Ocean reflect a shallower calcite compensation depth (see text) in this region; a general tendency can be seen for high concentrations of calcium carbonate to occur on the mid-ocean ridges, where the depth of water is usually less than the calcite compensation depth (A, calcareous sediment; B, deep-sea clay; C, glacial sediment; D, siliceous sediment; E, terrigenous sediment; ocean margin sediments occur in the areas left blank; from Davies and Gorsline[6], with permission from the authors and Academic Press).

should also be regarded as authigenic; these reactions account for the abundant zeolite mineral, phillipsite, and the clay mineral, montmorillonite, in sediments of the South Pacific basin. During some geological periods, large masses of evaporite sediments have been produced by the crystallisation of salts from sea water that has become concentrated by evaporation in shallow marginal areas. Such deposits are the starting materials for continental salt beds of economic importance. The major features in the distribution of sediment types in the deep sea are shown in *Figure 11.8* (see also *Figure 8.21*).

The Oceanic Particle Conspiracy

The dissolved constituents in the ocean can be removed, apart from those trapped in accumulating sediments as pore waters (see also *Figure 8.26*), by conversion into particles or by binding to particle surfaces. The solution composition is controlled by the balance between the input processes and the particle–solution interactions that occur in the water column. Turekian[22] summed up these interactions as follows: "The great particle conspiracy is active from land to sea to dominate the behaviour of dissolved species ... As these scrubbing agents operate around the world, they imprint their material burdens on accumulating sediments."

The particle conspiracy involves many processes and reactions. Dissolved ions undergo electrostatic exchanges at charged sites on particle surfaces. Potentially stronger associations arise by specific binding with chemical groupings, especially hydroxyl (OH^-) and carboxyl (COO^-), a process strongly encouraged by the ubiquity of coatings of organic matter on particle surfaces. Incorporation into organic material, and into skeletal minerals produced by organisms, involves the major structural elements, such as carbon, calcium, and silicon. Also, many of the trace elements become caught up in the formation processes, either because they have essential biochemical roles, like the metals which occur in enzymes, or adventitiously, because of chemical similarities to the major elements or a tendency to form complexes with organic substances. As a result of these various processes the concentrations of many elements are regulated at levels greatly below those at which actual precipitation occurs. The various processes by which elements in solution at low concentrations become associated with particles already present in the system are collectively referred to as 'scavenging'.

Where the permanent pycnocline forms a stratified water column, the return of the dissolved material to the surface by water transport is much less effective than is the down-column transport when the element is linked to a particle. There is, therefore, an efficient 'biological pump' for these constituents from surface to deep water, so they increase in concentration with depth over much of the ocean, an increase which may be very pronounced [*Figure 11.12(c)*], they are termed 'recycled elements'. A proportion of the sinking particles undergo deposition into the sediment, but the average atom of a recycled element travels with a particle into the deeper water and is released and returned to the surface layer many times before it is finally removed from the water column.

The organic matter produced by photosynthesizing organisms in the upper ocean is a complex mixture of compounds. Nevertheless, in the deeper waters, where recycling occurs, the increases in the concentrations of nitrate and phosphate and the decrease in the concentration of dissolved oxygen, which is used to oxidise the organic matter, occur in essentially constant proportions. This reflects the fact that, when averaged over time and space, the uptake of the main elements (carbon, nitrogen, and phosphorus) used to synthesize cellular material occurs in a common relationship, which reflects the mean composition of the mixture of compounds produced, such as carbohydrates, lipids, proteins, and nucleic acids. The atomic ratios in which carbon, nitrogen, and phosphorus are utilised are taken as 106:16:1; oxygen is utilised in the oxidation of organic matter in the ratio of 138 molecules per 106 atoms of carbon. These are known as Redfield ratios[18].

The types of distribution shown in *Figure 11.12* are idealised, but they also match quite closely the actual vertical profiles of constituents which show these behaviours to a marked extent (see *Figures 11.14* and *11.15*). All constituents show some particle reactivity and undergo recycling and scavenging, but for some the consequences, in terms of distribution, are less marked; in some cases they may be difficult or impossible to detect analytically. This may be because the down-column flux is low in absolute terms or because it is low relative to the dissolved concentration in the ocean reservoir. Only a slight vertical concentration gradient is detected for calcium in the ocean, despite the large flux of calcium carbonate from surface waters to the deep ocean, because of the high concentration of calcium in sea water. Few constituents of sea water are as abundant as calcium, however, and generally a major involvement in the particle conspiracy is reflected in clearly nonconservative behaviour.

The two main types of nonconservative behaviour are reflected in systematic horizontal variation in concentrations in the intermediate and deep waters of the major ocean basins[3]. A major part of the deep water in the ocean basins has a common origin in surface waters of the North Atlantic Ocean, particularly in the Norwegian Sea, where dense waters formed by surface cooling sink and flow south in the western Atlantic (*Figure 11.13*). The deep water, modified by additions from the Southern Ocean, flows northward in the Indian Ocean and in the Pacific Ocean. During this transit, the composition of the water is influenced by the particles which enter it from the surface. As the water becomes older, in the sense that more time has elapsed since it sank from the surface, more particles passing through on their way to the sediments scavenge the strongly particle-reactive constituents. The concentration of these constituents therefore decreases along the direction of the deep-water flow (*Figure 11.14*). In contrast, the recycled constituents increase in concentration (*Figure 11.15*), because the older waters have been subjected to more releases from biogenous particles sinking from the waters that overlie their route.

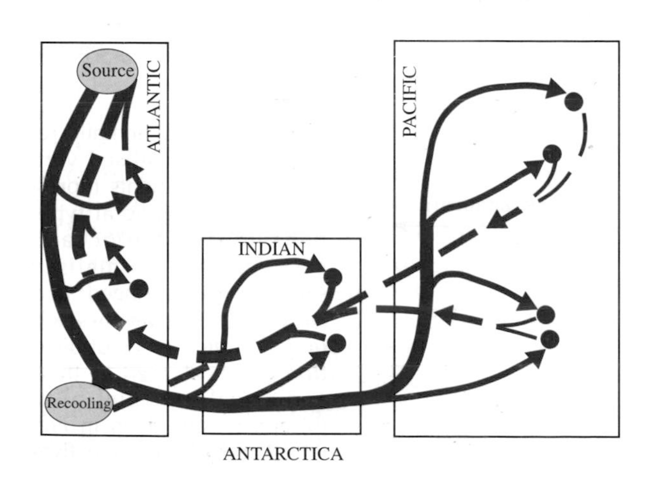

Figure 11.13 Schematic representation of the circulation of deep water in the major ocean basins (Broecker and Peng[3]; with permission from the authors). The solid lines show the flow of deep water, originating largely by sinking in the far North Atlantic Ocean. The dashed lines show the flow of surface water and the filled circles represent areas of localised upwelling.

Figure 11.14 Vertical profiles of some scavenged dissolved elements in the oceanic water column: (a) dissolved manganese and (b) dissolved aluminium in the northeastern Atlantic Ocean; (c) dissolved aluminium in the central North Pacific Ocean. Dissolved aluminium has a MORT of about 1000 years, and is therefore removed from the ocean on a timescale similar to the transit time of deep water from the North Atlantic Ocean to the North Pacific Ocean. There is thus a marked decrease in concentration between the two profiles. A strong sedimentary source shows up in the bottom waters in the North Pacific profile against the low deep-water concentrations. Surface input is dominated by dissolution of aeolian-transported mineral dust in profiles (a) and (c); this also adds significantly to the off-shelf transport of dissolved aluminium, which is important in the surface waters in profile (b). Dissolved manganese has a shorter MORT than aluminium and varies less in concentration in the deep ocean, except where *in situ* sources (e.g., hydrothermal) are significant. Compare these examples with the schematic representation in *Figure 11.12(b)*. Data are from (a) Statham *et al.*[19] (with permission from Macmillan Magazines Ltd), (b) Measures *et al.*[13] (with permission from the authors and Elsevier Science Ltd), and (c) Orians and Bruland[17] (with permission from the authors and Elsevier Science Ltd).

11.14

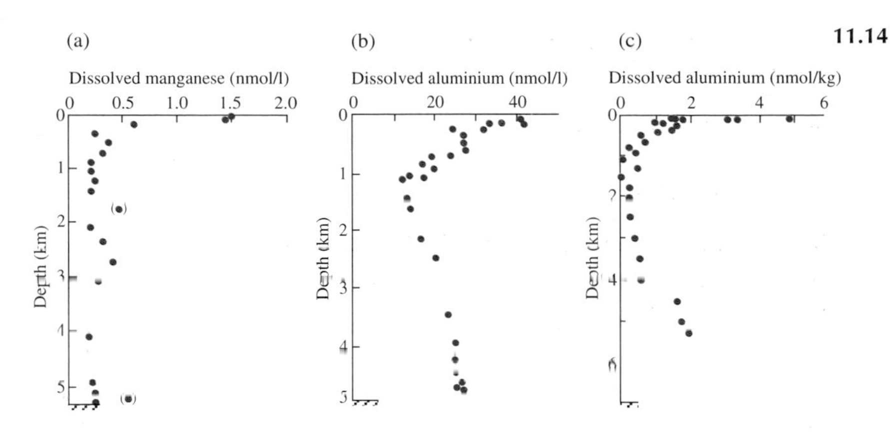

Figure 11.15 (a) Vertical profiles of some recycled dissolved constituents in three major ocean basins (squares, North Atlantic[4]; circles, southwestern Indian[16]; diamonds, central North Pacific[4]): the micronutrients are (i) silicon and (ii) phosphate, and the trace metals are (iii) cadmium and (iv) nickel. Within the overall feature of increasing concentration with increasing 'age' of the deep water, detailed differences in the profiles can be seen; in particular, nickel is less depleted in the surface waters than are the other constituents. A very close similarity in distribution is shown by phosphate and cadmium, reflecting the association of the metal, which has no known biological function, with organic matter sinking from the surface to deeper waters. The more gradual increase in concentration with depth for silicon, compared with phosphate, occurs because the skeletal material which carries silicon down the water column is recycled deeper in the ocean than are the soft organic tissues. Compare these examples with the schematic representation in *Figure 11.12(c)*. Figure from Morley *et al.*[16] (with permission from Elsevier Science Ltd). (b) The vertical distribution of concentrations of nitrate (μmol/kg) along a section through the western Atlantic, and Southern and Central Pacific Oceans (after Sharp[20], based on GEOSECS data; with permission from the author and Academic Press). The section illustrates more fully the pronounced increase in concentration with depth and the increase along the direction of the major deep-water circulation.

11.15a

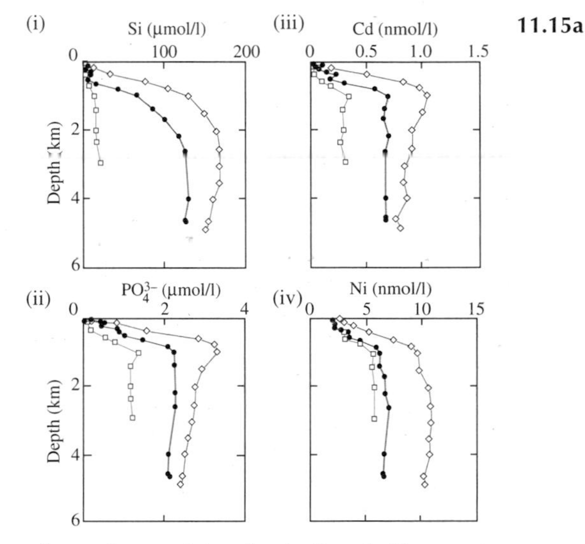

11.15b

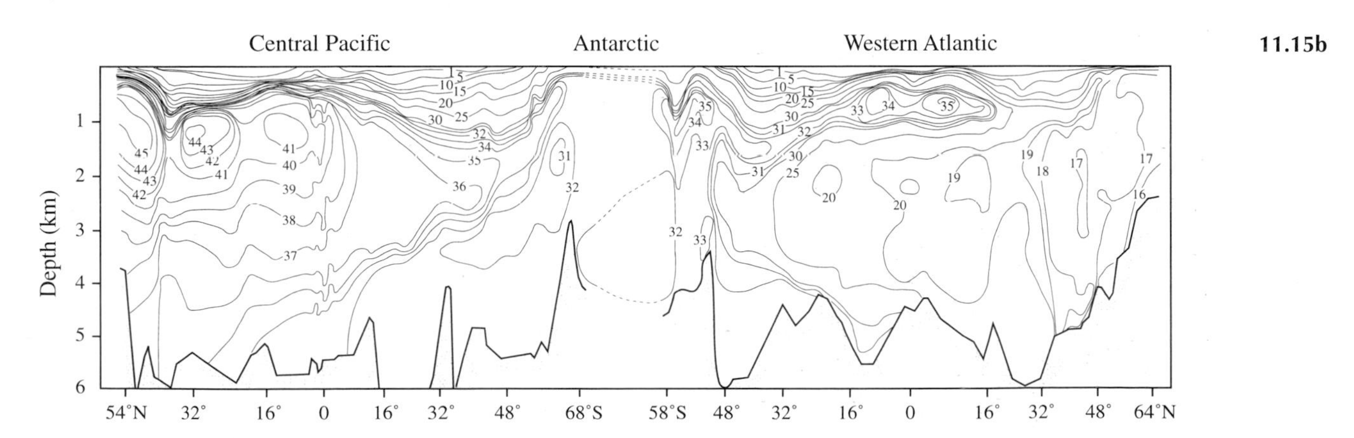

12.5

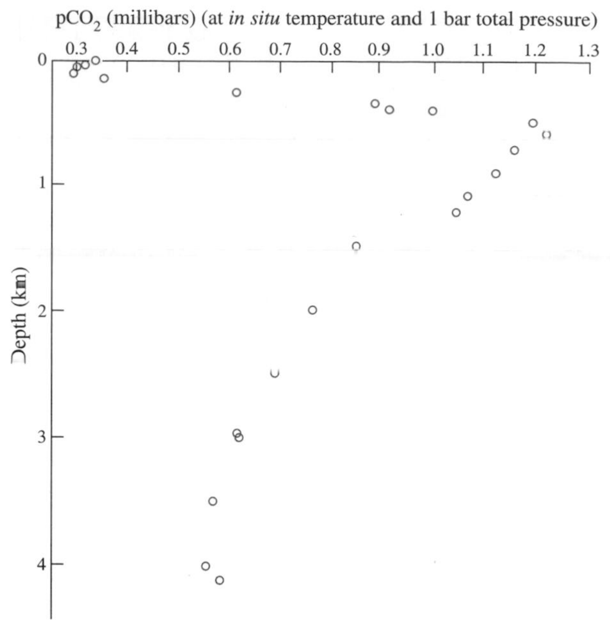

Figure 12.5 The typical variation of CO_2 concentration with depth. The profile shows a very sharp sub-surface increase to a maximum at approximately 500 m, due to planktonic respiration processes which release the previously fixed CO_2 back to the water column. The profile was taken in the Eastern Pacific (28°20′N 121°41′W) during the National Science Foundation GEOSECS (Geochemical Ocean Sections) program in 1969. (Adapted from Takahashi *et al.*[10])

Box 12.1 The buffering capacity of sea water

The pH of sea water varies over a surprisingly narrow range, centred at pH 8±0.5. The most important point is that the dissociation of carbonic acid (H_2CO_3) forms a *buffering system*, which may be summarised by the general weak acid–conjugate base equilibrium, equation (12.2), which results in a solution pH given by equation (12.3), where K_a is the equilibrium constant for the dissociation reaction. Small additions of acids or bases alter the ratio of anion (HCO_3^-) to acid (H_2CO_3) only slightly, and have little effect on the pH of the solution. The buffering capacity is the extent to which the pH is changed by a given addition of acid or base. The higher the concentration of carbonic acid, the greater the buffering capacity.

$$H_2CO_3 \rightleftharpoons H^+ + HCO_3^- \qquad (12.2)$$

$$pH = pK_a + \log([HCO_3^-]/[H_2CO_3]) \qquad (12.3)$$

sea water, after nitrogen and oxygen (see Chapter 11). Most gases dissolve in sea water in proportion to their atmospheric partial pressures, but CO_2 is an exception. It is an extremely reactive gas and has an elevated aquatic concentration relative to its atmospheric partial pressure. CO_2 is intimately involved in biological processes, being consumed by plankton, which photosynthesise in the surface waters, so leading to the production of organic matter tissue (equation 12.1):

$$CO_2 + H_2O \rightarrow \text{'}CH_2O\text{'} + O_2 \qquad (12.1)$$

where 'CH_2O' is a general term for organic plant material. A consequence of primary production (see Chapter 6) is that the upper waters of the ocean are generally undersaturated in CO_2 over large areas. In contrast, upwelling of deep waters in the equatorial region and along the west coast of the American continent, for instance (see Chapter 3), brings water supersaturated with CO_2 to the surface. These waters are rich in CO_2 because sinking organic matter has decomposed in the deeper waters of the ocean. The amount of the gas that is dissolved in sea water is determined by an interplay of chemical, physical, and biological factors. In turn, CO_2 helps to maintain the acidity of sea water in the range of pH 8.0±0.5 (see *Box 12.1*).

The capacity of the ocean to absorb CO_2 from the atmosphere appears great (*Figures 12.5–12.7*), and ultimately it can dissolve orders of magnitude more than is already present. Estimating the true capacity is, however, difficult since nothing is at equilibrium and the system is highly dynamic. CO_2 forms carbonic acid with water, which then dissociates to form hydrogen carbonate (HCO_3^-) and carbonate (CO_3^{2-}), which are the main forms of dissolved carbon in sea water. A simplistic view of the dissolution of CO_2 into sea water is given in equation (12.4) [details are discussed later – for example, see equations (12.6)–(12.9)].

$$CO_2(g) + H_2O \rightleftharpoons H_2CO_3 \rightleftharpoons H^+ + HCO_3^- \rightleftharpoons 2H^+ + CO_3^{2-} \qquad (12.4)$$

Reactions (12.1) and (12.4) illustrate the buffering ability of the marine carbonate system. An increase in atmospheric CO_2 increases the total amount of inorganic carbon within the sea. While this increases the buffering capacity, it also induces a slight increase in the ocean's acidity and thus acts to oppose further entry of the gas.

CO_2 concentrations increase with depth because CO_2 is used during photosynthesis and released again during respiration, and because the solubility of CO_2 increases with pressure (*Figure 12.5*). When

Box 12.2 Definitions c

The *alkalinity*, A_t, of sea water is
and carbonate ions (CO_3^{2-}), expre:
electrically neutral and is determi
alkalinity is the *amount of acid (h)*
tive un-ionised acids. The alkalini
negatively charged ions, equatio
[*SA*], the surplus alkalinity, is the
The second and third ionisation
contributor to A_t from the borate
so equation (12.14) may be sim
regions, when sulphide, ammonia

Typical values for the alkalinity
the contribution to alkalinity from
ly obtained by subtracting the b
obtained from the expressions for
tion, K_B, equations (12.17)–(12.19).
perature. The borate contribution i

The *total dissolved inorganic ca*
dissolved forms, and is given by e
total equilibrium concentrations a
(see *Box 12.3*). There are other
alkalinity of sea water, but under
their contributions are usually mu
(In the oceanographic literature, to
alkalinity as A_C; total carbonate as

At the pH of most ocean waters
and the H_2CO_3 concentration is onl

It is important to remember the
and carbonate ions. It is not a mea
it is easy to appreciate that sea w
DIC is high, so are alkalinity and
and acidity.

One of the few places where inc
Banks, where the sea is shallow
warmer and more saline the water
of CO_3^{2-} is also large, and often ri
$CaCO_3$, so that the inhibiting effec
carbonate (in the form of aragonit
This helps to explain the apparent
likely to dissolve, and vice versa.

$A_t + [H^+] = [HCO_3^-] + 2[CO_3^{2-}] + [OH^-]$

$A_t = [HCO_3^-] + 2[CO_3^{2-}] + ([OH^-] - [H^+$

$A_t = [HCO_3^-] + 2[CO_3^{2-}] + [B(OH)_4^-]$

$CA = [HCO_3^-] + 2[CO_3^{2-}]$

$B_T = [H_3BO_3] + [B(OH)_4^-]$

$K_B = \{[H^+][B(OH)_4^-]\}/[H_3BO_3]$

$[B(OH)_4^-] = K_B B_T/(K_B + [H^+])$

$DIC = [HCO_3^-] + [CO_3^{2-}] + [CO_2]$

12.6

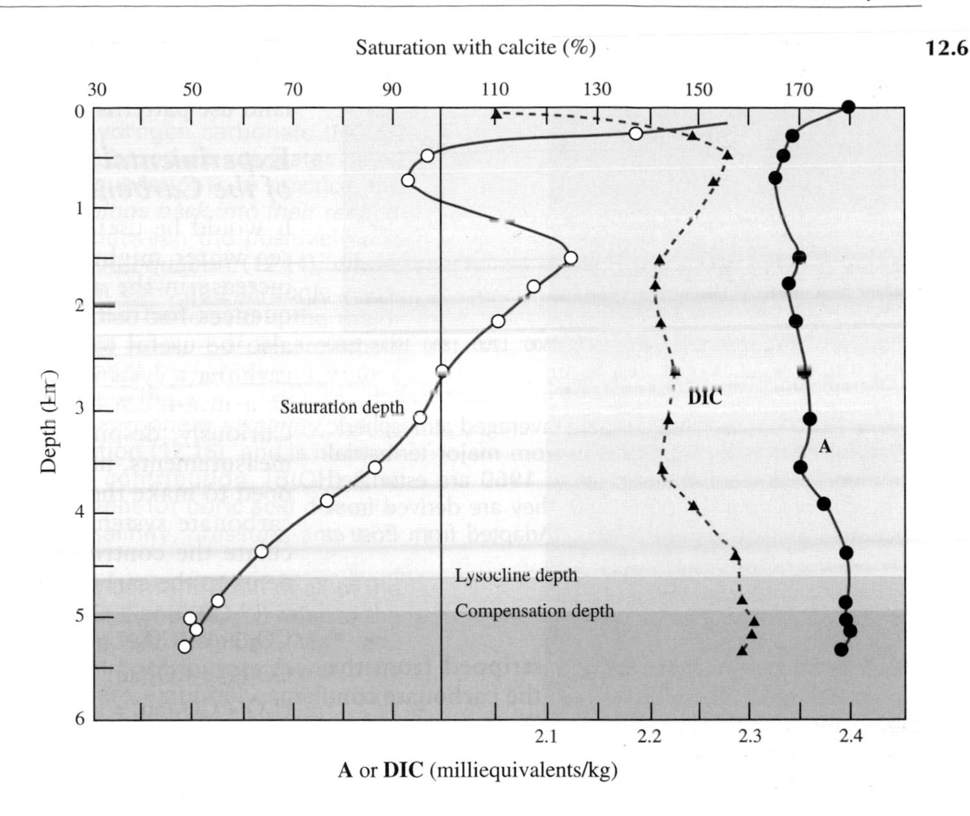

Figure 12.6 Total alkalinity, *A*, total dissolved inorganic carbon (*DIC*) concentration, and degree of calcite saturation as a function of depth in the equatorial Atlantic. Also indicated are the lysocline (see Chapter 11) at approximately 4600 m depth, where there is a perceptible amount of calcium carbonate dissolution, and the compensation depth at approximately 4900 m, below which all calcium carbonate should be dissolved. Since calcite dissolves at deeper depths than it should according to the calcite saturation index, this implies that plant and animal remains (which contain calcite) sink faster than they can be dissolved. (Adapted from Edmond and Gieskes[3].)

12.7

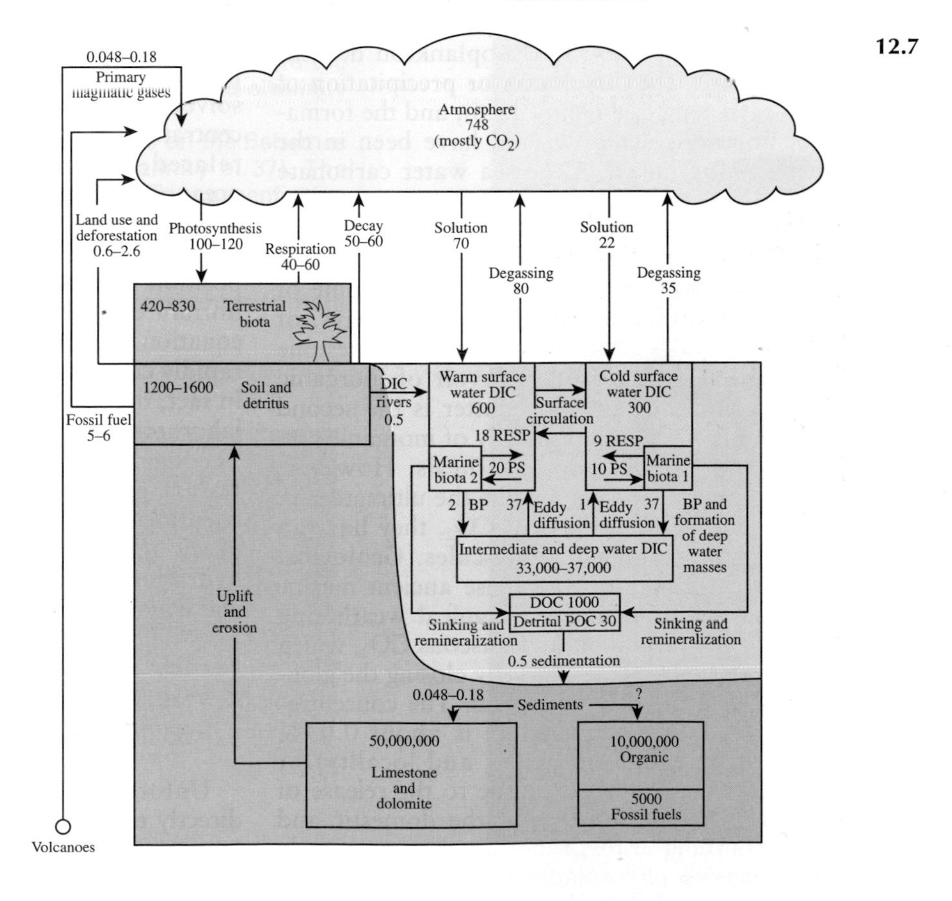

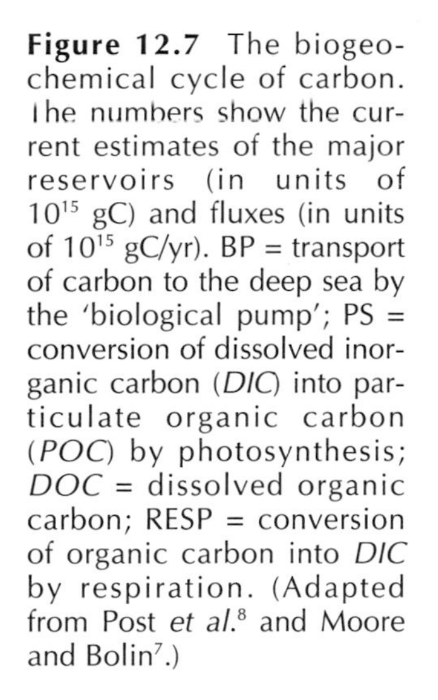

Figure 12.7 The biogeochemical cycle of carbon. The numbers show the current estimates of the major reservoirs (in units of 10^{15} gC) and fluxes (in units of 10^{15} gC/yr). BP = transport of carbon to the deep sea by the 'biological pump'; PS = conversion of dissolved inorganic carbon (*DIC*) into particulate organic carbon (*POC*) by photosynthesis; *DOC* = dissolved organic carbon; RESP = conversion of organic carbon into *DIC* by respiration. (Adapted from Post *et al.*[8] and Moore and Bolin[7].)

Box 12.5 The removal of atmospheric CO_2 by iron enrichment of surface waters

Little effort has been made to alter the rate of fossil-fuel consumption or land use – despite the social, economic, and political impacts. This inactivity is partly due to the level of uncertainty regarding the current predictions of environmental change that will be induced by the increasing atmospheric concentrations of CO_2. It is also due to the perceived prohibitive costs of developing alternative energy sources and systems. As a result, attention has focused on removing the excess of CO_2 from the atmosphere rather than cutting its production at source.

This has led to a proposal that marine phytoplankton growth might be stimulated by fertilising surface waters of the North East Pacific, the equatorial Pacific, and the Southern Ocean (an area of over 10% of the world's ocean). These waters contain abundant nitrate and phosphate, but support an unusually low biomass. The late John Martin believed that primary production in these nutrient-rich waters is limited by iron. Laboratory (microscale) experiments had already convincingly shown that nanomolar iron enrichments of high-nitrate, low-chlorophyll waters do, indeed, stimulate phytoplankton growth and biomass[12]. The 'bottle experiments' have also been successfully repeated in waters south of the Galapagos Islands (project 'IronEx' – see *Figure12.12*), although the results differed from those expected. The observed changes in the partial pressure of CO_2 and in nitrate, fluorescence, and chlorophyll levels were considerably less than theory predicted, presumably due to various unquantified loss terms (the grazing by zooplankton fortunate enough to benefit from the increase in phytoplankton numbers, the export of organic carbon, etc.). However, mesoscale field experiments are notoriously difficult to undertake and interpret, and contain many unexpected and unquantifiable factors.

If the 'iron hypothesis' is correct, adding sufficient soluble iron to these waters should stimulate enough primary productivity to consume one-third to one-half of the anthropogenic CO_2 flux. Supporters of this hypothesis have described this as a rapid method for recreating sedimentary organic matter, to counterbalance the rate of fossil fuel destruction. However, in the extreme, it is also possible that sustained elevated levels of photosynthesis might also remove sufficient CO_2 from the atmosphere to induce periods of glaciation.

12.12

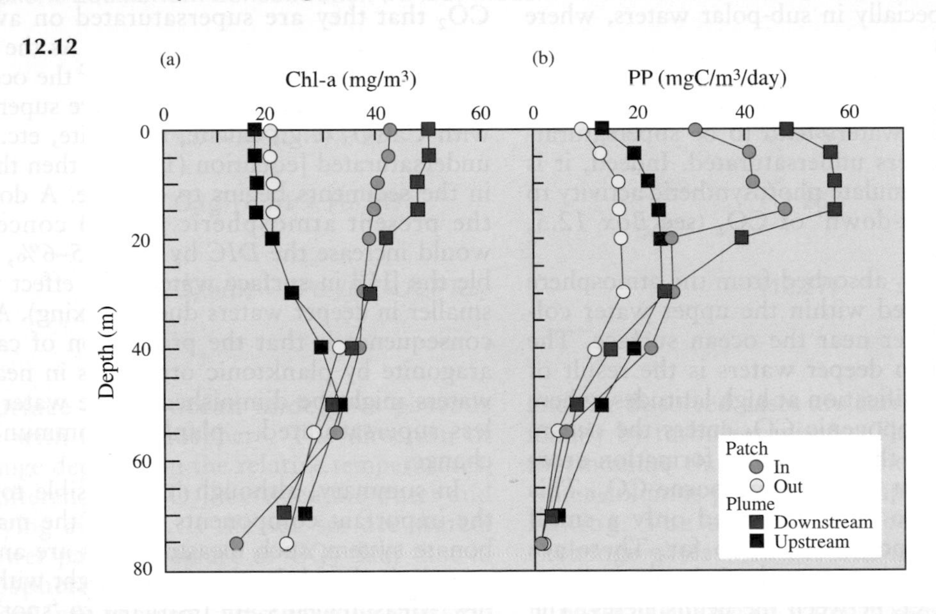

Figure 12.12 During project IronEx in November 1992, a patch of sea water was injected with iron, continuously tracked, and various chemical and biological parameters recorded. This was a particularly tricky operation involving aircraft overflights, radio-drogue buoys, and special chemical tracers (SF_6, for instance). The ship spent a total of 10 days sampling the patch of 'injected' iron. Here are shown comparisons of (a) chlorophyll concentration (Chl-a) and (b) primary production (PP), both in and out of the patch, and both upstream and downstream of the Galapagos Islands. There is a clear increase in both the chlorophyll concentrations (standing biomass) and the primary production (current activity of the plankton). (Data courtesy of the US JGOFS Steering Committee.)

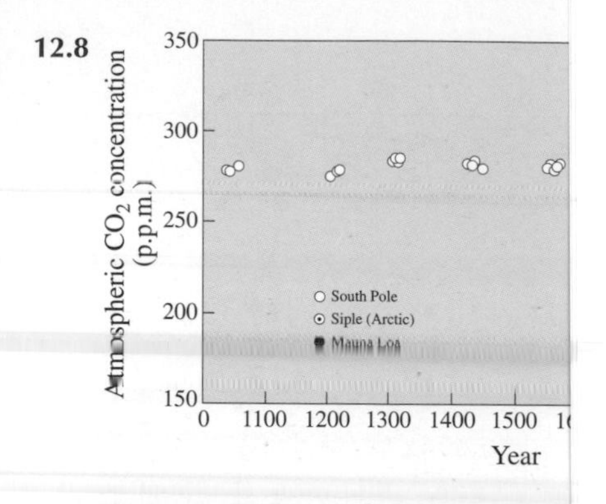

Box 12.6 Dynamic or stable – a question of scale!

What is the time taken for a mass of water to re-establish carbonate equilibrium if the temperature were to suddenly decrease 1°C from 21 to 20°C? At this temperature, the concentration of dissolved CO_2 drops by 4%. The original CO_2 in solution associates with CO_3^- to form hydrogen carbonate ions in order to accommodate the shift in the equilibrium concentrations of all species. CO_2 has to invade surface waters because the water is out of equilibrium with the atmospheric CO_2 concentration. Using the basic equations for CO_2 equilibria, it is possible to calculate that the total inorganic carbon concentration would have to increase by 0.4%, or 8×10^{-6} mol/kg, assuming the original CO_2 concentration was 2×10^{-3} mol/kg. If we assume that the thermocline (the main barrier to mixing with deeper waters) is at 100 m, then the immediate effect is to cause a CO_2 deficiency of approximately 0.8×10^{-3} mol/kg total inorganic carbon. Because of the physics of gas exchange and other meteorological factors, the time taken to re-establish equilibrium with the atmosphere will be slightly longer than 1 year.

Quite a long time! In reality, as CO_2 enters the ocean the deficit between the air and the sea becomes less and the net flux across the interface falls exponentially, so it is difficult to calculate exactly how long it would all take. The oceans change temperature seasonally (this 'drop' takes place every six months). The mixing of waters between polar and equatorial regions happens over a longer time scale (e.g., years), so the equilibration of air and sea never quite reaches completion – hence the term dynamic!

have to be fully utilised, and we must make use of developments elsewhere to help in our research. Future research in this area must concentrate on the development of sensors that can be deployed (over monthly and annual periods of time) on buoys or submersible platforms. The present evidence suggests that the marine environment interacts strongly with the atmosphere in such a way that the climatic consequences of increasing levels of atmospheric CO_2 may be accurately modelled – provided that we can find out more about the marine environment. This is a highly political and controversial area (see *Figure 12.13*, for example). Although the Earth may seem to be in some sort of homeostatic state of regulation (Gaia – *Box 12.7*) this should not lead us into a false sense of security. Short-term variations in atmospheric CO_2 may still create significant changes in meteorological conditions and global climate. This is an area that we have only just begun to research, and in which we have much to learn.

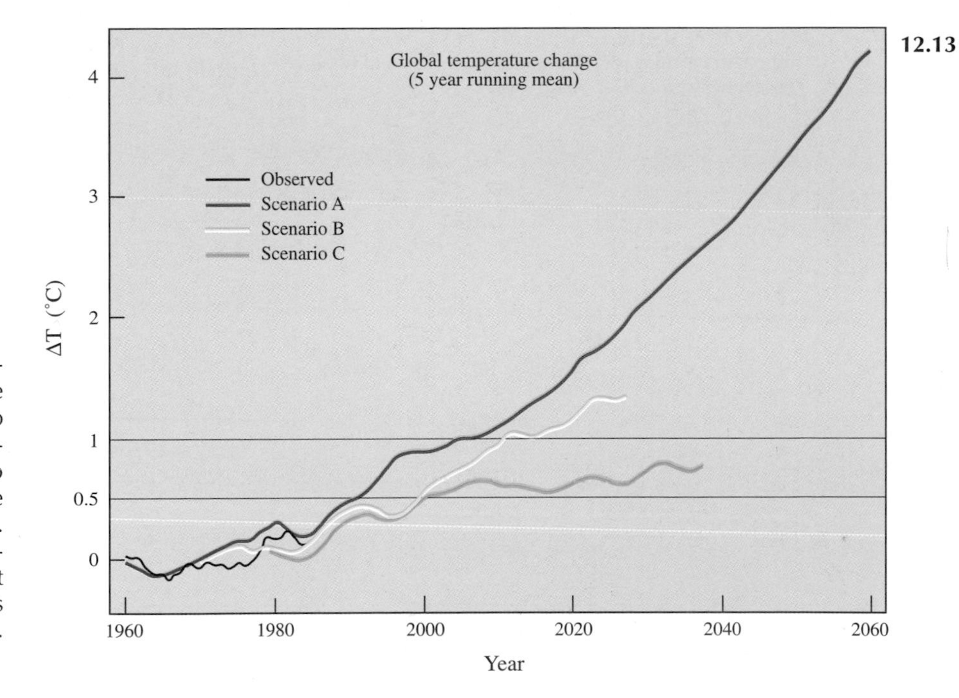

Figure 12.13 Predicted temperature increases, ΔT, for three different scenarios. In scenario A, the current rate of CO_2 emissions is continued. In scenario B, the emissions of CO_2 are held at their present values. Finally, in scenario C, the output emissions of CO_2 are cut such that atmospheric levels stabilise by the year 2000 AD. (Adapted from Libes[5].)

Box 12.7 The Gaia hypothesis

The biosphere appears to counteract naturally the artificial increase in atmospheric CO_2 by acting as a sink for it, and so buffering the greenhouse effect. The principal cause of the progressive fall in the ratio $[CO_2]/[O_2]$ is biological activity, removing CO_2 and releasing O_2 during photosynthesis. Relationships of this kind have led to the novel concept that the surface of our planet is actively maintained as a life-supporting environment by biological activity which acts as a feedback mechanism. This is James Lovelock's concept of the **Gaia** hypothesis, first proposed in the 1970s:

... without life's interference, CO_2 would accumulate in the air until dangerous levels might be reached.

It is the strong interaction of the geochemical cycles of the elements and the biosphere that *regulates* the environment, and Gaia's policy is always to turn existing conditions to its advantage. The biosphere actively maintains and controls the composition of the atmosphere so as to maintain an optimum environment for life and self-perpetuation. The geological fact that atmospheric conditions have remained practically the same over several tens or hundreds of thousands of years suggests that this dynamic system has been really quite stable!

General References

Broecker, W.S. and Peng, T.H. (1982), *Tracers in the Sea*, Eldigio Press, New York, 690 pp.

Butler, J.N. (1982), *Carbon Dioxide Equilibria and Their Applications*, Addison-Wesley, Reading, MA, 259 pp.

Lovelock, J. (1991), *Healing Gaia*, Harmony Books, New York, 1992 pp.

Skirrow, G. (1975), The dissolved gases – carbon dioxide, in *Chemical Oceanography*, Vol. 2, Riley, J.P. and Skirrow, G. (eds), Academic Press, London, Ch 9.

References

1. Bacastow, R.B. and Keeling, C.D. (1979), Models to predict atmospheric CO_2, in *Workshop on the Global Effects of CO_2 from Fossil* Fuels, Elliot, W.P. and Machta, L. (eds), Report CONF-770385, US Department of Energy, Washington, DC.
2. Coulson, K.L. (1975), *Solar and Terrestrial Radiation*, Academic Press, London.
3. Edmond, J.M. and Gieskes, J.M.T.M. (1970), On the calculation of the degree of saturation of sea water with respect to calcium carbonate under *in situ* conditions, *Geochim. Cosmochim. Acta*, **34**, 1261–1291.
4. Keeling, C.D. (1968), *J. Geophys. Res.*, **73**, 4547.
5. Libes, S. (1991), *Marine Biogeochemistry*, Academic Press, New York, 734 pp.
6. Marland, G.T., Boden, T.A., Griffin, R.C., Huan, S.F., Kancircuk, P., and Nelson, T.R. (1989), *Historical and Predicted Emmissions of Greenhouse Gases*, ORNL/CDIAC-25, Oak Ridge National Laboratory, Oak Ridge, TN.
7. Moore, B. and Bolin, B. (1987), The marine carbonate cycle, *Oceanus*, **29**, 11.
8. Post, W.M., Peng, T.H., Emmanuel, W.R., King, A.W., Dale, H., and DeAngelis, D.L. (1990), The biogeochemical cycle of carbon, *Amer. Sci.*, **78**, 314.
9. Schneider, S.H. (1989), The changing climate, *Sci. Amer.*, **261**, 38–47.
10. Takahashi, T. (1989), The effect of the marine carbonate system on climate, *Oceanus*, **32**, 29.
11. Takahashi, T., Weiss, R.F., Culberson, C.H., Edmond, J.M., Hammond, D.E., Wong, C.S., Li, Y.-H., and Bainbridge, A.E. (1970), Global effects of CO_2 from fossil fuels, *J. Geophys. Res.*, **75**, 7648–7666.
12. Wells, M. (1994), Pumping iron in the Pacific, *Nature*, **368**, 295.

CHAPTER 13:

A Walk on the Deep Side: Animals in the Deep Sea

P.A. Tyler, A.L. Rice, C.M. Young, and A. Gebruk

And now, how can I retrace the impression left upon me by that walk under the waters? Words are impotent to relate such wonders.

Professor Pierre Aronnax, in Jules Verne's *Twenty Thousand Leagues under the Sea*

Introduction

Water is a totally alien environment for us humans. Few of us do more than enter the sea and, as we swim, dip our heads briefly beneath the surface to take a momentary glimpse of a blurred and strange world. Even the best-equipped Self-Contained Underwater Breathing Apparatus (SCUBA) divers can penetrate no more than a few tens of metres, for to go deeper requires highly specialised equipment or very sophisticated and expensive submersibles. But imagine that, like Captain Nemo and the crew of Jules Verne's *Nautilus*, we could walk freely across the ocean floor and let us review what we might experience if we took the long journey across the Atlantic Ocean from Britain to the Bahamas.

If an extra-terrestrial visitor to the Earth wanted to take back a collection of animals from its most typical environment, he could do no better than sample the abyssal deep-sea floor across which our route takes us. For the world ocean covers 70% of the planet's surface and over 80% of it is more than 3000 m deep. Yet our own knowledge of what lives there is very recent. Although a few samples had been taken in relatively deep water in the first half of the nineteenth century, serious study of the deep-sea fauna began with the cruises of HMS *Porcupine* in 1868 and 1869[6], and received a further boost from the circumnavigation of HMS *Challenger* from 1872–1876 (see Chapter 1). The reports on the data collected from HMS *Challenger* filled 50 large volumes and included descriptions of hundreds of new species. But because of the coarse nets routinely used from HMS *Challenger*, and for the next 80 years or so, most of the animals collected from the sea floor were relatively large and lived on or just beneath the sediment surface, just the sort that we are likely to see. But we should be aware that this is just a fraction of the deep-sea fauna, for the vast majority is made up of tiny animals measuring millimetres at most and hidden from our eyes within the bottom muds beneath our feet. The richness of this tiny fauna became apparent only in the 1960s, when the Americans, Howard Sanders and Robert Hessler, started sampling the deep-sea floor with much finer nets and recovered an amazing variety of small invertebrates[5]. A consequence of studies using sampling apparatus to collect the smaller organisms is the recent suggestion[2] that the deep sea may have a species diversity equivalent to, if not greater than, that of tropical rain forests (see also Chapter 15). With this proviso in mind, let us plan our expedition. Although nets collect animals from the sea floor, the use of the deep-sea camera has allowed us to view this environment undisturbed. The photographs presented in this chapter come either from cameras mounted on trawls, from 'Bathysnap' (see Chapter 7), or from cameras mounted on submersibles, the last of which allows the operator to select the target.

Route Planning

Our route takes us from the intertidal (see Chapter 16), across the continental shelf to the west of the British Isles (*Figure 13.1*). As we cross the edge of the shelf at about 200 m depth, the slope of the sea floor steepens perceptibly and the already dim downwelling light diminishes rapidly. By the time we reach a depth of 300–400 m we are no longer able to detect any daylight from the surface and we enter a vast zone of perpetual darkness.

Our path takes us down the gentle northeastern slope of the Porcupine Seabight, named after HMS *Porcupine*, for this was one of the very first areas of deep sea to be sampled. We head first south and then west where, at a depth of about 3500 m, the Seabight opens out onto the Porcupine Abyssal Plain. A more direct route would have been down the eastern flank of the Seabight, but this area is riven with deep channels known as submarine canyons. The fauna here is different from that on our route, but the difficulty of the terrain makes sampling with surface-deployed gear hazardous so that this area, together with similar rough topography in the world's ocean, is best sampled from sub-

13.8

Figure 13.8 Fresh phytodetritus collecting in depressions (about 20 cm in diameter), some of them feeding marks left by sea stars, at a depth of about 2000 m in the Porcupine Seabight in May 1982. Phytodetritus consists largely of dead and dying phytoplankton cells which picks up a variety of other particles as it sinks through the water column (Chapter 7). Much of it is recycled during its downward journey, but a significant proportion reaches the deep-sea bed as a seasonal pulse and forms an important food source for the bottom-living animals, in this case the sea-urchin *Echinus affinis*.

13.9

Figure 13.9 'Old' phytodetritus concentrated in biogenic depressions (about 50 cm diameter) on the Porcupine Abyssal Plain at a depth of 4850 m in September 1989. The patchiness produced by this phenomenon may at least partly explain the high biodiversity of the deep-sea floor communities.

include sponges such as *Euplectella* sp and the pompom-shaped sponge *Crateromorpha* (*Figure 13.10*). The cnidarians are richly represented here by gorgonians, the most dramatic being the bottle-brush *Thouarella* sp., *Iridiogorgia* sp., and the fan-shaped Paramuriceidae (*Figure 13.11*).

The last of these orientates normal to the prevailing current, allowing the maximum cross-section for particle filtration. Filter feeding is also found in the crinoids, the dominant echinoderm group in this habitat, of which the bright yellow *Anachalyspicrinus nefertiti* (*Figure 13.12*) and *Porphyrocrinus thallassae* are the most spectacular examples. A similar life-style is adopted by the brisingid sea stars, such as *Brisingella multicostata*, which sit atop rocks in this zone to benefit from the accelerated flow over topographical highs (*Figure 13.13*).

The fauna of this part of the slope can be used to determine current direction (*Figure 13.14*). Non-random orientation of the crinoid *Porphyrocrinus thallassae* and the downstream winnowing of sediment behind a glacial erratic indicate the flow of Northeast Atlantic Deep Water along this slope[8].

13.10

Figure 13.10 The sponge *Crateromorpha* at about 2120 m depth to the west of the Porcupine Bank. The main filtering apparatus is the pompom at the distal end of the stalk, some 20 cm above the rock surface.

13.11

Figure 13.11 A branched gorgonian (ca. 1 m high) belonging to the family Paramureicidae at a depth of 2800 m on the western slope of the Porcupine Bank. The small anemone-like polyps found along each branch filter particles from the water column. These colonies, which may be a metre high, are orientated at right angles to the current so the broadest face of the colony faces the current, thus maximising particle capture.

13.12

Figure 13.12 The brilliant yellow crinoid *Anachalypsicrinus nefertiti* at 2480 m depth to the west of the Porcupine Bank. The star-shaped part of the animal filters particles from the water column, the stalk keeping this filtering apparatus well above the sea bed. The animal can be up to 25 cm high.

13.13

Figure 13.13 Three individuals of the multi-armed sea star *Brisingella multicostata* resting on a small rocky hillock at 2890 m depth to the west of the Porcupine Bank. The arms are extended into the water column to trap small animals and particles brought past the rock by the current.

13.14

Figure 13.14 Individuals of the bright red crinoid *Porphyrocrinus thallassae* attached to the side of a block of rock at 2310 m depth to the west of the Porcupine Bank. The crinoids are all orientated in the same way, showing that the current is from the right.

13.26

Figure 13.26 The multi-armed sea star *Novodinea antillensis* in a typical feeding posture at 660 m depth off the Bahamas. Food, such as small crustaceans, is trapped by tiny pincers on the arms and moved to the mouth by the tube feet. The brilliant colour is typical of brisingid sea stars (see also *Figure 13.13*).

leathery. If we reach out and touch this animal, we find the body is soft and delicate. It can be damaged beyond recognition by only a moderate poke of the finger.

On a nearby rock outcrop there is a brilliant red starfish resembling *Brisingella,* which we saw on the slope west of Ireland. *Novodinea antillensis* (*Figure 13.26*), the Bahamian version, is perched on the highest available portion of the bottom, where it has access to the currents that pass by. Its posture is that of a filter-feeder, but close examination of its long arms reveals it to be a formidable predator of euphausiids and other swimming crustaceans. The entire body is covered with tiny pincers which form a velcro-like surface capable of entrapping any small objects unfortunate enough to encounter it. When we brush the starfish arm with our own, it immediately grasps our hairs in hundreds of places. We have no trouble escaping, but a small crustacean whose legs are trapped is in a much worse position. Once captured, the prey is enclosed in a loop of the arm, grasped with the tube feet, and moved directly to the gaping mouth on the central underside of the disk.

Novodinea is not the only animal here that gives us a feeling of *déjà vu.* As we look at the stalked crinoids, the gorgonians, the hexactinellid sponges, and the other animals populating this rock outcrop,

13.27

Figure 13.27 The crowns of six individuals of the crinoid *Endoxocrinus* at 640 m depth off Egg Island in the Bahamas. Some of the arms are raised to be flicked rapidly downward, a mechanism which apparently they use to dislodge annoying crustaceans. Compare the number of arms seen on these individuals with the five seen on the species in *Figure 13.12*.

it is apparent that, given the same substratum and depth, faunas may be very similar in widely separated parts of the world ocean. The names of the players may be different on the two sides of the Atlantic, but the roles they play are very much the same.

Like flowers struggling toward the light in a shaded garden, stalked crinoids cluster along the ridges where currents are faster, thereby bringing more food to the waiting arms and pinnules (*Figure 13.27*). Each crinoid bends with the currents in exactly the same way as its neighbours; all have arms poised in a parabolic fan to entrap tiny particles in thick mucus. Occasionally, we see one of these arms abandon its normal posture to move up and down rapidly. It appears that the crinoid is waving to us, but a closer look reveals the fallacy of this interpretation. Small crustaceans occasionally dart toward the feeding groove on the oral side of the arm, possibly to steal some of the concentrated food that the crinoids have accumulated in their mucus. The crinoid responds to this irritation by flicking the offender away, much like a horse swishing flies with its tail.

On the soft sediment near this rock outcrop is one of the most common sea-urchins of this region, *Stylocidaris lineata* (*Figure 13.28*). If it is springtime, probably each individual will be paired with another of its kind. Sea-urchins do not mate, but instead cast their eggs and sperm into the sea, where fertilisation occurs. Normally, these deep-sea animals are so far apart that egg–sperm encounters are unlikely. They pair in the spring to increase the odds that sperm will find eggs to fertilise. A pure white snail living on this same slope (*Tugurium caribaeum*) carries odd bits of broken shell and other refuse around the margin of its shell.

13.28

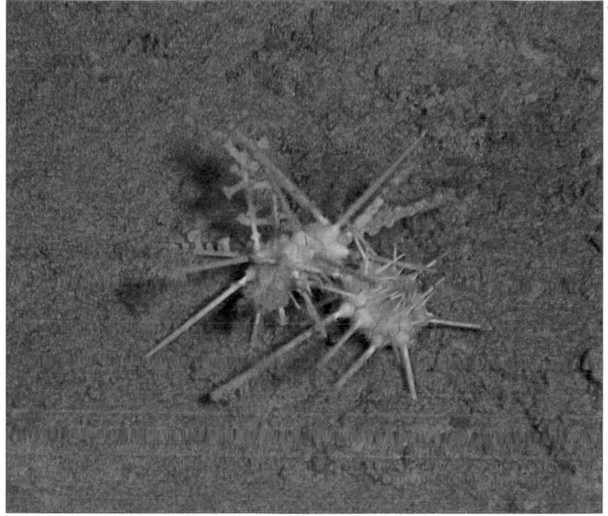

Figure 13.28 The sea-urchin *Stylocidaris lineata* on a rocky outcrop at 500 m depth off the Bahamas. During their reproductive season individuals of this species come together in pairs for spawning to aid their fertilisation success.

By 450 m depth, our dark-adapted eyes start to perceive a greyish glow of light from the surface. Two large sea-urchins catch our eye. The first, *Calocidaris micans* [*Figure 13.29(a)*], is the largest sea-urchin we have seen anywhere on the trip and the second, *Coelopleurus floridanus* [*Figure 13.29(b)*], is the most beautiful. The spines of *Calocidaris* are perfectly straight and have the consistency and lustre of ivory elephant tusks.

3.29a

13.29b

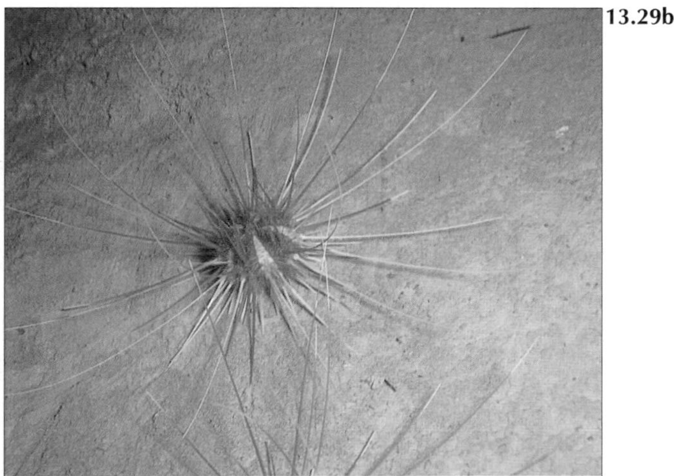

Figure 13.29 Protection in the deep sea. (a) The sea-urchin *Calocidaris micans* at 260 m on sediment off the Bahamas; the straight spines can be up to 20 cm long. (b) The striped sea-urchin *Coelopleurus floridanus*, at a depth of 450 m on sediment in the Tongue of the Ocean. The black, white, and red shell, and the elegant curve of the spines, make this a most striking animal.

14.2

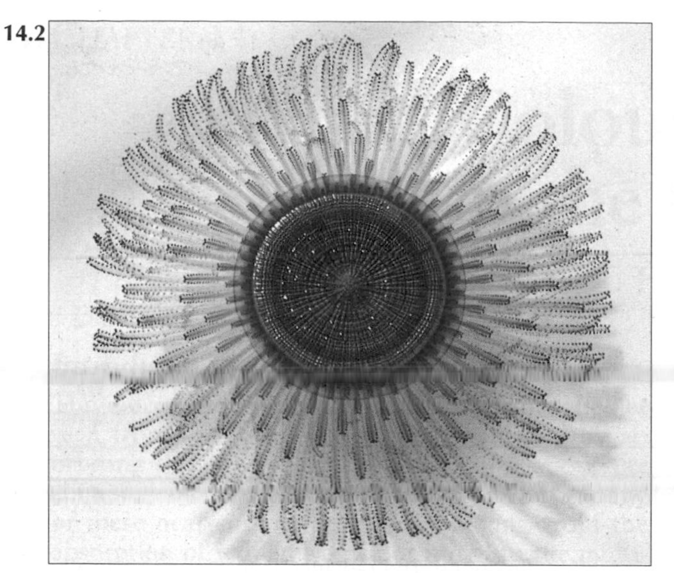

Figure 14.2 The 40 mm diameter coelenterate *Porpita* floats at the surface, where it is camouflaged by a blue carotenoprotein pigment.

14.3

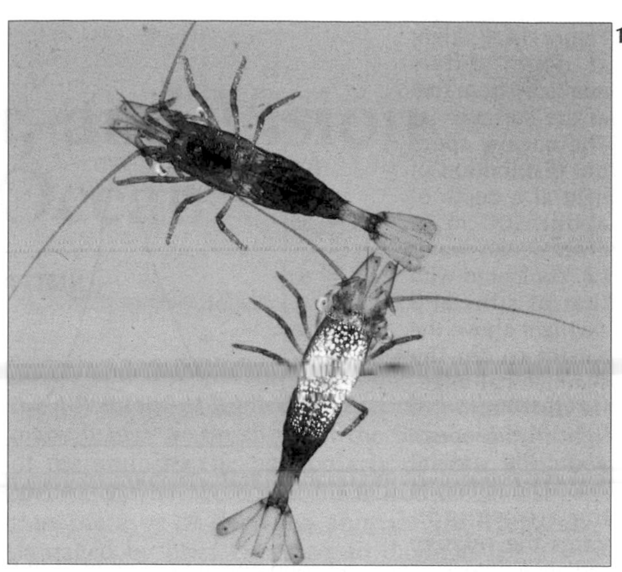

Figure 14.3 This 10 mm long surface living shrimp, *Hippolyte coerulescens*, is camouflaged by a blue carotenoprotein pigment and white reflective dorsal chromatophores. (Courtesy of the Southampton Oceanography Centre, England.)

14.4

Figure 14.4 Blue colour can be achieved without pigment. In the isopod *Idothea metallica* (12 mm long), tiny reflective particles in the upper structure reflect blue light much better than red. The transmitted red is then absorbed by a dark pigment beneath the reflective layer. (Courtesy of the Southampton Oceanography Centre, England.)

upward scattered radiance of clear ocean waters. The colour is achieved in different ways by different species. Blue carotenoprotein or biliprotein pigments are commonly used (*Figures 14.2*, *14.3*). Blue structural colours (selective diffuse or specular reflection) are other means of achieving the same result (*Figure 14.4*). The deep blue colour with which many upper ocean fish camouflage their dorsal surfaces has a similar structural basis.

Transparency and silvering

For smaller organisms in the upper waters, camouflage can be achieved by transparency. Many planktonic species, particularly the gelatinous forms, rely on this for their protection (*Figure 14.5*). For larger animals the complexity of the body tissues renders transparency impracticable. However, in the particular light environment of open water (with the brightest light from vertically above and a symmetrical radiance distribution about this axis), a mirror stood vertically in the water becomes invisible from any angle of side view. Many animals have taken advantage of this to mimic transparency by turning themselves into the equivalent of vertical mirrors. This is best

14.5a

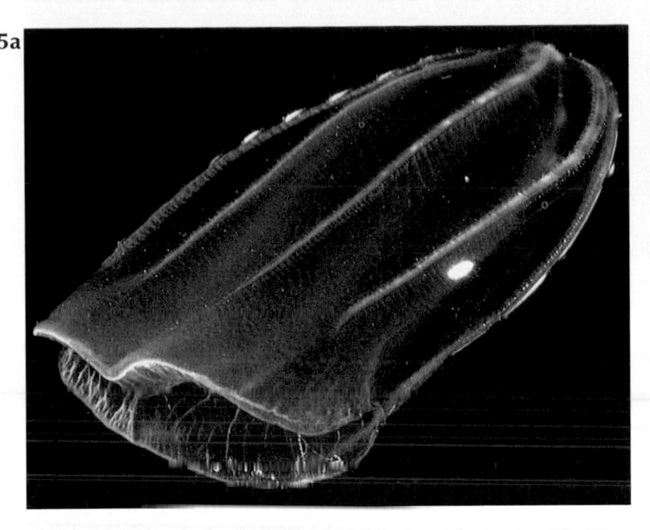

14.5b

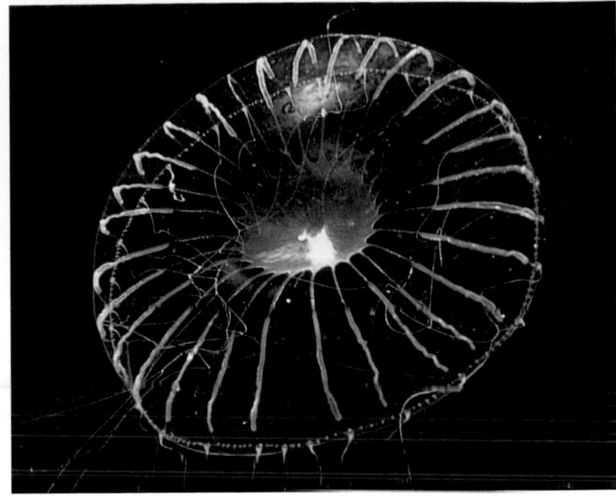

Figure 14.5 Many jellyfish are often almost invisible by virtue of their transparency. Typical examples are (a) the comb jelly *Beroe* (60 mm diameter) and (b) the medusa *Aequorea* (25 mm diameter). (Courtesy of Image-Quest 3-D.)

achieved by flattening the body, so that the flanks are vertical, and covering these with reflective material. Fish are consummate examples of this strategy, none more so than the hatchet fish (*Figure 14.6*). For their reflective material they use tiny crystals of guanine (an excretory product derived from nucleic acids), which are aligned parallel to the body surface. The crystals are arranged in multiple stacks, alternating crystal and cytoplasm (the watery matrix of the cell), whose spacing is such as to achieve constructive interference reflection. For a particular wavelength of light, λ, viewed at right angles to the stack of crystals, ideal interference reflection occurs[1] when the optical thickness of each layer is 0.25λ. With such a system, almost 100% reflectance is achieved with only 5–10 crystals. Without such spacing, five crystals would have a reflectance of only 20%.

One potential drawback of the system is that the best-reflected wavelengths shift toward the blue end of the spectrum as the angle of viewing becomes more oblique.

Effective silveriness requires reflection of all wavelengths, at all angles of view. To achieve this, the spacing of the stacks is adjusted so that either different colours are reflected from different stacks (which may be adjacent or superimposed on one another), or the spacing within the stacks varies in a regular way. Vertical flanks are not compatible with a muscular stream-lined body, so most fish

14.6

Figure 14.6 In the radially symmetrical light distribution in the ocean (*Figure 14.1*), a vertical mirror is invisible from the side. These hatchet fish (*Argyropelecus*, 25–50 mm long) have turned themselves into mirrors by flattening their sides and silvering them, using stacks of guanine platelets as interference reflectors.

have a more elliptical cross-section. If the reflecting stacks in the skin or scales remain vertically oriented, despite the curved body surface, the effect of a vertical mirror is retained[1].

14.7

Figure 14.7 Where some organs remain opaque they can still be camouflaged by silvering them separately, as is the case for the liver and eyes of the 40 mm long squid *Cranchia scabra*.

Even when much of the body is transparent, there may be particular tissues that remain opaque (e.g., eyes, red muscle, or digestive organs). These organs can still be silvered individually to achieve effective camouflage. Squid use the same strategy, but employ reflective platelets of protein rather than guanine crystals (*Figure 14.7*).

The concealment value of vertical reflective surfaces rapidly disappears as the surfaces are tilted. This property can be used to good effect to distract predators (e.g., the flashing of a twisting school of fleeing sardines or the eponymous silversides). Changes in body orientation can also be used to send optical signals to nearby members of the school.

The dramatic colour changes visible in some oceanic fish (e.g., the coruscating colours of a captured dolphin fish, *Coryphaena*) are brought about by very rapid changes in the spacing of the crystals in the reflecting cells. Each stack of crystals behaves like venetian blinds as contractile elements in the cells tilt the individual crystals, altering their distance apart and hence their reflected colour. Rapid colour changes, under similar control, take place in the reflective stripe of the freshwater neon tetra and in some damselfish.

Deep-water colours

If reflection is the saviour of animals in reasonably well-lit open water, it spells potential disaster for those in the dark of the ocean depths. Here, the reflection of a bioluminescent flash or glow could break the cover of an animal hitherto invisible

14.8

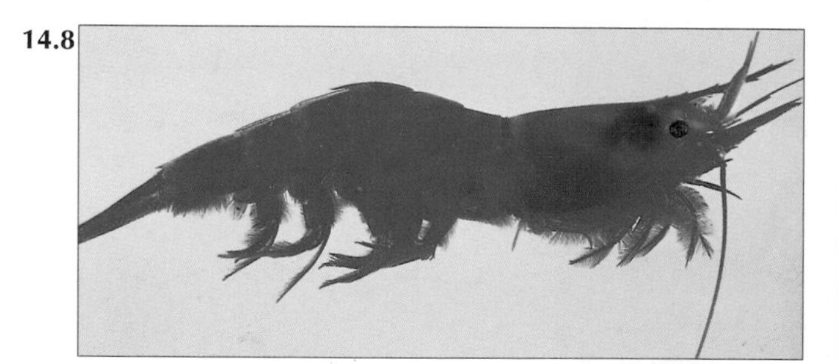

Figure 14.8 The uniform scarlet colour of the 70 mm long deep-living shrimp *Acanthephyra purpurea* is due to a carotenoid pigment that absorbs any incident blue light. We see it as scarlet in the white light of the camera flash; it is invisible in the deep sea, where blue bioluminescence is the norm and most predators only have blue-sensitive eyes. (Courtesy of the Southampton Oceanography Centre, England.)

14.9

Figure 14.9 The black melanin pigment of this typical deep-sea fish (*Gonostoma bathyphilum*, 110 mm long) plays the same camouflage role as does the scarlet pigment of the shrimp. (Courtesy of the Southampton Oceanography Centre, England.)

14.10

Figure 14.10 In midwater, where light from the surface is still important and day–night changes are substantial, the limited colouring of the 50 mm long shrimp *Sergestes* is mostly distributed in large dorsal chromatophores. This allows the animal to change its appearance according to the light environment.

against the black background. Thus, silvering is an anathema in deeper water, and is replaced by uniform matt colours of brown, purple, black, or scarlet (*Figures 14.8, 14.9*). This is a particularly clear example of the functional equivalence of subjectively very different colours. Dramatically different as these colours may appear to us in the sunlight on deck, they are all equally effective at preventing the reflection of any residual dim blue light filtering down from the surface, or of stray flashes of blue bioluminescent light. As the bioluminescence may come from any direction, the colouring is spread over the whole of the animal.

At intermediate depths, where daylight from the surface is dim but still significant, and day–night intensity changes are still important, a compromise is reached in which animals have some element of colour and some of silvering or transparency. This enables them to adjust their colouring quite markedly in response to changes in light intensity. Fish silvered in daylight have mobile dark pigment cells which disguise their silveriness at night. Shrimps have a 'half-red' appearance, in which the red pigment is present in large pigment cells, and are able to disperse or aggregate the colour as appropriate to the light conditions (*Figure 14.10*). In both cases, the pigment distribution is primarily dorsal, in response to the continued dominance of downwelling light.

As the bottom is reached, quite marked changes occur in the colours of the animals. Many animals are now grey, pale, or even white. There is no obvious rationale for this change, except that in any bioluminescent light the paler species present a lesser contrast when seen against the lighter sediment than do their heavily pigmented pelagic relatives only a few tens of metres above (*Figure 14.11*). Many of the animals on the bottom have large sensitive eyes, so vision clearly still plays an important role in this environment.

Many oceanic animals undertake substantial vertical migrations during their lifetimes. In general, the larvae and juveniles live at shallower depths than the adults, and thus experience gradual changes in the light conditions as they move deeper in the water column. Their appearance at any given stage of development reflects the depth at which they are living. Early shrimp larvae near the surface may be transparent, the juveniles at mid-depths half-red, and the adults at depth uniformly scarlet. Colour is clearly a key feature in the life of oceanic animals, but its appearance to the denizens of the deep is not always quite what it seems to the human eye in sunlight.

14.11

Figure 14.11 Many bottom and near-bottom animals are very pale, like this 220 mm long rat-tail *Nezumia*, at a depth of 1100 m off southwest Ireland. In the light of the flash these animals are not easily distinguishable from the pale sediment on the bottom; this may also be the case in whatever dim bioluminescence exists on the sea floor. Also here (just above and to the left of the fish) is a shrimp with a reflective eye (see also *Figure 14.18*) and (above and to the right) two large pot-like glass sponges. (Courtesy of the Southampton Oceanography Centre, England.)

14.12

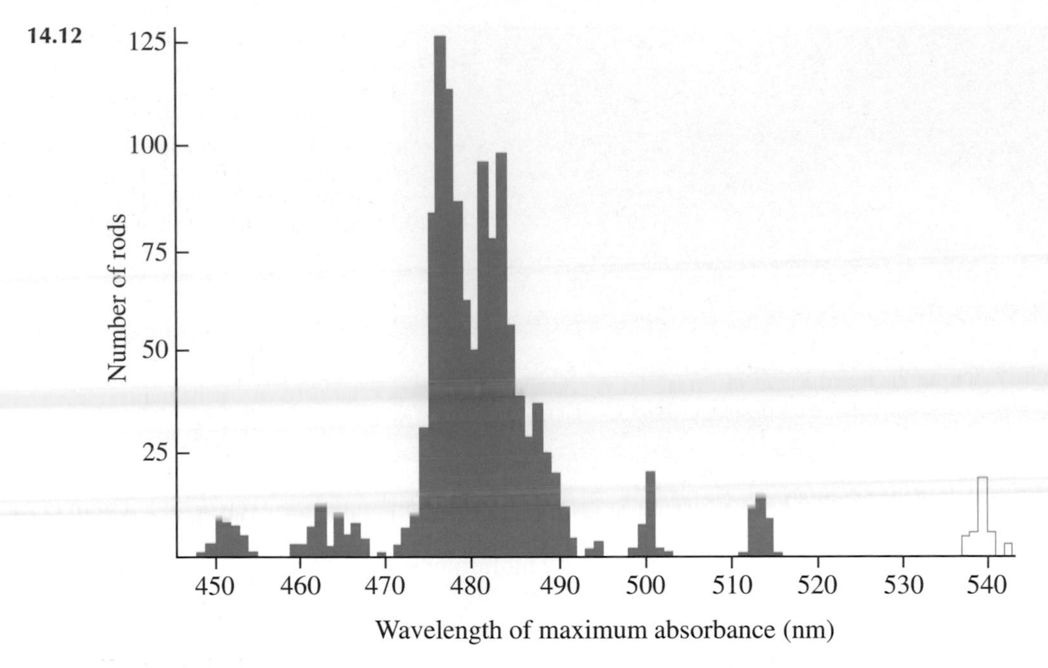

Figure 14.12 The distribution of 1370 measurements of the visual pigment maxima in the rods of 57 species of deep-sea fish shows a good match with the blue–green light in their environment (*Figure 14.1*). Most pigments are rhodopsins; the few longer wavelength porphyropsins are shown as open blocks (from Partridge *et al.*[13]).

Vision

The light environment of the ocean is matched by the visual adaptations of its inhabitants. Vision depends on the absorption of photon energy by the visual pigments and its transduction into a neural signal. The spectral sensitivity of the eye is determined by the absorption characteristics of the visual pigments in the retinal receptors. In vertebrates these receptors are single cells (rods and cones); in invertebrates, they are units (rhabdoms) formed from several cells.

Visual pigments

In general, the absorption maximum of the main visual pigment is a good match to the spectral characteristics of the environment. Thus, deep-sea fish usually have rod visual pigments with absorption maxima in the blue wavelengths around 480 nm (*Figure 14.12*), while shallow coastal species have maxima at longer wavelengths[11,13]. Near the surface the high intensity and broad spectral range of ambient light provide the opportunity for both colour vision and high acuity. The dominant visual task is to maximise the contrast present in the target area. When a *dark* object, or silhouette, is seen against the background of downwelling light, or horizontally against an infinite background of scattered light, the contrast is maximised by having a visual pigment which matches the spectral transmission of the water. When, on the other hand, a *bright reflective* object is viewed horizontally, maximum contrast can be achieved by exploiting the spectral differences between the reflected and background light, using a visual pigment whose absorption maximum is offset from that of the background.

Visual pigments are formed by linking a protein, one of the opsins, to a vitamin A_1 or A_2 derivative (forming a rhodopsin or a porphyropsin, respectively). Porphyropsins absorb at longer wavelengths than do their rhodopsin partners. Although a few marine fish do have this pair of pigments, many of them lack the porphyropsin, but have more than one rhodopsin (i.e., vitamin A_1 with different opsins) and thus retain the potential for colour vision. Additional visual pigments may also be present, usually in different types of cone cell, and coloured filters or oil droplets may further differentiate the spectral sensitivity of individual receptors. In the most extreme cases (some mantis shrimps) there may be up to eight kinds of receptors, each with different spectral sensitivities. Recent work has shown that the shrimp *Systellaspis debilis* has a visual pigment which is sensitive to near-ultraviolet light, as well as one sensitive to blue–green light[4] (*Figure 14.13*).

Colour vision is also theoretically possible for fish with only one visual pigment, but with a retina containing multiple banks of rods. Each layer modifies the spectral nature of the light transmitted to

4 14.13

Figure 14.13 The shrimp *Systellaspis debilis* (60 mm long) has two visual pigments, one absorbing in the blue–green, the other in the near ultraviolet. Since these wavelengths have different transmission characteristics, their ratio could give the animal an indication of its depth. The dark spots on the thorax and abdomen are light-emitting organs.

the next layer, giving them, in effect, different spectral sensitivities.

Major changes in the light environment of fish occur during the lifetime of those, like the eel, which have a marine and a freshwater phase or, like the pollock, migrate into deeper water as an adult. These changes are compensated by visual pigment changes, either between rhodopsin and porphyropsin pairs or by opsin shifts between different rhodopsins. In either case the new suite of pigments is more appropriate to the visual tasks of the new environment.

The upward view

Light in the deep-water environment is dimmer, bluer, and highly directional. Animals in the upper few hundred metres are likely to sight prey or detrital particles within an angle of 35° from the vertical; here, the downward radiance does not drop below 50% of its maximum value (*Figure 14.1*). This limited, but brighter cone of view (70°), dominates the visual environment. Many animals at these depths have responded by evolving upwardly pointing eyes[9]. Every stage between fully lateral and fully upward eyes can be found in one or other species of mesopelagic fish, culminating in the extreme cases of *Opisthoproctus*, *Benthalbella* (*Figure 14.14*), and the hatchet fish *Argyropelecus*. Visual acuity (resolution) needs to be maximised in this direction and both amphipods and euphausiids have a gradation of forms whose eyes range from round, with a uniform acuity over the whole visual

14.14

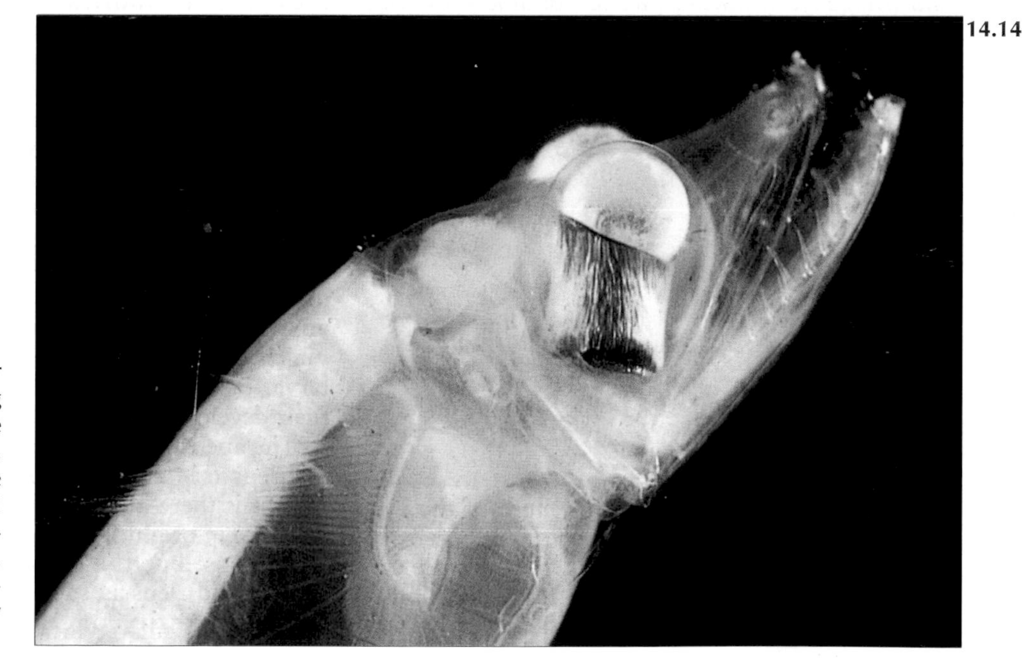

Figure 14.14 The tubular eyes of the 100 mm long fish *Benthalbella* provide a binocular overlap, allowing it to determine the range of prey as well as providing a large aperture for high sensitivity. (Courtesy of the Southampton Oceanography Centre, England.)

14.21

Figure 14.21 *Malacosteus* (175 mm long) has both blue- and red-sensitive visual pigments. The red sensitivity is further enhanced by a red reflector (tapetum) behind the retina, providing the red colour to the eye visible here in daylight.(Courtesy of Dr N.A. Locket, Adelaide University, Australia.)

14.22a

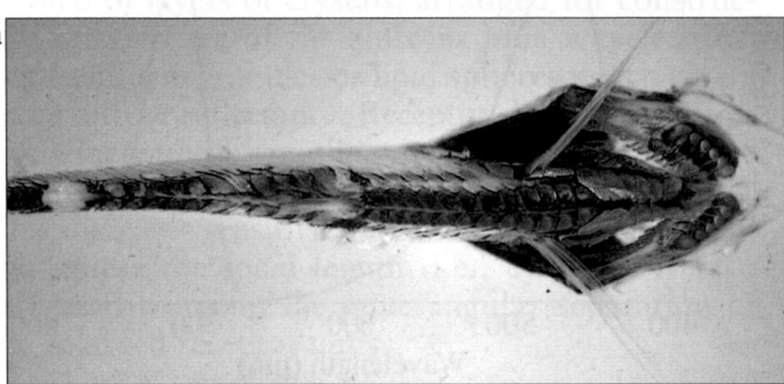

14.22

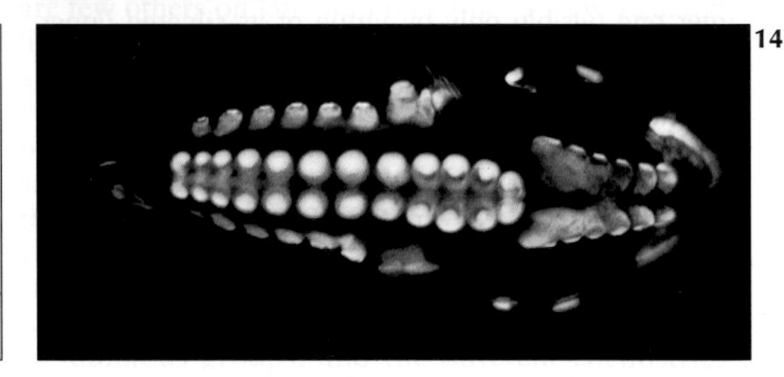

Figure 14.22 (a) The hatchet fish *Argyropelecus* (*Figure 14.6*), seen from below, showing the photophores arranged along its ventral margin, each containing a magenta-coloured filter. This results in the emitted bioluminescence being a clear blue, exactly matching the colour of light in the sea (courtesy of P.M. David, Southampton, England). (b) A luminescing specimen by its own light.

The long wavelength light must be very important to the fish, but it would not normally be able to see it if it had only a typical blue–green sensitive visual pigment. *Malacosteus* turns out to have a red-sensitive visual pigment as well. It also has a scarlet tapetum (see earlier) to maximise its sensitivity to these long wavelengths (*Figure 14.21*). It thus has a 'private' wavelength, which could be used either as a secure communication with others of the same species or to break the camouflage of red shrimps, whose colour only works in blue illumination and whose blue-sensitive eyes would not detect the red illumination. Red light of these wavelengths is rapidly absorbed by sea water, so it can only be effective over short visual ranges[3].

A few other marine animals produce light of more than one colour. The stalked sea-pen *Umbellula*, for example, has green luminescence on its stalk, but blue emission from the polyps at the top. Some squid produce green and blue light from different light organs, and can also change the emission spectrum from a single light organ. Another coelenterate, a sea anemone-like zoanthid, has colonies in which some individuals have green light and some yellow.

Luminous camouflage

The ecological value of most of the exceptions noted above is not yet clear, but the value of precisely controlling the colour has been well-established in the hatchet fish. This fish lives at depths where daylight is still important and, as already described, has mirror-like camouflage and tubular eyes. Despite being very laterally flattened, it cannot altogether avoid being seen in silhouette from below. Like many other fish at mid-depths, it eliminates this silhouette by having rows of light organs along its ventral surface [*Figure 14.22(a)*]. All its light organs point downward, except one, which points *into* the eye. By adjusting the luminescence shining into the eye to match the downwelling daylight, it simultaneously matches all its ventral lights to the surrounding light – and vanishes. However, the broad bandwidth light produced *within* the light organ is not quite the

Figure 14.23 Mature females of the pelagic octopod *Japetella* (100 mm long) develop a yellow-coloured bioluminescent oral ring which degenerates after they spawn. Males have no luminous organ, so it is assumed that this provides a sexual signal.

same colour as light in the sea. The aperture of each light organ contains a purple filter pigment which corrects this spectral mismatch. After passage through the filter the luminescence has a narrow bandwidth blue emission with a maximum at about 475 nm, corresponding exactly to the spectrum of light in the sea [*Figure 14.22(b)*]. To complete the camouflage, the design of the reflectors in the light organs ensures that the angular distribution of the luminescence also matches that of submarine daylight[3], as illustrated in *Figure 14.1(b)*.

Other luminous defences

Camouflage is an example of a passive defensive function. The most common use of light in the sea is as an active defence, which can take the form of a single, short, bright flash (dinoflagellate), a volley of flashes (some fish), a wave of flashes moving over the body (sea-pens and medusae), or the discarding of sacrificial parts of the body which flash independently to distract a predator (scale worms and brittle stars). Another widely employed active defence is that of a squirted luminescence. Many shrimps, worms, a few squid and fish, and some medusae and ctenophores produce copious amounts of light in this form. Particularly among the ctenophores and medusae, it is not simply a cloud, but is composed of separately scintillating particles. Many of these gelatinous animals hang passively in the water fishing for prey; for any larger animal in the area, they collectively form a luminous minefield.

Figure 14.24 Female angler-fish, such as this 45 mm long *Chaenophryne,* maintain a culture of luminous bacteria in the very elaborate lure. Although regarded primarily as a means of attracting prey, it may also be a means of identifying the female to the non-luminous male.

Sexual light signals

Small crustaceans, such as copepods and ostracods, have luminous glands whose secretions the animals kick away defensively as they swim. They can also be used for sexual displays. The pattern, timing, and trajectory of the luminous gobbets produced in the mating displays of males of the *Vargula* group of ostracods identify them to the waiting females, who swim up to join them. Syllid worms have analogous displays in many parts of the world (e.g., the Bermuda fireworm). Sexual differences in the size or position of the light organs of male and female lantern fish, stomiatoid fish, and some cephalopods also suggest that they have a sexual function (*Figure 14.23*). Female angler-fish have lures containing luminous bacterial symbionts, but the males do not. It is assumed that the lure attracts prey, but perhaps it also sends a specific signal to the males (*Figure 14.24*).

CHAPTER 15:

Ocean Diversity

M.V. Angel

Biodiversity is the term used to describe the rich variety of life found on Earth. It was the subject of the UNCED conference in Rio in 1992, which resulted in the signing of the International Convention on Biodiversity and the adoption of Agenda 21, which lays down the guidelines under which the Convention will operate. However, biodiversity is used to express this variety over a great range of levels of organisation. It can be applied to different types of ecosystems, or to express the number of species found locally or globally, or even to the amount of genetic variation within individual species. This can lead to confused thinking, unless the term's precise meaning is explicitly stated in each context. Our discussion concentrates on ecosystem diversity, and species richness and dominance within local and regional areas.

Diversity at these levels of organisation is the product of evolution. New species almost always evolve as a result of a subpopulation becoming isolated from its parent population for thousands of years, and being subjected to different selective pressures. Generally, the stock of species of the world has increased linearly over geological time (based on the fossil record), although several mass extinctions have interrupted this trend (*Figure 15.1*). The causes of these mass extinctions are still a subject of vigorous debate[17]. However, since the beginning of the Mesozoic era the increase has been almost linear. If we are to understand the present patterns of biodiversity in our seas, we need to appreciate, first, how and why evolution has given the world such a rich compendium of species and, second, how ecological processes continue to maintain this richness.

The Origins of Disparity

The earliest traces of life that have been found so far are imprints of single-celled micro-organisms in sedimentary rocks laid down in the prototype ocean about three and a half billion years ago. The first to appear were prokaryotic cells or bacteria, which have no nucleus and are termed prokaryotes. Since they have the most ancient lineage of any living thing so far, these apparently simple cells show remarkable variations in their physiology, cell chemistry, and genetics. As we learn more about them, there are proving to be greater differences

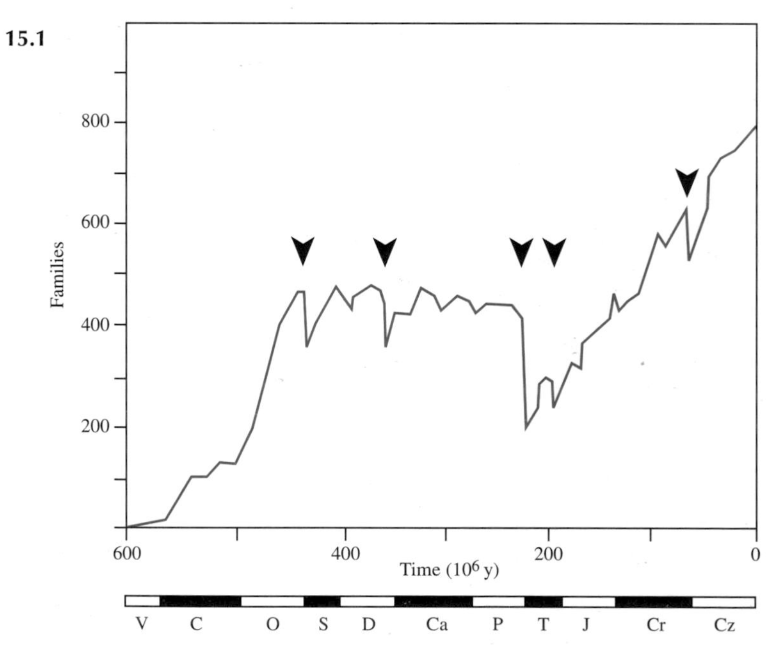

Figure 15.1 Diversity of marine families through the geological record since metazoan organisms first appeared. Note the steady increase until the end of the Ordovician (O in the lower scale), when the first of the mass extinctions occurred. Then the numbers of families remained roughly constant until the mass extinctions of the Triassic (T) and the beginnning of the fragmentation of the supercontinent Pangea (see *Figure 15.4*). Compare the steady increase in numbers of families since the Triassic with the patterns of continental drift shown in *Figure 15.4*. The one interruption was the mass extinction at the end of the Cretaceous (Cr), which more-or-less coincided with the split developing between Australasia and Antarctica, and resulted in the start of circumpolar circulation in the Southern Ocean (redrawn from Sepkowski[17]). (V = Vendian; C = Cambrian; O = Ordovician; S = Silurian; D = Devonian; Ca = Carboniferous; P = Permian; T = Triassic; J = Jurassic; Cr = Cretaceous; Cz = Cenozoic.)

15.2

Figure 15.2 Specimens of the large deep-living copepod species *Megacalanus princeps*. Copepods numerically dominate the vast majority of plankton samples, no matter at what depth they are collected. They outnumber all other animals of comparable size in any other ecosystem, including all insects, and yet there are only just over 1900 known species compared with perhaps a million insects. (© Heather Angel.)

within the prokaryotes than within all other living organisms[9].

It took another two billion years before the first appearance of single-celled organisms *with* nuclei – the eukaryotes. Both the nucleus, which contains the genetic information, and other organelles within the cells – the mitochondria, in which the reactions occur that provide the cell with energy – are thought to have originated as a result of different prokaryotes forming symbiotic associations, which through time became a permanent and obligate relationship.

It was almost another billion years before the first multicelled organisms made an appearance, just before the beginning of the Cambrian era some 670 million years ago, and it was yet another 155 million years before multicellular forms began to invade the land during the late Silurian.

Considering its much greater geological age, it is hardly surprising that the fauna of the oceans is far more disparate than the terrestrial fauna, being made up of 28 phyla – the name given to the basic types of animals (*Table 15.1*), compared with just the 11 phyla represented in terrestrial faunas. However, is this greater disparity also repeated in a greater species richness in the oceans compared to the richness on land?

Terrestrial Versus Marine Species Richness

For pelagic faunas the answer appears to be no; terrestrial faunas appear to be far richer in species. For example, the most abundant and species-rich group of plankton are the copepods (*Figure 15.2*), of which there are just over 1900 species known from oceanic and brackish waters. Compare this with the most diverse inhabitants of terrestrial habitats – the insects (a group virtually absent from the oceans), in which just one order, the beetles, includes several hundreds of thousands of known species, and whose true global richness is estimated to be in excess of a million. Much the same applies to the plants; there are an estimated 250,000 species of terrestrial green plants, but merely 3500–4500 in the oceans (*Figure 15.3*). Why are there such major differences in species richness?

15.3

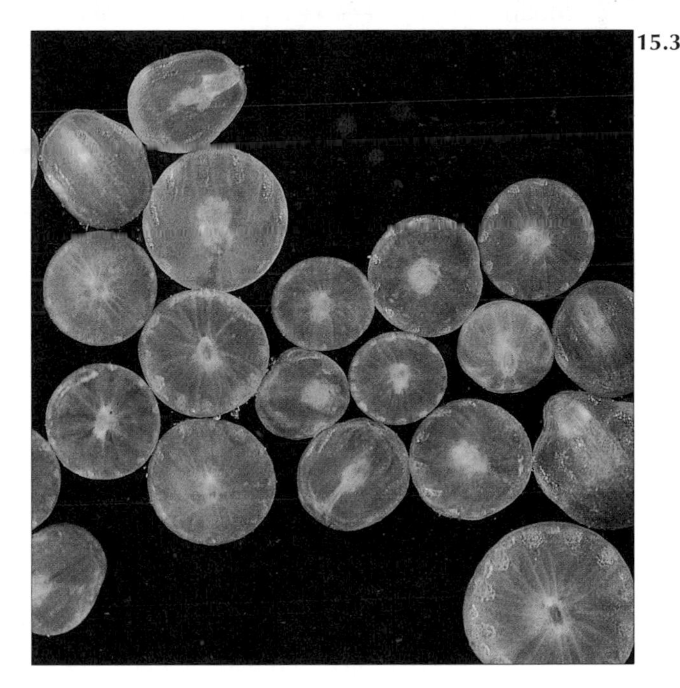

Figure 15.3 Giant phytoplankton cells (1–2 mm in diameter) of the monad *Halosphaera viridis*. Apart from the large algae that grow in shallow coastal waters, virtually all life in the ocean is dependent on the photosynthetic production of the 3500–4500 species of phytoplankton. Which should receive the greatest priority for conservation, one of the few phytoplankton species or one of the quarter of a million terrestrial green plants? (© Heather Angel.)

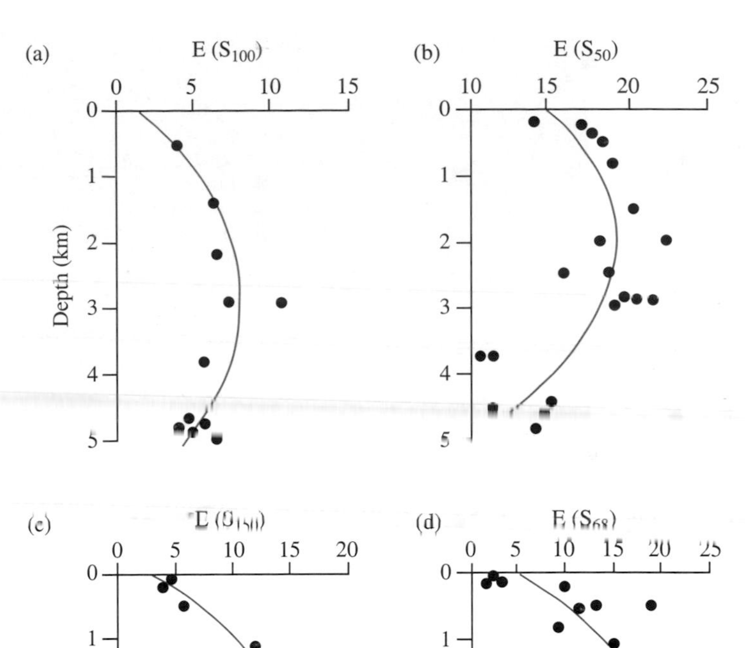

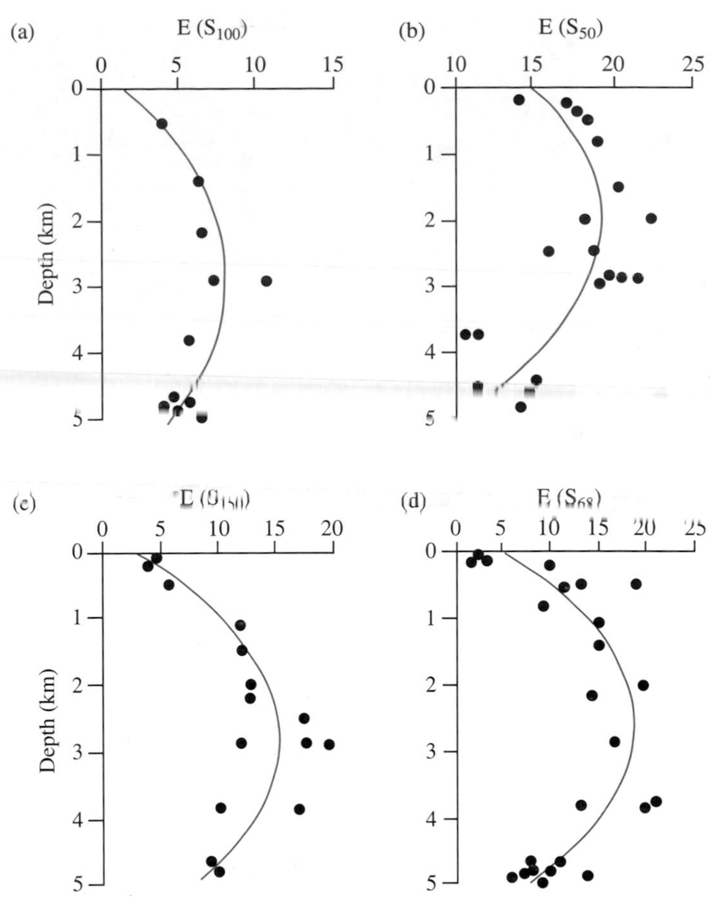

Figure 15.17 Depth profiles of the expected numbers of species found in a sample of *S* specimens, randomly selected from dredge and sledge samples down-slope in the northwestern Atlantic, of (a) protobranch molluscs, (b) polychaetes, (c) cumaceans (crustaceans), and (d) gastropod molluscs (redrawn from Rex[14]).

2000–3000 m (*Figure 15.17*). As for pelagic communities, both benthic biomass and the density of organisms decrease almost exponentially with depth (*Figure 15.18*), with a 10% reduction in biomass between 200 m and 2000 m; so, once again, the link between productivity and diversity is not straightforward, since the peak in species richness occurs where the biomass of the benthic community has fallen to nearly a tenth of that at the shelf-break.

Some recent evidence suggests that the species richness of benthic assemblages may match the rich disparity and may even be on a par with that of tropical rain forests. An intensive programme of sampling the benthic communities living on the sea bed at depths of 2000 m off the east coast of the US, involving the analysis of over 200 grab samples, revealed the presence of nearly 800 different species, of which over 50% were new to science[5]. When the sampling was extended both northward and southward along the slope, an extra species was added for each extra square kilometre that the sampling programme covered. If that rate is maintained throughout the global ocean, then the numbers of benthic species would be approximately equivalent to the area in square kilometres of the ocean floor – over 300 million. Moreover, this study did not include the meiobenthos, which contains the two most speciose groups in the oceans – the nematodes and foraminifers. Over 100 species of each of these groups were found in a cubic centimetre of mud scraped off the surface of the sea bed at a depth of just over 1000 m off the south-west of Ireland.

No one really knows how to extrapolate from the results of a few very localised samples to give a credible estimate of how many species occur on Earth, or even if this is possible. Many benthic species appear to have very extensive geographical distributions, so estimates that they exceed a million might seem excessive, but are they? It is worth noting that the huge estimates of the numbers of species inhabiting tropical rain forests are based on similar extrapolations from data obtained by misting just six trees with insecticides, and so are equally lacking in credibility! But perhaps, in this case, the extrapolations are not so wildly excessive given the very large numbers of plants, especially trees, that occur in rain forests and the strong ecological links between the plants and specialist insects.

Effects of the Seasonality on Benthic Diversity

Are changes in the production cycle reflected in changes in benthic diversity? We know that at temperate latitudes, where the production cycle is highly seasonally pulsed, the amounts of detrital material reaching the sea bed vary seasonally (Chapter 7). So, once again, we expect there to be substantial change in the benthic communities at latitudes of around 40°. Each year the deep-sea bed to the west of Britain, at depths of 1000 m, becomes carpeted with detrital material 6–8 weeks after the onset of the spring bloom. This provides a food bonanza for the bottom-living animals, some of whom appear to specialise in exploiting this detritus. No such bonanza is seen further south, either at 30°N on the Madeiran Abyssal Plain or at 20°N on the Cape Verde Rise. The animals which specifically feed on these detrital falls are missing at these latitudes, and there appears to be insufficient food available to support some of the larger animals.

So each region is inhabited by a different assortment of animals, with much fewer large species at the more southerly sites. For example, sea cucumbers – the holothurians – which are animals that feed on the enriched surficial sediment layer, become rare where the sedimentary flux is too low. So their species richness runs counter to the expected latitudinal trend seen, to some extent, in other groups (*Figure 15.19*). Fish abundances are also sharply reduced at the lower latitudes, but the specimens there belong to an unexpectedly rich assortment of species. The numbers of species taken in the limited sampling that has been done are lower than those taken at temperate latitudes. However, comparisons of the species counts in samples of a standard size indicate that sample diversity is as high as, if not higher than, those at temperate latitudes.

The difficulties of drawing generalisations are further illustrated by recent results comparing macrobenthic communities from undisturbed soft muddy sediments at depths of 30–80 m off Spitzsbergen (78°N), in the North Sea (55°N), and off Java (7°N), all sampled and analysed using exactly the same methods[7]. The results show that the diversity profiles at all three sites were indistinguishable, and gave no indication of a latitudinal trend (*Figure 15.20*). Is this observation merely a quirk of history? In other words, were the investigators unfortunate in choosing sites where, by chance, evolutionary history had resulted in the diversity being identical. This solution seems far-fetched, but is testable by repeating the comparison elsewhere. An alternative explanation might be that the processes that maintain the diversity at these three locations are not influenced by latitudinal forcing. For example, in such muddy environments, is there always an unlimited amount of organic matter around, irrespective of the production cycle?

15.18

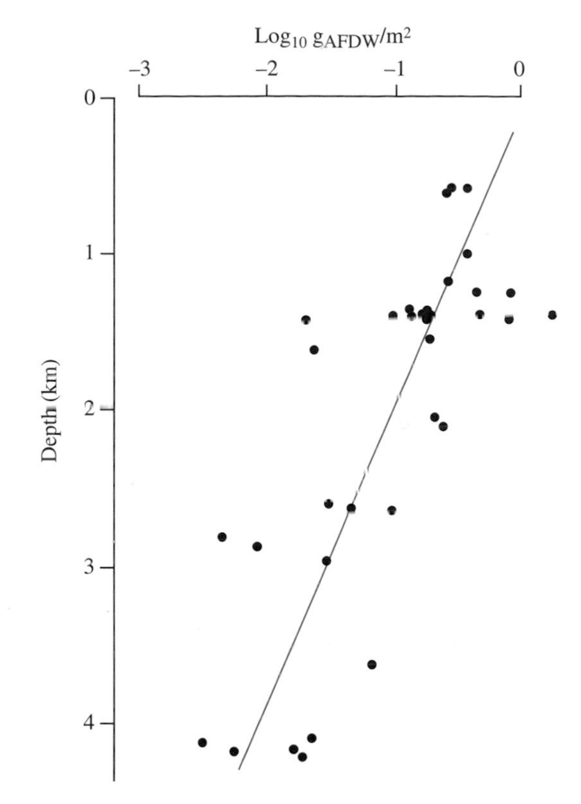

Figure 15.18 Profile of benthic biomass expressed as ash-free dry weight (AFDW) per square metre versus depth in the Porcupine Seabight region off southwest Eire[8]. The scatter is the result of the patchiness of the benthic communities, especially at depths of around 1250 m, where some samples contained several specimens of the sponge *Pheronema*.

Diversity and Productivity

There is an apparent paradox in the trends in pelagic diversity with depth and geographically, which run counter to the trends in productivity and biomass (standing crop) of animals, certainly in the water column and maybe on the sea floor. Populations in low-productivity regions seem to be characterised by being rich in species-richness without any one or two species being overwhelmingly dominant. In contrast, where productivity is high (and often more variable seasonally) far fewer species occur and one or two species tend to be numerically dominant. Does this explain why, when we fertilise the seas with our sewage and with the agricultural run-off of dissolved nitrates and phosphates in our rivers, the local productivity goes up, but the numbers of species goes down (a process called eutrophication)?

Chapter 16:

Life in Estuaries, Salt Marshes, Lagoons, and Coastal Waters

A.P.M. Lockwood, M. Sheader, and J.A. Williams

Introduction

Any one location offshore generally presents relatively little change in its physical and chemical conditions over a short time-scale. The same cannot be said of near shore waters, estuaries, and coastal lagoons, which provide some of the more varied and unstable environments on Earth. The organisms inhabiting such areas may experience wave action to differing degrees, tidal change, salinity, temperature, and oxygen variations, high sediment loads, and tidal currents of varying velocities on a daily or seasonal basis. Locations such as salt-marsh pools (*Figure 16.1*), estuarine creeks (*Figure 16.2*), and shallow sloping beaches (*Figure 16.3*), can experience quite wide ranges in one or more of their physical characteristics. Tidal progression (*Figure 16.4*) may further compound the difficulties for colonisers by imposing rapidity in the changes of factors such as salinity and temperature. Add to these effects in the water column the diversity of types of substratum available for attachment or burrowing (*Figure 16.5*) and the impact of man (*Figures 16.1, 16.3–16.5*) and it is apparent that such waters can be inhabited only by species capable of responding to physical, mechanical, and physiological challenges.

Response to environmental fluctuations cannot, however, be restricted to mere passive tolerance; adaptive features must incorporate a whole gamut of positive measures, including biochemical, physiological, and morphological features.

Reproductive responses, too, must be involved; it is not much use having adult stages adapted to a particular environment if the motile young lack the facility to locate and colonise suitable habitats.

The conditions, then, are harsh, but few of the challenges presented by coastal waters have not been met. True, sites such as mobile shingle banks

16.1

16.2

16.3

Figure 16.1 Estuarine raised marsh on the River Test, Hampshire, England, showing tidal creeks and saline pools. The creeks experience substantial salinity variation during the tidal cycle, while the pools are exposed to sudden changes when the surrounding grass is submerged on high spring tides.

Figure 16.2 Closer view of the tidal creek shown in the middle distance in *Figure 16.1*. This is the site of the salinity measurements illustrated in *Figure 16.8*.

Figure 16.3 Estuarine beach areas are often impacted by anthropogenic influences, as well as experiencing physical changes.

16.4a

16.4b

Figure 16.4 Views of an estuarine site (a) at low water and (b) at high water on a neap tide. On high water spring tides, the raised marsh to the left would generally be submerged.

16.5

Figure 16.5 The variable habitats presented by a tidal creek. Rocks with fucoid algae provide refuge at low water for non-burrowing forms, such as the gammarid amphipods; the mud banks are burrowed by the annelid *Nereis diversicolor*, the amphipod *Cyathura carinata*, and by the non-feeding adults of the isopod *Paragnathia formica*. The prawn *Palaemonetes varians* and the mysid *Neomysis integer* may be found where deeper pools are left at low water.

16.6

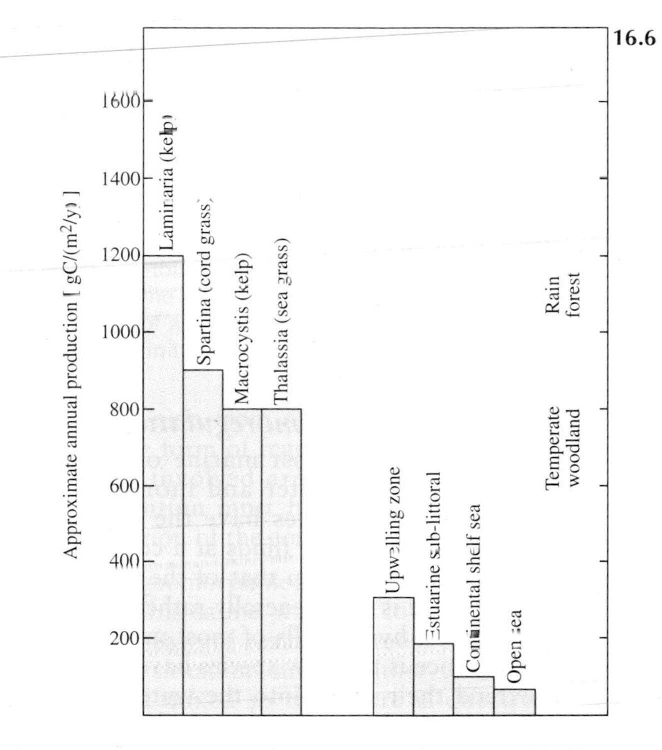

Figure 16.6 Estimates of primary production (gC/m²/yr) for various marine and terrestrial ecotypes. Note that values vary widely with both conditions and plant density. The figures indicate, first, that the productivity of seagrasses and the fixed algae of the coastal edge can be as high as that of tropical rain forest and, second, that the unit area productivity of offshore waters, even in the nutrient-rich upwelling zones, tends to be considerably lower.

and salt-marsh pools have a restricted fauna, but no coastal environment is without any colonisers. So successful, indeed, are some of the inhabitants that shore and near-shore regions have the highest primary productivity per square metre of any marine area (*Figure 16.6*).

There is, of course, no single answer to the problems posed for organisms in the coastal and shelf seas, nor is there any single simple lifestyle. In this chapter, therefore, we select three diverse topics to illustrate the amazing versatility of estuarine and coastal forms. These include:

- The physiology of osmoregulation – the regulation of water and ion balance by organisms – to indicate the variety of solutions to the problems generated for living forms by variation in the chemistry (salinity) of the habitat.
- Animal–sediment interactions, to outline the importance of the geological and hydrodynamical regime for benthic animals.
- Patterns in community structure of zooplankton, to interrelate the changing populations with features of the physical environment and biotic factors.

17.2

Figure 17.2 Wrecks also provide settlement substrate for epibiota, as seen by a diver looking at growth on the railings of a wrecked freighter, near Sardinia, Italy.

Artificial reefs are incidentally created by engineering works, such as harbour breakwaters (*Figures 17.4–17.6*) and supports for bridges (which are often built on muddy or sandy sea beds). Such structures provide habitats for species that could not live on the open sea bed, and attract mobile benthic species, such as cryptic fish, lobsters, and crabs. Oil production platforms also attract numerous fish species and provide a good settlement surface for animals and plants. In the Gulf of Mexico, the states of Louisiana and Texas are involved in an active artificial-reef creation programme utilising obsolete platforms. These are either relocated to shallow water and sunk, or toppled *in situ*. The money saved by the oil companies in disposal costs is ploughed back into the management of these structures for recreational fishing – over 4000 structures are available to anglers. In the UK there is interest in using North Sea oil production platforms as artificial reefs, now that they are beginning to reach the end of their useful working life (25–30 years).

17.3

Figure 17.3 The excavation of the historic wreck of the Mary Rose (sunk 1545) provided a range of new habitats in the muds and silts of the Solent (southern England). Large shoals of pouting congregated around the excavated timbers.

17.4

Figure 17.4 Coastal defence structures, such as these concrete tetrapods off Funchal, Madeira, provide new surfaces for colonisation by algae and encrusting animals.

17.5

Figure 17.5 This breakwater in Monterey, California, provides a convenient spot for Californian sea-lions to bask in the sun and groom. (Courtesy of Jane Jensen.)

17.6

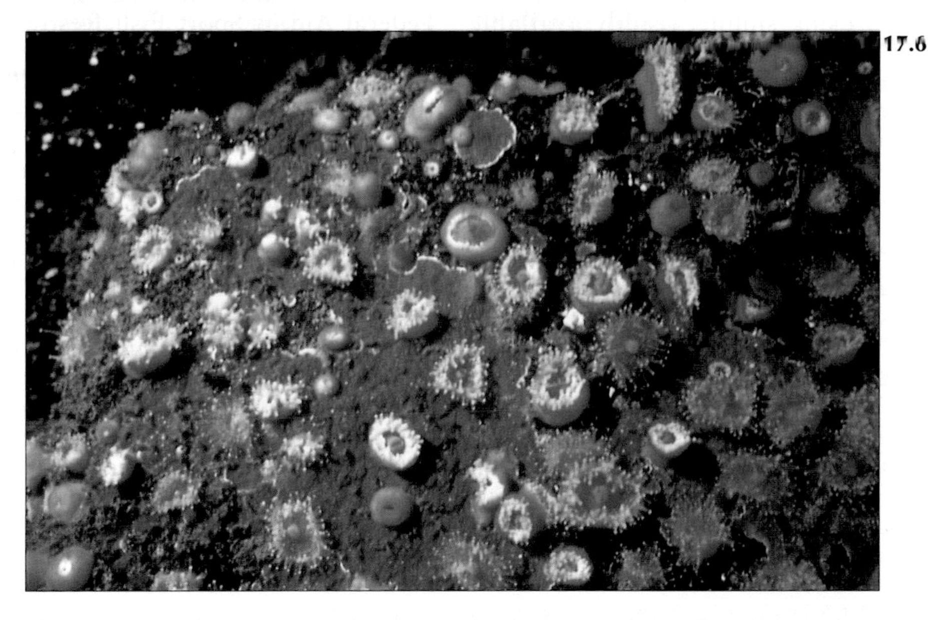

Figure 17.6 Below water the breakwater boulders provide good anchorage for a giant kelp forest (Monterey, California). The boulders are also settled by a profusion of sessile animals, such as these colourful solitary corals.

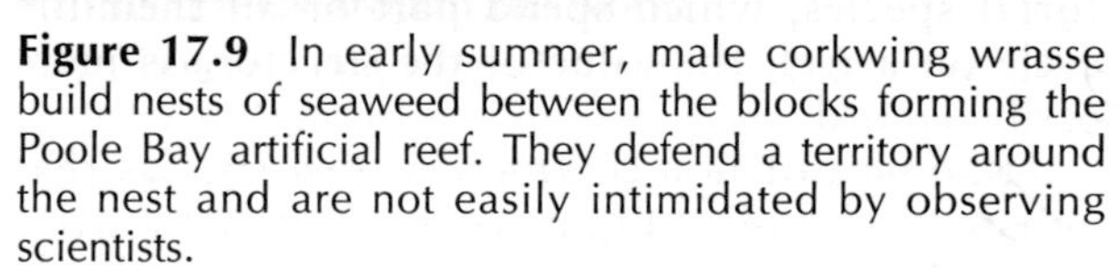

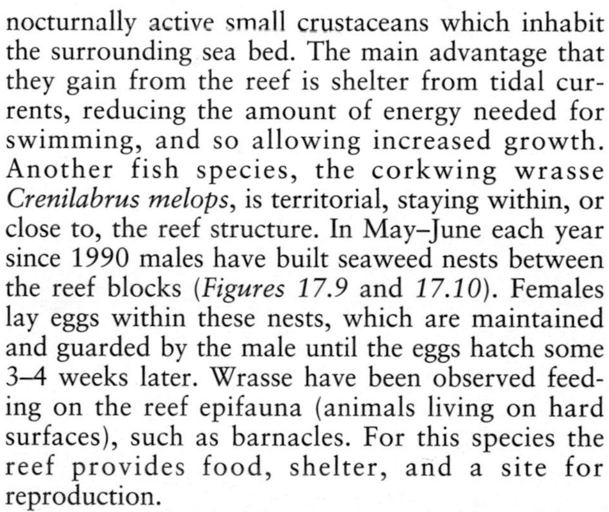

Figure 17.9 In early summer, male corkwing wrasse build nests of seaweed between the blocks forming the Poole Bay artificial reef. They defend a territory around the nest and are not easily intimidated by observing scientists.

Figure 17.10 The corkwing wrasse constructs a complex nest from seaweeds, seen here in its mouth, taken from algae growing on the reef and also drifting past in the current.

nocturnally active small crustaceans which inhabit the surrounding sea bed. The main advantage that they gain from the reef is shelter from tidal currents, reducing the amount of energy needed for swimming, and so allowing increased growth. Another fish species, the corkwing wrasse *Crenilabrus melops*, is territorial, staying within, or close to, the reef structure. In May–June each year since 1990 males have built seaweed nests between the reef blocks (*Figures 17.9* and *17.10*). Females lay eggs within these nests, which are maintained and guarded by the male until the eggs hatch some 3–4 weeks later. Wrasse have been observed feeding on the reef epifauna (animals living on hard surfaces), such as barnacles. For this species the reef provides food, shelter, and a site for reproduction.

The potential fishery value of artificial reefs is more easily demonstrated with less mobile, but valuable commercial species, such as molluscs and crustacea. Artificial reefs in the Adriatic Sea provide settlement sites for large numbers of mussels, an important commercial species throughout Europe. Because mussels produce such large numbers of larvae, the artificial reefs are not being colonised at the expense of other rocky areas, but in addition to them. In the Adriatic, this fisheries enhancement case appears to be well-proven; mussel harvests boost the commercial catches from artificial reefs, giving a three-fold return on deployment costs over 7 years. Net proceeds for fishermen operating within the artificial reefs have been shown to be 2.5 times that of activity outside the reefs. Elsewhere in Europe, attention has focused on lobsters as a commercial species under increasing fishing pressure that may benefit from habitat provision in the form of artificial reefs.

Case study: artificial reefs and lobsters

Man-made shelters ('pesqueros' in Cuba, 'casitas cubanas' in Mexico), have been used to provide small, temporary refuges for spiny lobsters (*Panulirus argus*). These function as a focus for fishing effort, rather like a FAD for fin fish. Canada, Israel, and the UK, interested in other species of lobsters, have focused attention on artificial reefs as a specific lobster habitat. Canada built the first artificial reef specifically for lobster research in 1965. Over the following 8 years the lobster population of the artificial reef was monitored by diving scientists. The reef was initially colonised by large specimens of the American clawed lobster (*Homarus americanus*), which were thought to have outgrown their burrows, so being forced to roam to seek new shelter. By 1973 the size-frequency distribution of the artificial reef population was similar to that on natural reefs in the area. It was concluded that the standing crop on the reef might be increased by a different pattern of rocks. However, a cheaper source of reef material or a multiple-use reef was required before an artificial reef could be considered an economically viable proposition, so the research was halted.

In Israel, efforts focused on the non-clawed slipper lobster, *Scyllarides latus*, an important commercial species found off the Mediterranean coast. Research showed that slipper lobsters preferred horizontal shelters with two narrow entrances on the lower portion of the reef. Using shelters to hide from predators is believed to be an important defense mechanism for these animals, so the presence of the artificial reef provided a new and suitable habitat for colonisation. Slipper lobsters migrate into deeper water as the inshore water temperature rises in summer, but tagged individuals were seen to return to a coastal tyre reef during spring (*Figures 17.11* and *17.12*) over the project period of 3 years. In the long-term, populations of these heavily exploited animals could be protected against fishing effort by building appropriately designed artificial reefs in protected areas, such as underwater parks and reserves.

17.11

Figure 17.11 Scrap tyres provide a resilient material for constructing artificial reefs. However the tyres must be securely bound together and weighted, as in this example off the Israeli Mediterranean coast at Haifa. (Courtesy of Ehud Spanier, Centre for Martime Studies, University of Haifa, Israel.)

17.12

Figure 17.12 The spaces within tyres provide ideal shelter for fish such as these squirrel fish. This Red Sea species has migrated through the Suez canal into the Mediterranean Sea. (Courtesy of Ehud Spanier, Centre for Martime Studies, University of Haifa, Israel.)

Work continued since 1989 on the experimental reef in Poole Bay, England, found that lobsters (*Homarus gammarus*; *Figure 17.13*) appeared on the reef within 3 weeks of its deployment. Tagging studies were initiated in 1990, and data to June 1994 show that lobsters have found the artificial reef a suitable long-term habitat, the longest period of residence standing at 4 years. This can be compared to a maximum age in the range of 10–20 years, sexual maturity being reached at 4–5 years of age. Conventional tagging of lobsters below the fishery minimum landing size of 85 mm carapace length (ca 250 mm total length) in a nearby fishery revealed that these lobsters do not undertake any seasonal migration, and the range of most movements is less than 4 km in magnitude. The use of a novel electromagnetic telemetry system has started to reveal complex local movement behaviour. These data reveal that lobsters are mostly active at night and frequently change their daytime shelter. The internal galleries and tunnels of the conical reef units (1 m high, 4 m diameter) made from randomly stacked, cement-stabilised, pulverised fuel ash blocks (40x40x20 cm) were often occupied by more than one lobster. An animal was also monitored leaving the reef site for up to 3 weeks and then returning. Only the larval stage of the lobsters life cycle is planktonic; the juvenile and adult lobster live either in burrows in the sediment or in shelters in rocky sea bed for the rest of their lives. Diver observations and evidence from pot-caught lobsters suggests that the reef can support all aspects of the benthic life cycle; berried (egg carrying) females utilise the shelters and release their larvae from them, some reproducing more than once on the reef. Lobster larvae have been taken from the waters above the artificial reef and a wide size-range of juvenile and adult animals has been captured and/or observed by diving scientists.

17.13

Figure 17.13 The randomly stacked blocks of the Poole Bay artificial reef provide a wide variety of crevices and tunnels. Here a lobster emerges from the entrance to a gallery within the reef unit where it spends most of the day.

Artificial reefs have been shown to support effectively three species of commercially important lobster. Research in the UK has shown good survival of hatchery-reared juvenile lobsters released into the wild and subsequently recruited into the fishery. It seems feasible that an artificial reef could be 'seeded' with hatchery-reared juveniles and that they would live to become part of the fishery. At present, the maximum densities of lobster population that can be achieved are not established, but data for *H. americanus* suggests that the Canadian quarry rock reef supported one lobster per 6 m^2 while the Poole Bay reef is thought to hold one *H. gammarus* per 2 m^2. Since neither structure was designed to maximise lobster habitat there is a potential for improvement. The artificial reef densities can be favourably compared to results (in the order of one lobster per 30 m^2) of diver surveys of natural reefs. Animal density is strongly correlated to the number of suitable shelters available. Lobster territorial behaviour also influences the usage of habitat. It seems more than possible that in the future a reef could be designed to provide lobster shelters in various sizes to minimise 'off-reef' movement, caused by the need to seek a new shelter after increasing in size following moulting. Already predictions have been made of the number and size of shelters in a reef made up of spherical boulders (a starting point calculation for more realistic material shapes), which can be linked to results describing the habitat requirements of lobsters.

Aquaculture

Artificial reefs can serve several roles in fisheries management, from the enhancement of wild fisheries to more intensive aquaculture systems.

Settlement of mussels on reef structures in the Adriatic is described above. Stock density can be intensified by adapting the well-established 'suspended rope' system of mussel cultivation, practised throughout the world, stringing seeded ropes of mussels between artificial reef units. Suspended rope is also used for kelp culture in Japan.

In Japan, specially designed artificial reefs are used for the culture of kelp, urchins, and abalone.

Combined systems for all three have been constructed where the kelp supplies food for the urchins and abalone. There is interest in aquaculture systems to develop abalone as a cultured species in Europe; artificial reefs may play a part in this.

Many salmon farms in northwest Europe have started to utilise the 'cleaning' capabilities of species of wrasse, such as the corkwing and goldsinny, to remove sea lice from the bodies of the captive salmon as an 'environmentally friendly' alternative to treating the fish with chemicals. Currently, most of the wrasse are caught in the wild, so localised populations are depleted. As wrasse are territorial rock-fish and spawn and nest in rocky habitats, there is a potential to artificially enhance wrasse habitat near salmon cages. This would provide a self-sustaining source of 'cleaning' fish near the site of salmon farms. The potential of artificial reefs and their epifaunal filter-feeding community to act as 'biofilters' to remove waste (uneaten food and faecal material) close to aquaculture facilities in the Baltic Sea is being investigated. Mussels, acting as filters, could play an important role in lessening the impact of aquaculture on the environment.

Biodiversity Management

The contribution that artificial reefs can make in biodiversity (see Chapter 15) management is that of habitat manipulation. This has the potential to increase the number of species in an area and provide purpose-designed habitats for target species.

Provision of new habitat

Artificial reefs are usually constructed to provide elevated hard substrata where formerly there was none. Artificial reefs mimic natural reefs, but can be built to provide greater surface area, elevation, current shadow/disturbance, or niches/crevices to favour target species.

An early European artificial reef study was that by the Association Monégasque pour la Protection de la Nature off Monaco in the 1970s. Artificial reefs (2 m^3), made from hollow blocks or tiles cemented together, were laid on a muddy sea bed within a marine reserve. These attracted a good settlement of epifauna and provided a habitat and refuge for spiny lobster. Following on from this work, specifically designed cave habitats for rare red coral have been successfully developed by the Institut Océanographique at Monaco.

The placing of artificial reefs on a mud–sand sea bed smothers and kills infauna (animals living in the sediment) directly under the reef (see also Chapter 22, references to the Garrock Head waste-dumping ground). Some 100 infaunal species were replaced by more than 250 epifaunal species within two years in the Poole Bay artificial reef. The infaunal populations in undisturbed sediment around the reef were unaffected by reef deployment. The lowest estimates of epifaunal biomass per unit area were equivalent to that of the previous infaunal biomass. However, the greater surface area available on the reef (2.5–3.0 times ground area lost) gave a higher biomass estimate than from the sea bed that the reef covered. This new habitat led to a rapid increase in species numbers and diversity. After 4 years the epibiota (*Figures 17.14* and *17.15*) provided a food source for molluscs, lobsters, crabs, and fish; these, in turn, provided food for cuttlefish and predatory fish. The reef is providing a valuable site for reproduction, such as nesting

17.14

Figure 17.14 After 4 years underwater, the Poole Bay artificial reef density and variety of colonising animals and plants closely resemble the biological communities seen on local natural reefs.

17.15

Figure 17.15 Colonisation of the Poole Bay artificial reef was rapid. Within 1 year the surfaces were entirely covered by a variety of hydroids, bryozoans, and sponges.

17.16

Figure 17.16 Artisan fishing communities in Kerala, Southern India, have used locally available materials, such as bamboo, to construct artificial reefs which are deployed on their village fishing grounds immediately offshore. (Courtesy of Steve Creech, Hampshire, England.)

17.17

Figure 17.17 Smaller concrete slabs, cast in the sand on the beach, have been assembled into artificial reef units and are being paddled out to sea balanced on canoes made from three tree trunks lashed together (CWARP, Coal Waste Artificial Reef Program). (Courtesy of Steve Creech, Hampshire, England.)

by corkwing wrasse or the laying of egg masses by whelks (*Buccinum undatum*).

Restoration of damaged habitat

Modern techniques have increased the environmental impacts of fishing; overexploitation of stocks by modern, powerful vessels and heavier fishing gears (such as 'rockhopper' trawls; it is estimated that every square metre of the North Sea is trawled 3–5 times annually), have led to the destruction of some sea-bed habitats. As an example, in southwest India development aid was provided to equip fishermen with trawlers to increase fish harvests. The coastline of Kerala province is one of the most densely populated areas of India. Artisan fishing communities depend on catches immediately off the beach on which they live, up to 3 km offshore, the range of their log canoes. The trawlers, based in the northern part of the province, decimated the artisan fisheries further south, which, as catches declined, were no longer economic to operate. In an attempt to restore damaged fishing grounds the coastal communities have used a variety of local materials (stones, cast concrete and bamboo, *Figures 17.16* and *17.17*), with some success, to restore bottom-habitat diversity and fish catches.

In the Maldives, the lack of construction materials has led to the use of coral (as blocks and aggregate) for building on the low-lying islands. However, the loss of the coral reefs that provide a living coastal defence barrier ultimately threatens the entire archipelago. In an experiment, concrete structures normally used for coastal defence in Europe have been laid, and living corals transplanted onto them in an attempt to re-establish coral reefs. Corals have grown and fish have been attracted back to what had become a coral rubble desert. It seems ironic that concrete should be used to restore the reef destroyed in the quest for building materials.

Around the UK coast there are several offshore sites where waste could be dumped under licence (see also Chapter 22). Off Blythe (northeast coast of England), coal fly-ash from coal-burning power stations has been dumped for many years. This has resulted in the smothering of rocky outcrops with a sterile blanket of ash, covering ideal lobster and crab habitat, and fish-feeding grounds. While dumping is legislated to cease in 1995, it will be many years before this area will return to full productivity. One restoration method would be to continue dumping coal ash as stabilised ash blocks (tested in the Poole Bay reef) to restore habitat diversity and accelerate the recolonisation process.

Protection of existing habitat

In the Mediterranean Sea, most artificial reefs have been placed as nature conservation and/or habitat protection structures. At least 150 artificial reefs have been deployed for habitat protection by Israel, Italy, France, and Spain. Reef complexes range in size from a few hundreds of square metres to several square kilometres. Their prime role is to prevent the destruction of sea-grass meadows by trawling (*Figure 17.18*) – sea-grass is a valuable habitat for many commercial species of fish.

One example of such a protective reef or barrier is found off the new port of Loano (northwest Italy). Here, *Posidonia* sea-grass has been protected by an artificial barrier some 3 km^2 in area, deployed in 1986. Most of the anti-trawling barrier consists of 350 1.2 m^3 cubes placed on the sea bed. In the centre of the artificial barrier a number of concrete block 'pyramids' (*Figures 17.19–17.21*),

17.18

17.19

Figure 17.18 Illegal trawling of sensitive marine areas in Spain is being counteracted by these anti-trawling units. The steel railway lines protruding from the solid concrete blocks snag and destroy nets. (Courtesy of Technologia Ambiental, SA, Madrid, Spain.)

Figure 17.19 Another anti-trawling reef, off Loano, northwest Italy, uses pyramids of 2 m concrete cubes with holes to encourage cryptic animals.

17.20

Figure 17.20 The Loano artificial reef was deployed to protect an area of sea-grasses, which are important nursery grounds for juvenile fish.

17.21

Figure 17.21 The Loano artificial reef units provide a variety of habitat types. This massive bryozoan colony being inspected by the diver is typical of animals found on shaded overhangs on natural rock outcrops.

19.5

Figure 19.5 The multicorer takes eight short cores for biological and geochemical purposes (see Chapter 11). The stand rests on the sea floor and the corer is lowered into the sediments. (Courtesy of SOC.)

interface needed to be sampled, a need that led to the design of multicorers (*Figure 19.5*), with a grid of short tubes mechanically pushed into the sediment, retaining the sediment–water interface and causing minimal disturbance.

Towed sensors and samplers

The traditional oceanographic measurement station, where the ship is stopped and wires are lowered to make measurements, is costly in ship time. To save time, engineers and scientists have devised instruments and vehicles that may be towed behind a ship to give continuous coverage while underway. Following tows of instruments at constant depth came the need for vehicles which could sample the water column more completely.

Canada led the way with the Batfish™ in the early 1970s, a vehicle that could be towed at 5 m/s while undulating between the surface and 400 m. With a payload of a CTD and other instruments, it

19.6a

19.6b

Figure 19.6 Towed vehicles do away with the need for the vessel to stop and lower instruments. The SeaSoar undulator travels from the surface to 500 m depth in a horizontal distance of 2 km at a ship speed of 9 knots (about 4.5 m/s). It carries CTDs and other physical, chemical, and biological instruments to measure the structure of the upper ocean (see, for example, *Figure 4.10*). (a) Deployment through the stern 'A' frame of RRS *Charles Darwin* is straightforward in calm seas, but can be difficult in a gale (a spare vehicle is in the right foreground). (b) The vehicle being deployed – SeaSoar's ability to carry several instruments makes it especially valuable for research spanning more than one discipline. Increasingly, the vehicle is used to make measurements for biological oceanographers, including ocean colour, phytoplankton pigments, and zooplankton counts using optical and acoustic sensors.

19.7

Figure 19.7 Introduced in the early 1970s, the UOR – known commercially as the Aquashuttle – can be towed from ships of opportunity at speeds of up to 10 m/s. The vehicle can either be self-contained, following pre-set undulations and storing data internally, or it can be controlled from the ship through a conducting cable to give real-time data. It can be launched by one person and can record its data internally. It is ideally suited for deployment by non-experts on merchant ships, so extending the geographical coverage of data collection beyond the tracks of research vessels in a most cost-effective way. (Courtesy of Plymouth Marine Laboratory, Plymouth, UK.)

19.8

Figure 19.8 The Rectangular Midwater Trawl (RMT) 1+8 acoustically controlled mid-water net system consists of a pair of rectangular trawl nets within a single frame. The outer net has an opening area of 8 m^2 with a mesh of 4.5 mm to catch micronekton, the inner net an opening area of 1 m^2 with a finer mesh (320 μm) is for plankton. Acoustic commands from the ship control a mechanical release gear to open and close the nets. Data on temperature, depth, speed and distance travelled, and net position are telemetered acoustically to the ship. (Courtesy of SOC.)

19.9

Figure 19.9 The epibenthic sledge is designed to skim over the surface of the sea bed and collect the organisms living on the sea floor, immediately above it, and in the upper few centimetres of the sediment. An acoustic pinger indicates when the sledge has reached bottom, when the opening–closing mechanism has operated, and when the sledge leaves the bottom. Cameras and electronic flash may be mounted on the sledge to provide images of the nature of the bottom and to give some indication of the efficiency of the net in capturing animals. (Courtesy of SOC.)

gave new insights into the structure of the upper ocean at 1 m vertical and 1 km horizontal resolution. This idea was developed into the UK SeaSoar vehicle, which can undulate from the surface to 500 m in a distance of less than 2 km. Several expeditions have used SeaSoar for tows of over 10,000 km, with the greatest danger being during recovery or deployment (*Figure 19.6*).

Supporting SeaSoar at sea requires a team of electronic and mechanical engineers, as well as data-handling specialists, to enable the wealth of information to be processed aboard ship in real-time.

For many experiments this is too heavy a burden (too large a team to support financially or to accommodate onboard ship); for these, simpler vehicles are available to carry out similar measurements. The Undulating Oceanographic Recorder (UOR; *Figure 19.7*) is one such device for covering the upper 200 m.

Research laboratories have developed many other types of specialist towed instruments for sampling, many of which are unique, serving the needs of individual researchers or small teams – examples are shown in *Figures 19.8–19.11*.

19.10a

19.10

Figure 19.10 A thermistor chain (a) laid out under cover before deployment, and (b) being deployed. This chain of thermistors to measure the thermal structure of the upper ocean is 400 m long and consists of 100 sensing 'pods', each providing a measure of temperature, and 28 measures of pressure so that the depth of the measurements are well-established. A 2 tonne sinker weight keeps the chain near-vertical under tow at a speed of 4 knots (about 2 m/s). The chain uses digital data communication and contains only six wires, two for power supply, two for data, and two for control. Sampling the 100 pods takes only 0.9 s. This particular chain has been used on 14 ocean measurement surveys, and more than 30 weeks of continuous data have been collected, comprising some 38,000 km of track. Much of the work has been directed toward the detection and tracking of fronts and eddies in the area between Iceland and the Faeroes, where waters of the North Atlantic meet those of the Norwegian Sea. (Courtesy of Dr J. Scott, Defence Research Agency, Dorset, UK.)

19.11

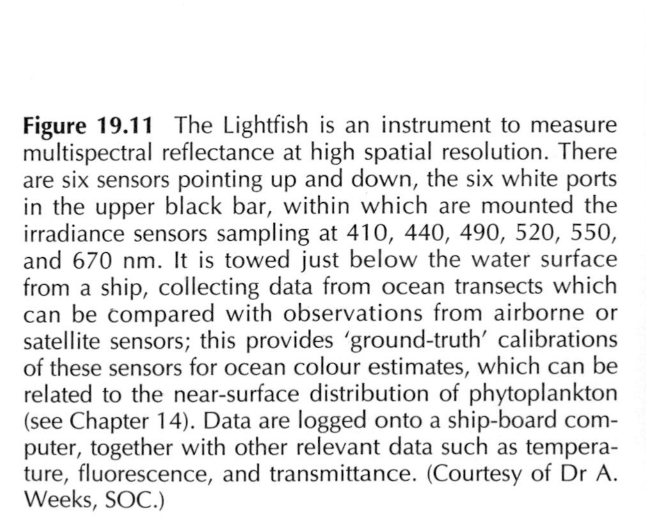

Figure 19.11 The Lightfish is an instrument to measure multispectral reflectance at high spatial resolution. There are six sensors pointing up and down, the six white ports in the upper black bar, within which are mounted the irradiance sensors sampling at 410, 440, 490, 520, 550, and 670 nm. It is towed just below the water surface from a ship, collecting data from ocean transects which can be compared with observations from airborne or satellite sensors; this provides 'ground-truth' calibrations of these sensors for ocean colour estimates, which can be related to the near-surface distribution of phytoplankton (see Chapter 14). Data are logged onto a ship-board computer, together with other relevant data such as temperature, fluorescence, and transmittance. (Courtesy of Dr A. Weeks, SOC.)

19.12a

Figure 19.12 (a) Deployment of an Aanderaa™ current meter mooring from the foredeck of RRS *Discovery* (courtesy of SOC). (Inset) A recovery beacon that enables the ARGOS satellite system to track and locate the mooring if it surfaces prematurely (courtesy of M. Conquer, SOC). (b) A release mechanism with biofouling after recovery from a 12-month deployment. (Courtesy of M. Conquer, SOC.)

Instruments fixed in position

Current meter moorings

Obtaining information on the behaviour of the ocean over time-scales of more than a few days requires scientists to resort to self-recording instruments moored to the sea floor, or left drifting. Ingenious non-electronic self-recording instruments were designed before the arrival of the transistor, but the rapid advances of recent years owe much to the microprocessor and to high-energy density batteries.

The highly successful Aanderaa™ current meter was designed in the early 1960s. Using mechanical encoding of the rotation of a Savonius rotor and the direction of a large vane, it owed a great deal to the weather vane and anemometer.

Nevertheless, the instrument became the standard for measuring deep-ocean currents, and in a solid-state form is still popular today. Capable of being deployed for periods of a year and more, its simplicity, reliability, and relative cheapness make it an almost ideal oceanographic tool.

In depths of up to some tens of metres, typical of waters close to shore in the shelf seas, equipment may be moored to the sea bed and to a surface buoy, enabling simple recovery.

Such a simple mooring is not practical in the deep ocean, where the buoyancy is usually placed beneath the surface, well away from the influence of surface waves and shipping or fishing (see *Figure 19.12*).

The problems are then, first, to find the mooring and, second, to retrieve it. A single instrument, the acoustic release, gives the answer. When the continually listening unit hears a coded sound-pulse from a ship it sends out a reply signal that indicates the range from the ship to the mooring.

As the ship homes in, another signal activates the release mechanism to separate the mooring from its anchor, so the mooring returns to the sea surface.

When rising the release provides a beacon signal, often augmented on the surface by radio transmitters, flashing lights, or radar reflectors.

As moorings contain increasingly more valuable equipment, and have a small probability of early failure, satellite position-indicating transmitters may be fitted to the buoyancy. These alert the laboratory to moorings that have surfaced prematurely, giving position and drift information that aids their recovery.

19.13

Figure 19.13 A surface meteorology buoy, with anemometers (for mean wind speed and direction), sea and air temperature sensors, a 3 m path acoustic current meter, a buoy motion package (to give wave height and directional spectrum), and radio and satellite data telemetry (see Chapter 2 for a discussion of the way the atmosphere affects the ocean). (Courtesy of A. Hall, SOC.)

19.14

Figure 19.14 A Wavecrest buoy, developed by the Netherlands company Datawell, being deployed. The buoy follows the sea surface and sensors measure the components of acceleration, which, integrated twice, give the wave-height variations. The buoy is moored to the sea bed using a compliant tether – usually a length of thick rubber line – to avoid mooring forces affecting the record. (Courtesy of C. Griffiths, Dunstaffnage Marine Laboratory, Oban, Scotland.)

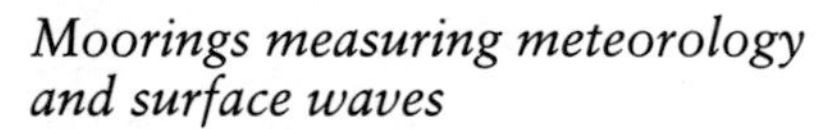

Moorings measuring meteorology and surface waves

Surface meteorological buoys (*Figure 19.13*) combine accurate sensors, replicated for reliability, with data telemetry via satellite and terrestrial radio links (see also *Figure 3.24*). Additional sensors that measure the motion of the buoy give estimates of wave period, height, and direction. An example of an instrument designed purely for wave measurement is shown in *Figure 19.14*.

Bottom-mounted instruments

Benthic landers are instrument packages that provide observations near the sea floor. They may be deployed from the ship's warp, when the experiment duration is short, or may be autonomous to give measurements over long periods.

Geophysical experiments make extensive use of sea-bottom seismographs that record the signal associated with natural earthquakes or from induced sound sources, such as explosives and air guns (*Figure 19.15*; also see Chapter 8).

Scientists studying the geochemistry of the sediment–water boundary require water samples from within the sediment (the pore waters). Bottom landers have been built that contain hydraulically driven syringes to penetrate the sediments, take samples at a range of depths, and retract in readiness for recovery. Other sensors, such as those to measure pH and oxygen concentration within the sediments,

19.15

Figure 19.15 The Digital Ocean Bottom Seismometer (DOBS) is a self-recording listening station with an in-water hydrophone and geophones in contact with the sea floor. Sound generated from ship-towed air guns or from dropped explosive charges reaches the DOBS directly through the water and also through the ocean floor, through which the sound speed is much higher. From the characteristics of the different propagation paths, the nature of the sea floor can be inferred. (Courtesy of SOC.)

can be included on these landers (*Figure 19.16*).

Movement of sediment along the sea floor helps to shape much of the coastline, so several instrument platforms have been designed to study sediment concentration and transport close to the sea bed. Rapid turbulent motions, critical in causing fine sediment to become suspended, need to be measured along with the concentration or mass of the suspended matter. Measurements are made several times a second. Instruments based on the principle of electromagnetic induction are suitable and can be made sufficiently robust. A voltage sensed by electrodes on the instrument is generated by the movement of the conducting sea water through a magnetic field produced by an internal solenoid. These current meters can be combined with acoustic probes to measure the sound scattered from the suspended particles in bottom landers (*Figure 19.17*).

Increasing emphasis is being given to obtaining data over many years from the deep ocean to monitor natural and man-made change. Some locations lend themselves to regular visits by research ships, others are either too remote or visited too infrequently. New bottom-mounted instruments have been designed to solve the problem of obtaining data over several years from remote locations. The MYRTLE (Multi Year Tide and Sea Level Equipment) package releases data podules from the sea bed that, on the surface, return their data via satellite to the laboratory (*Figure 19.18*).

Figure 19.16 This sampler contains hydraulically driven syringes that penetrate the sea floor to take sediment and pore-water samples at a range of depths. Other sensors, such as pH and oxygen probes, may be fitted to the lander. (Courtesy of SOC.)

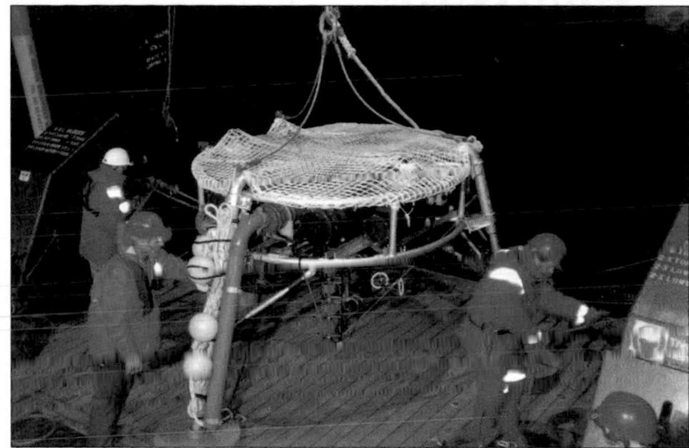

Figure 19.17 Work on research vessels proceeds through the night – here deploying the Sediment Transport and Boundary Layer Equipment (STABLE) benthic instrument platform with current and high-frequency acoustic sediment transport sensors. (Courtesy of J Humphrey, Proudman Oceanographic Laboratory, Bidston, UK.)

Figure 19.18 Sea-level changes of centimetres can be observed from the deep-ocean floor by the MYRTLE package. Designed for operation over 5 years, MYRTLE releases data 'podules' to the surface, where, as they drift, they telemeter data to the laboratory via ARGOS satellites. The package shown here is being deployed on the continental slope off the Antarctic Peninsula, the southern boundary of Drake Passage. With another MYRTLE package on the northern slope off South America, the slope of the sea level between the two sites provides a measure of the transport of water through the Drake Passage by the Antarctic Circumpolar Current, and of its variability (see Chapter 4). (Courtesy of Proudman Oceanographic Laboratory, Bidston, UK.)

Box 20.1 What is an electromagnetic wave?

An electromagnetic (EM) wave is a combination of oscillating electric and magnetic fields which transport EM energy. EM waves are able to propagate in a vacuum at a speed of approximately 299,776 km/s, and cover a wide frequency band which includes visible light (*Table 20.1*). EM waves are reflected or absorbed by objects and refracted by changes in the refractive index of the medium through which they travel. Waves undergo a velocity change when refracted as they travel through different parts of the Earth's atmosphere. This results in a potential source of error in measuring the ranges from satellites, which becomes important in the calculation of accurate positions.

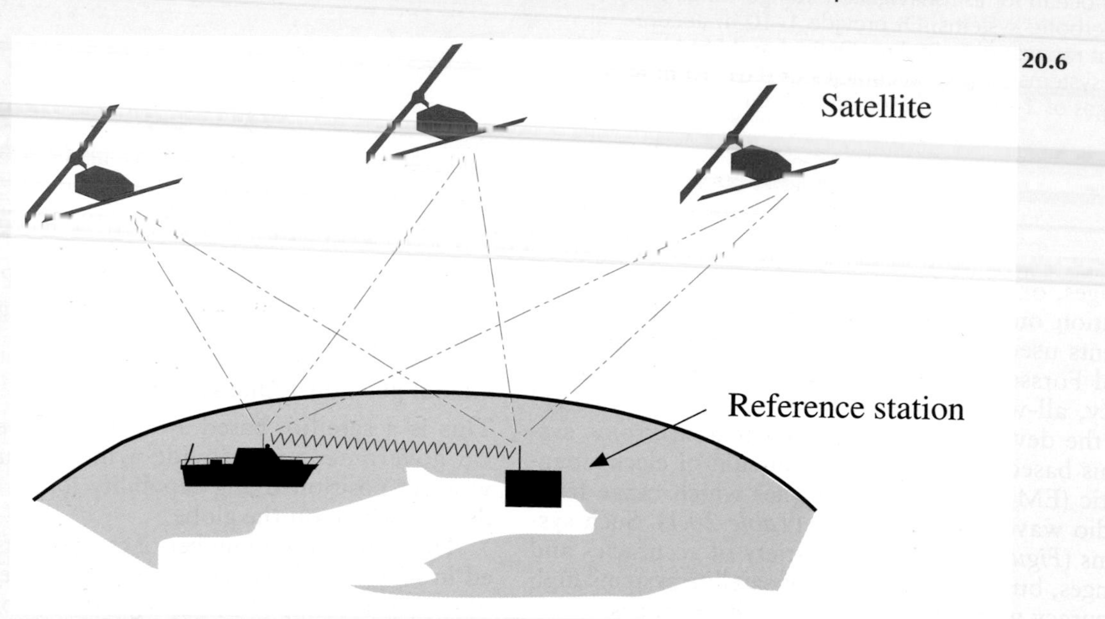

Figure 20.6 Differential GPS (DGPS) involves corrections based on the known position of a reference station being calculated and transmitted to the vessel. The vessel uses the correction information to recalculate and correct its known position, which has been calculated from directly received satellite signals. This provides positional accuracy of less than a metre.

Differential GPS (DGPS)

This development of GPS uses a reference station with knowledge of its own position to calculate the errors in the satellite signals caused by SA and by atmospheric conditions (*Box 20.1*, *Figure 20.6*). The errors, or an error correction, are then transmitted in real-time from the reference station to the vessel, where they are applied to give much greater accuracy than is provided by GPS. The accuracy of DGPS partly depends on how close the reference station is to the survey. Where the reference station is within 30 km of the survey vessel, 'instantaneous' accuracies of less than 1 m are consistently obtained. When used in a special geodetic mode, centimetre accuracies can be obtained. Some companies have set up networks of reference stations to routinely provide corrections that allow a positional accuracy of around 5 m over an area extending hundreds of kilometres from the base stations.

The position of the sensor

Some survey sensors are towed at the end of a cable, which can be several kilometres in length (*Figure 20.7*). If the position of the sensor is not well-known, it is impossible to produce an accurate sea-bed map of the property being measured. In practice, acoustic devices on board or placed on the sea bed (e.g., an array of accurately positioned transponders or navigation beacons) may be used to measure the relative position of a sensor package.

The Sea Floor

What are we looking for? It may be the size, orientation, shape, and composition of natural features (or natural obstacles), or the location of manufactured objects, such as sunken vessels and mines. Surveys of shallow-water environments tend to be more detailed and to find wide local variations of features, sediment type, and their distribution. In

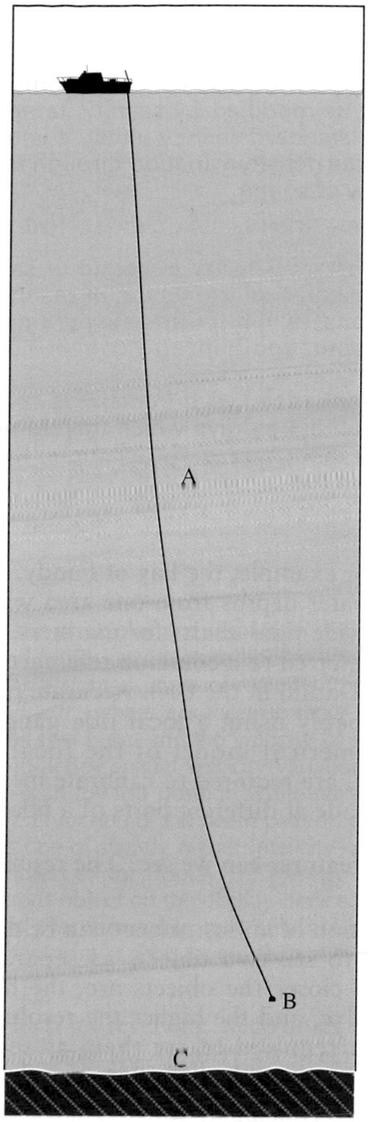

Figure 20.7 Surveying at depth in the ocean involves a tow cable (A) and a sensing package (B), which is held near to the sea bed (C). Scaled to a vessel length of 225 m, the water depth here is 3150 m, but sensors can be deployed at 5000 m or more.

the deep-water environments, which cover a substantial area of the Earth's surface, there is generally more interest in the larger scale changes in sediments and features (see Chapters 8 and 9). In either environment, the minimum expected size of the feature of interest usually determines the type of system used, the survey line spacing, and, consequently, the type of position-fixing system. Sedimentary features vary in size from sand ripples of a few centimetres in wavelength and height, to the massive slumps and turbidite features described in Chapter 9. Small features are often found superimposed on larger features, which may be important in understanding the general evolution of an area. Changes in sediment type can be detected by changes in the related reflective properties of the sea floor, as can be seen, for example, in *Figure 9.13*. Deep underwater canyons are easy to identify, but difficult to survey accurately. The ocean liner *Titanic* remained hidden in 5 km of water for over 40 years before Bob Ballard[1,2] used sophisticated sonar systems and submersibles with cameras to find and explore the wreck.

Principles of Surveying Equipment

The echo-sounder

The use of instruments which remotely sense the sea floor is well-illustrated by the echo-sounder, which is used for measuring water depth (*Figure 20.8*). A short pulse of sound at a given frequency is transmitted into the water column by an acoustic transducer, either on the vessel's hull or towed at a known depth beneath the surface. The transducer can both transmit and receive sound pulses. The pulse reflected back from the bed is collected, its travel time to reach and return from the sea bed and its amplitude are measured. The amplitude of the returning pulse relative to that transmitted depends on the depth of water (since water attenuates sound energy) and the reflection characteristic (or reflectivity) of the sea bed. The size, shape, reflectivity, and orientation of the sea-bed features to the arriving sound pulse also affect the returning pulse amplitude.

Some systems use the characteristics of reflected pulses to obtain other information, such as the type of sediment which makes up the sea bed. The relationship between a sound *pulse* and a sound *wave*

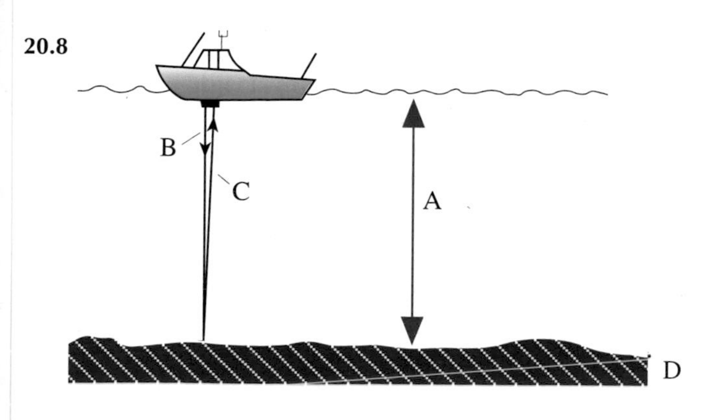

Figure 20.8 In echo-sounding, the time a pulse of sound takes to travel to the sea bed and back is measured; since the speed of sound in water is known to be about 1500 m/s, it can be converted into depth. The amplitude (strength) of the returning, reflected signal depends on the depth and absorbence of the water and the type of sea bed – e.g., rock will give a stronger reflection than sand. (A) Depth of water, (B) path of outgoing pulse, (C) path of reflected pulse, and (D) sea bed.

Table 20.2 Surveying instrumentation. The data given can only serve as a guide, since the parameters depend on the configuration of each system, and the type of sea bed. The compromise between system resolution and depth of penetration is clear. Chirp systems are a relatively recent development in profiling technology, and employ a swept-frequency pulse and sophisticated processing of the return signal to achieve high resolution, while giving good penetration of the sea bed.

System	*Frequency* (kHz)	*System Resolution* (cm)	*Penetration* (m)
Echo-sounder	200	5	0–0.5
Echo-sounder	35	20	0–2
Pinger	3.5	20	20–75
Boomer	1–10	40	50–100
Chirp systems	**0.5–12.5**	**10**	**40**
Sparker	0.1–1	200	100–200
Air gun	0.05–0.1	200–5000m	100–tens of km

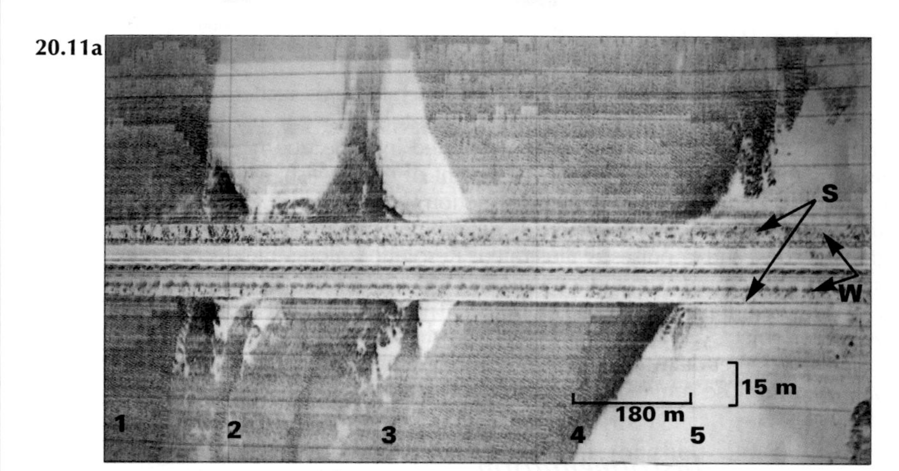

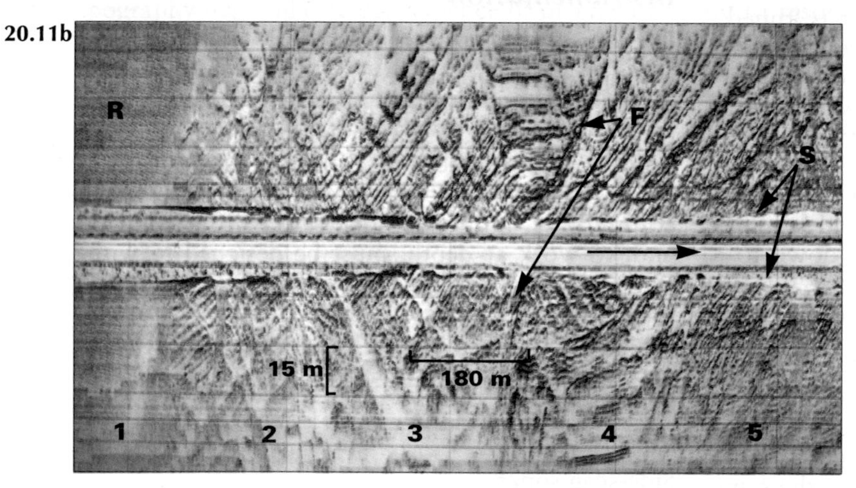

Figure 20.11 Two annotated sections of side-scan traces recorded in shallow water. The grey scale is calibrated for the strength of reflections, from black (maximum) to white (minimum), and is typical of this type of system. Both traces were recorded during the same survey with consistent system settings. (a) This is from an area of soft sediment. The numbers 1–5 are fixed reference points marked on the trace at one minute intervals. Note the distance scales marked on the trace and the consequent scale distortion of the features – along track distances are proportionally larger than across track distances. The bed has distinct zones which are rippled (darker areas) and unrippled, respectively, interspersed with other dark areas, ambiguous because of the lack of ripples. The sea bed in this area is predominantly sand with varying quantities of shell fragments. The sand ripples are of small amplitude, typically less than 0.2 m, and the apparently narrow boundary between rippled and unrippled areas extends over several metres. (S, sea-bed reflection; W, water-column noise.) (b) The trace shows a rocky area which has undergone considerable folding and faulting, and is directly adjacent to an area of rippled sand (R). Note the fault indicated by the strong reflection (F) running diagonally across the centre of the trace. The layers of rock are being viewed from 'end on', dipping at a steep angle down into the page.

20.12a

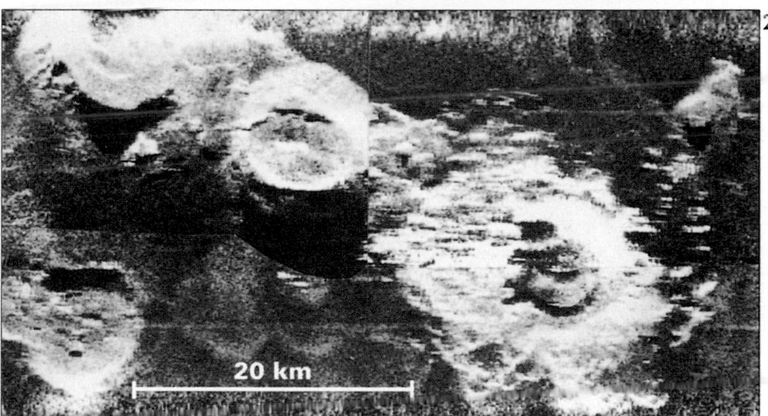

20.12b

Figure 20.12 (a) The GLORIA side-scan vehicle during deployment from its dedicated launch cradle. The vehicle is 7.75 m in length and weighs approximately 2 tonnes in air, but is neutrally buoyant in sea water. It is towed at a speed of between 8–10 knots. (b) GLORIA 6.5 kHz side-scan image showing a linear chain of submarine volcanoes in the Pacific Ocean, west of San Francisco (scale bar is 20 km) – strong sonar targets are white, acoustic shadows are black (note this is opposite to those in *Figure 20.11*).

20.10(b) and *20.10(c)*. In the interpretation of side-scan pictures, a number of other factors influenced by the range of operation have to be kept in mind.

The images built up, as illustrated in *Figure 20.10(a)*, are named sonagraphs and represent the **intensity of reflection** plotted in the two space co-ordinates, range and distance, along the vessel's track. The intensity depends on the reflective nature of the sea floor as well as on the orientation of the features [*Figure 20.10(a)*].

Features may be compressed or expanded in the along-track direction unless a correction is applied to allow for the speed of the vessel (*Figure 20.11*). Other distortions of the image can be caused by electronic noise and interference, and the movement of the towfish in the water[5].

Sonagraphs provide much information about the nature of the sea bed, (e.g., texture, composition, and the orientation of features). Their interpretation requires experience and some knowledge of an area in order to decide what features are likely to be real. Natural features, for example, tend to follow particular patterns or shapes and have a reflectivity which is often different to human artefacts.

GLORIA (Geological Long Range Inclined Asdic)

GLORIA, or Geological Long Range Inclined Asdic (*Figures 20.12* and *20.2*), is a unique side-scan sonar system, developed at the UK Institute of Oceanographic Sciences, for rapid imaging of large areas of sea floor. It operates at a frequency of 6.5 kHz and has a depth-dependent swath width with a practical maximum of 45 km. Towable at speeds of up to 5 m/s, up to 20,000 km^2 of sea floor can be covered in a single day (an area the size of Wales or Massachusetts). Over a 25-year period it has evolved from a large and difficult-to-deploy single-sided sonar vehicle into a much more compact vehicle which produces both side-scan imagery and swath bathymetry (see later) on both sides of the vehicle track. The neutrally buoyant vehicle has its own handling system and is configured with buoyancy at the top and transducers below to give good roll stability. It is towed from the nose at a depth of 40–80 m, depending on the ship's speed and its tow-cable length (usually about 400 m). The resolution of GLORIA, as with other side-scans, varies with range, being largely dependant on the horizontal beam width (about 2° for

20.13

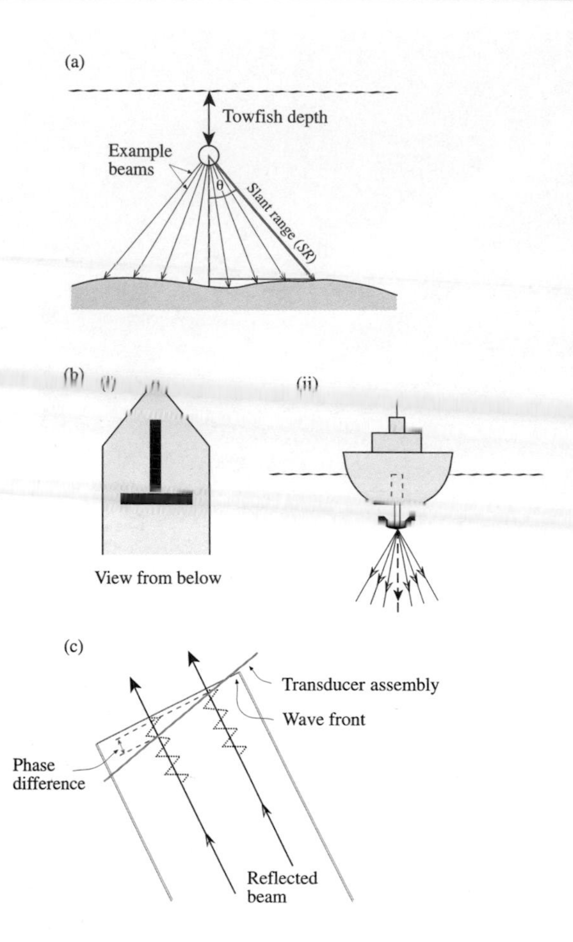

Figure 20.13 The more common types of deeper water *multibeam* systems (e.g., Simrad EM12, Seabeam, Krupp Atlas Hydrosweep) use a cross-shaped array of transducers to form a series of beams fanning out perpendicular to the ship's track. In these systems, the beam pattern is controlled by transmitting from the along-ship arm of the cross and receiving on the across-ship arm. Multibeam systems are generally more accurate by a factor of 5–10 than the interferometer systems, but give lower swath widths. (a) The travel paths of a number of example beams – continuous depth measurements are available over the full width of the swath. Simplistically, depths are calculated by converting the beam travel time into a slant range (*SR*) and using the equation $D = SR\cos\theta + T$, where D is depth, T is the towfish depth, and θ, the beam angle to the vertical, is known or measured. (b) Examples of multibeam transducer array configurations. (i) line array (discussed in the text); (ii) cylindrical array, which can be directly hull-mounted, or placed on a retractable unit to avoid turbulence associated with the hull. (c) In interferometer measurements, a reflected wave-front strikes the transducer assembly at an angle. By measuring the change in phase as the wave moves across the transducer(s), its angle of incidence can be measured, allowing the beam angle, θ, to be calculated. The angle of incidence of the wave front is related to the distance of its reflecting origin from the ship's track.

GLORIA). At mid-range, it can resolve features about the size of a football pitch in a water depth of 5 km, i.e., an area around 120 x 60 m.

Swath-sounding systems

Echo-sounders provide depth along a profile; side-scans provide both depth beneath the vessel from a straight down–up return and a picture of an area based on the variation in the intensity of reflections. It would be more efficient if one system could provide depth and intensity measurements at several ranges (not only below the vessel). This is achievable by swath sounding.

A development of the echo-sounder/side-scan technology, swath sounders measure depth along a swath which extends each side of the vessel's track [*Figure 20.13(a)*]; depths are calculated for each point along the swath. The shallow-water versions of these systems can provide full coverage and an overall accuracy of better than 10 cm in depth and 5 cm in range over the swath. The configuration of a system determines the depth range in which it is used, with maximum swath widths (i.e., maximum ranges) of shallow-water systems being typically 5–8 times the depth of water beneath the transducers. Most multibeam systems do not provide side-scan data, although these are available with some systems.

There are two main types of swath sounders, multibeam and interferometer, both of which provide line-by-line measurement of depth along the swath. Transducers can be hull- or towfish-mounted and are operated in a number of configurations,

20.14

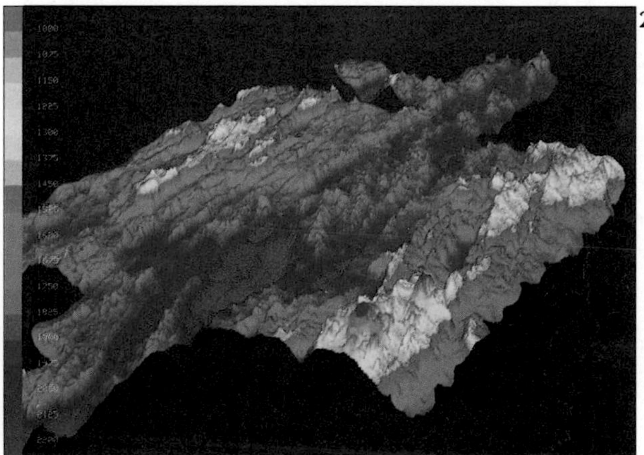

Figure 20.14 Three-dimensional visualisation of a section of the mid-Atlantic Ridge based on data from a Hydrosweep multibeam swath system. The image shows a series of deep basins (dark blues and purples) within the central mid-ocean ridge rift valley, which crosses the image from bottom left to top right. The higher rift valley shoulders are seen toward the edges of the image (yellows and greens).

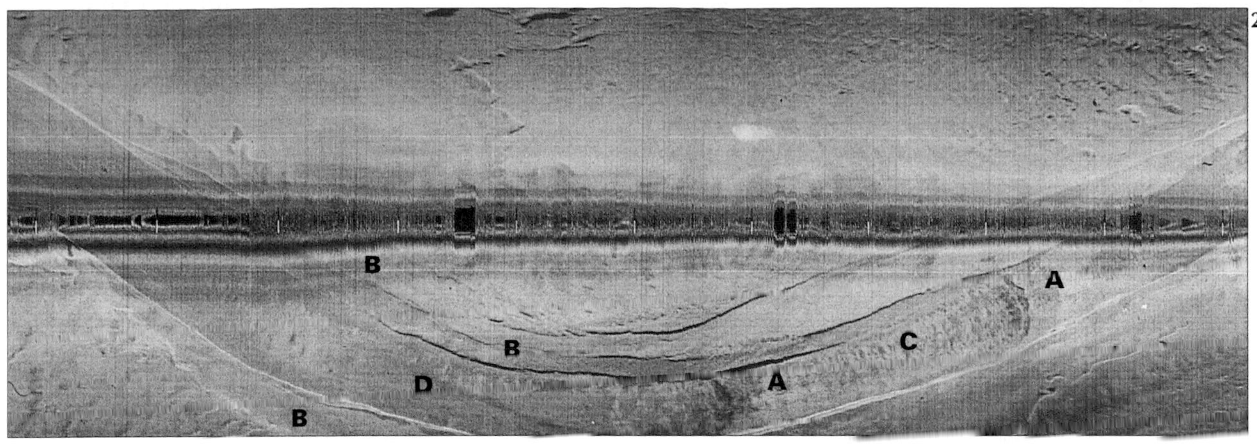

Figure 20.15 TOBI 30 kHz side-scan sonar image of a deep-sea channel on Monterey Fan, off the western US – strong sonar targets are white, acoustic shadows are black. The channel, here at a depth of about 4100 m, is between 1–2 km in width and up to 50 m deep. Note the terraced walls (B) and 'waterfalls' (A) within the channel (D), showing strong similarities with similar features associated with subaerial rivers (C is a sediment wave field).

of which two examples are illustrated in *Figure 20.13(b)*. A number of overlapping beams are electronically formed from the same set of transducers to produce the *multibeam* scan necessary for swath bathymetry (e.g., one system produces 52 beams from four transducers). The technology is changing rapidly[3], and more recent developments have produced compact, relatively portable systems which still function in depths of <1 m.

Essentially a development of side-scan systems, *interferometer* systems (e.g., GLORIA, SeaMARC II) use a set of transducers to measure the interference pattern of reflected sound [*Figure 20.13(c)*]. This sound 'interferometer' allows the accurate measurement of the travel time of sound waves reflected from small, adjacent areas of sea floor. Travel time is again converted into distance and, knowing the beam characteristics and direction, the depth can be calculated. The intensity of reflection can also be used to provide more information about sediment type or sea-floor orientation.

A three-dimensional relief map, based on swath bathymetry, is shown in *Figure 20.14*. Swath bathymetry provides an 'ordnance survey' or topographic-type map, in contrast to the 'aerial photograph' provided by side-scan sonar. Near geological-quality interpretation can be achieved when the results of both are combined using computational tools, such as a Geographical Information System (GIS).

The Towed Ocean Bottom Instrument (TOBI)

TOBI is one of a family of deep-towed instrument platforms, which includes, among others, the original Scripps deep-tow, the American SeaMARC 1, and the French SAR system. By operating within a few hundred metres of the sea floor, such vehicles can provide resolution far higher than that of surface-towed vehicles such as GLORIA. The penalty is that this mode of operation can require the use of many kilometres of tow cable, resulting in very slow tow speeds. TOBI (*Figure 19.26*), has a swath width of 6 km, can cover 400–600 km^2 per day, and can operate at depths of 6000 m, allowing its use over all but a few percent of the sea floor. *Figure 20.15* shows a TOBI sonagraph.

Able to carry a wide range of instrumentation, TOBI currently carries a side-scan sonar operating at 30 kHz with a beam width of about 0.8°, giving a resolution on the scale of a few metres. Other instruments carried include a sub-bottom profiler (7 kHz), and a transmissiometer to allow the measurement of temperature and depth.

Sub-bottom profiling

Sub-bottom profilers are used to explore the even more invisible world beneath the sea floor. Still using reflected sound waves, the frequency of operation is chosen such that the sound can penetrate the sea floor and be reflected from interfaces between different types of rock or sediment [see *Table 20.2* and *Figure 20.16(a)*]. As in the conventional echo-sounder, the system records a profile of data along the vessel's track; a three-dimensional picture of the subsurface structure requires the collection of many closely spaced profiles. The speed at which sound waves travel varies in rocks of different type and density (see *Table 20.3*). This wide range of velocity increases the potential for errors in the conversion of travel time into depth of a reflecting layer beneath the sea bed (depth = travel time times velocity). The variation of velocity with-

Table 21.6 Whale catches (thousands)[19].

	1985	1986	1987	1988	1989	1990	1991
Blue and fin whales	7.9	6.5	6.3	0.68	0.61	0.65	0.66
Sperm and pilot whales	83.2	153.8	135.6	132.7	168.4	105.8	50.0

important in determining stock levels from place to place[11].

Clearly, we need to understand and to be able numerically to forecast the natural variability in the system in order to exploit the living world more effectively. By itself, however, such an understanding will be inadequate; we also need to prevent overfishing. The difficulties in policing fishing operations make effective control unlikely. This is another argument in favour of adopting the methods of the farmer, rather than those of the hunter, to ensure the true potential of the ocean for feeding the world's growing population is achieved.

When technology was primitive, fishing was indiscriminate; early man took what he could get. Advances in technology brought greater control, and an ability to target particular species systematically, no matter where they might be found. Specialist fisheries arose, like the cod fishery of the Grand Banks, or the herring fishery of the North Sea. Whaling (*Figure 21.10*) provides a good example of the evolution of a hunting-based fishery[4]. It can be done by primitive tribes in coastal waters, and in Europe it began to develop into an industry by the twelfth century in northern Spain. Local overfishing drove whalers further afield, to Newfoundland and Spitzbergen, where whaling had become big business by the mid-seventeenth century. The business boomed in the nineteenth century, when technology made whaling easier and turned it into a global industry. It reached its peak in the 1930s and 1940s, but the combination of rapid technological advance, making the job easier, and the limits of the whale population lead to decline by the 1960s. Large-scale commercial whaling is now a thing of the past, although some

21.11a

21.11b

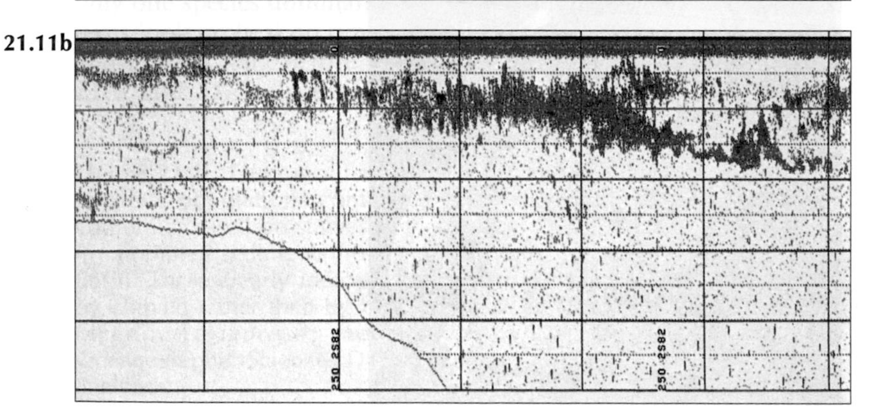

Figure 21.11 (a) Antarctic krill, *Euphausia superba* (size approx. 2 cm), plays a key role in carbon cycling in the Southern Ocean. (b) The echo-sounding trace, with horizontal scale lines spaced 25 m apart, shows a swarm of krill as a large widespread acoustic signal (in blue and red) centred at about 50 m (the red horizontal line) over the edge of the Antarctic continental shelf (shown at the lower left). Whales feed on these dense layers of krill and other zooplankton. (© British Antarctic Survey, England.)

21.12a

21.12b

Figure 21.12 (a) Development by the Japanese Marine Science and Technology Centre (JAMSTEC) of an artificial sea floor (20 x 20 m), which can be raised to the surface (right) and lowered below the surface (left) to and from its working depth of 4 m; it forms a platform for growing abalones, and its tanks contain black rock-fish. (b) On the prototype in Ryohri Bay, Japan, the abalone are fed with kelp and measured at the surface once a week. (© Mineo Okamoto, JAMSTEC, Japan.)

nations, notably Norway and Japan, continue to catch large numbers of whales[33] – *Table 21.6* shows the decline in catch in recent years. The creation of a whaling reserve around Antarctica in 1994 should help to renew and maintain whale populations.

In the case of herring, the drift net was replaced by the purse-seine net in the 1960s in Europe, which led to overfishing and bans on fishing in the North Sea at various times. Perhaps inevitably, the response of the local industry to a decline in one stock is to go for another. Declines in the herring catch by European fisheries were offset by a tremendous rise in industrial fishing for sand-eels and the like for fish-meal and fertiliser. The consequent decline in sand-eels is thought to be the cause of larger than usual numbers of deaths of sea-birds, which feed on them[32]. New fisheries are already developing to harvest new species of fish, like the Grenadier, from deep water, or different creatures, like squid and krill[42] (*Figure 21.11* and *Table 21.3*).

The lessons from the whale, cod, and herring fisheries show that, without protection, living resources cannot survive the technological shift from primitive to industrial fisheries. Bigger and better ships, with more sophisticated, expensive, and durable equipment, are depleting a supply that once seemed boundless. With the help of satellites, sonars, computers, and refrigeration, combined with global-range ships, humans will sweep the seas more or less clean of fish if they continue unchecked.

Mariculture

One answer is mariculture, farming the sea (*Figure 21.12*). Oysters have been cultivated in Asia for over 2000 years. The Chinese developed freshwater aquaculture over 1000 years ago, and wrote the first fish-farming textbook in 475BC; it described the farming of carp (fresh water), milkfish (brackish), and mullet (marine)[3]. Mariculture has been common for centuries in Southeast Asia, where fish farmers seed ponds with the eggs of fish and crustaceans, transferring the larvae and juvenile forms to larger ponds and feeding them on algae and plankton[4]. World mariculture production is currently around 10 million tonnes, excluding freshwater environments – fin fish account for about 45%, shellfish for about 25%, and seaweeds for the rest. The Chinese are the largest producers. Japan derives about 15% of its total ocean produce from mariculture, including such exotic items as sea-urchin roes and seaweed (which is grown on nets[35]); note the growth of seaweed shown in *Table 21.3*, which also shows the growth in fish taken in inland waters (much of it from mariculture). The UN Food and Agriculture Organisation (FAO) estimates that by 2000AD aquaculture (including freshwater fisheries) will account for 20–25% by weight of the world fisheries production, and about 50% of the value[3].

21.13a

21.13b

Figure 21.13 (a) Long-line mussel farm in Loch Etive, Scotland. The black objects are plastic floats from which hang the ropes on which mussel spats settle (© Jim McLachlan, McLachlan Shellfish and Fish Farming Equipment). (b) At harvest time, after around 3 years, there may be 250 kg of mussels per rope (© Nicki Holmyard, Association of Scottish Shellfish Growers).

Around Europe, oysters and mussels have been cultivated in coastal areas for centuries (*Figure 21.13*; note the growth in mussels shown in *Table 21.3*). Since the late 1960s, salmon farms have grown apace in the fjords of Scotland and Norway (*Figure 21.14*); the salmon industry in Scotland now equals the size of the beef and lamb industries there[3]. The market is expanding and the outlook is good – note the growth in salmon shown in *Table 21.3*. Interest is growing in farming other species, especially halibut and turbot, and in developing salmon and other farms in the open sea rather than in lochs[3]. The advantage of the open sea is that the farms could be larger, and are not at such great risk from local pollution.

One variety of mariculture that does not use farms is ocean ranching, in which juvenile fish are released into the sea to be caught at a later stage. Salmon ranching is a commercial success in Japan and Alaska[28]. It has been shown that plaice ranching could work on the Dogger Bank in the central North Sea, where plaice hatchlings transplanted from the Dutch coast grew more rapidly and successfully[42]. Unfortunately, this operation has not been taken up commercially.

Legislation

Fishing can be controlled by legislation, although not without difficulty. It is a global business (*Table 21.5*), so the ability of the large fishing countries, like Russia, South Korea, and Japan, to fish just about anywhere means that the stocks of fish off countries with less fishing capacity could be depleted without any recompense. To counter this problem, most fishing countries agreed in 1977 to a 370 km (200 mile) limit, to keep to themselves the right to fish their own waters and to prevent overfishing[35]. It was a declaration of extended geographic limits around Iceland that kept British fishing vessels out of Icelandic waters and led to the so-

21.14

Figure 21.14 Salmon farm in Loch Creran on the west coast of Scotland. Young salmon, bred in fresh water, are transferred to the octagonal, 4.5 m (15 foot) diameter sea cages at 18 months and fed on high-protein fish food for about 2 years. Netbags (4–6 m deep) are suspended from the floating cages; tides help to flush effluents away. The farmer here is putting wild wrasse (a 'cleaner' fish) into the cages to keep the salmon free of lice. Total production from this one farm is around 40 tonnes/year. (© Jim Buchanan, Association of Scottish Shellfish Growers.)

Figure 21.15 A desalination plant – removing salt from sea water provides drinking water for people in arid countries. (© Zefa Pictures.)

called cod wars between the two countries in the mid 1970s[3,35]. The European deep-water fishery was severely affected by the extension of the Canadian fishery exclusive zone to 200 miles over the Grand Banks. Factory ships from other nations can circumvent the problem by 'setting up shop' seaward of the limit and paying local vessels for their catches from within the zone[35].

In some countries Governments have set quotas on the size of fish or amount of catch in order to control fishing and preserve stocks. Unfortunately, controls work poorly as they are virtually impossible to police effectively. In protest against quotas, French fishermen burned down the Town Hall of Rennes, in Brittany, in 1993. Tempers will continue to run high as the hunting culture clashes with the implementation of the principles of sustainable development.

We are at the beginning of what Borgese[4] has called the Blue Revolution, in which sustainable techniques, such as farming, will take over from hunting as the method of harvesting food from the sea. Farming in the coastal zone will grow in volume and expand in area to become common everywhere, with more and more species being farmed. Farms will move offshore to take advantage of space, on the one hand, and local supplies of nutrients (pumped from cold subsurface waters), on the other hand.

Chemicals and Medicines from the Sea

The sea is a vast storehouse of dissolved minerals. Unfortunately, most of the dissolved constituents are disseminated in such tiny amounts that extraction will never be profitable. Only a few are abundant enough to be extracted on a commercial basis, the most common of which is sodium chloride, or common table salt. What many people do not realise, unless they live in arid coastal areas, is that one of the principal extracts of salt water is fresh water. Distilling plants to produce fresh water from salt water have been common on ocean-going ships for well over a century. Desalination plants have become increasingly common on land, especially in arid coastal regions like the Red Sea and the Persian Gulf (*Figure 21.15*). At present, they make commercially unattractive the once-popular idea of obtaining fresh water from icebergs towed north from Antarctica[9].

Chemicals

About 3.5% of the weight of salt water is dissolved solids; sodium chloride accounts for 71% of this (see Chapter 11). It has been a prime ingredient for cooking and a principal article of trade for well over 5000 years. Roman soldiers were part paid in salt (salarium argentium, from which the word salary derives). Humans probably first came across natural salt in dried-up lakes and coastal ponds. It was not a great leap of imagination to create artificial coastal ponds, so bringing the process of evaporation under control, and the basic approach has not changed to the present day (*Figure 21.16*). However, raw sea-salt derived in this way can be impure and rather bitter, containing iron, calcium, and magnesium compounds as well as sodium chloride (to obtain pure salt, several stages of crystallisation must be used). Evaporation accounts for around one-third of the world supply of salt, most of the world's sea-salt being produced by India, Mexico, France, Spain, and Italy[16].

The only two elements commercially extracted from the sea on a large scale are magnesium and bromine[16]. After oxygen, hydrogen, chlorine, and sodium, magnesium is the next most common element in sea water. In recent years, some 18% of the world magnesium production of 1.8 million tonnes has come from sea water, mostly produced

Figure 21.16 Harvesting salt from salt-pans in Portugal. (© Centro de Coridade Nossa Senhora.)

21.21a

21.21b

Figure 21.21 (a) A schematic cross-sectional cutaway of the prototype shore line wave energy plant on the Scottish island of Islay, demonstrating the operating principle: (1) waves oscillate the water column; (2) upward water motion in the chamber forces air through the turbine, driving the generator to produce electricity. (3) the turbine converts reciprocal air flow into high speed uni-directional rotation. The plant has a 75 kW capacity, an energy cost of 7 p/kWh, and an energy output of 300 MWh/yr. It indicates a future potential of 1000 kW, with an energy cost of 3–4 p/kWh. This amount of production would make wave energy plants economic for the islands (courtesy of AEA Technology, Harwell, England). (b) The Islay Wave Power Station showing the wave chamber (foreground) and turbine housing (rear) (courtesy of Don Lennard, England and Australia).

21.22a

21.22b

Figure 21.22 (a) Salter's ducks are hollow floats made of reinforced concrete, with a cone-shaped cross-section. Each duck is about 33 m long and 20 m across. Named after their inventor, Professor Stephen Salter of the University of Edinburgh, several would be connected in a line by a hollow shaft attached to land or a fixed platform at one end. Incoming waves tilt each float, or duck, upward, absorbing the waves' energy and leaving calm water in its lee. Within the oscillating duck, gyroscopes (inset) activated by the rocking motion, drive a high-pressure hydraulic system; the circulating oil drives a turbine that generates electricity. Power from each duck would flow to shore through a central cable. Compared with other such plants, the ducks are extremely effective, providing the highest output per metre of sea. A pilot plant has been tested and research continues. (© ETSU, Harwell, England.) (b) A model duck at about 1/150th scale in a narrow tank, being tested for response to freak waves. The duck (centre, in water) is being subjected to a 50-year design wave coming from the wavemaker (offstage to the right). The rig above water level, to which the duck is attached by angled struts, constrains the axis or 'spine' of the single duck to move as if it were part of a 'duck string' of 50 or more such devices restrained by a compliant mooring system, while allowing the duck–spine surge and heave reaction forces to be studied (© J. Taylor, EUWP, Edinburgh University, Scotland).

21.32a

21

Figure 21.32 Manganese nodules carpet the
photo sledge during RV *Sonne* cruise S079; th
90°42.5'W, showing the sea bed densely cov
pushed the nodules apart in its search for foo
less dense coverage by larger nodules, with the
a star fish near the centre. Most nodules look l
cauliflower heads. (Courtesy of Dr von Stackelb

water is 100–1000 m deep, so they are p
accessible with the right mining gear.
deposits on land are mined for most of th
phosphate fertiliser. Where the economics
offshore deposits could provide the basis fc
phosphate fertiliser industry. But land res
abundant and cheap to mine, so it seems
that offshore phosphorites will become c
cially viable before 2000AD. A likely fi
tender is actually an east-coast deposit,
Chatham Rise off New Zealand. There a
deposits also off the east coast of the south
US, from Florida to the Carolinas.

Elsewhere on the sea bed, iron–manganes
minerals are deposited in the form of nod
crusts[10]. In certain places on the deep-ocea
(4000–5000 m deep), where the rate of sedi
tion is extremely slow, the sea bed is carpete
small potentially mineable manganese nodul
size and shape of potatoes or tennis balls (
21.32). Nobody is very interested in their man
content; the attractive feature is their high cont
combined copper, nickel, and cobalt (aver
2.4% in places[18]). In shallower waters (ar
1000–2000 m), the flanks of volcanic island

Figure 21.33 Manganese encrustation on pillow b
from Nod Hill, an abyssal hill rising from the wester
of the Madeira Abyssal Plain close to the eastern fla
the Mid-Atlantic Ridge. (1) Basalt core; (2) oxidised g
surface of pillow basalt; (3) 1.6 cm thick manga
encrustation. Crust has been removed to show
smooth oxidised surfaces in the glassy sect
Encrustations[41] can reach 25 cm. (Photo, M. Conque
IOSDL.)

Waves

Where tides are not powerful enough to warrant the construction of barrages to generate electricity, waves may be an alternative source of energy[36]. The power potential of an average wave per kilometre of beach is around 40 MW. It has been estimated that a substantial part of Britain's energy needs could be met by putting wave energy to work[3].

Wave energy can be harnessed by fixed or floating devices. A fixed device is operational on the Isle of Islay, on the western coast of Scotland, where the wave energy offshore is equivalent to between 50–77 kW per metre of shore-line per year[3], decreasing to 20 kW/m at the device. Here, Britain's first shore-line wave power station, delivering 75 kW, was constructed by a team from the Queen's University of Belfast in 1988 and commissioned in 1991 (*Figure 21.21*). It works on the principle of an oscillating water column, with waves pushing air through a turbine[3]. The construction of this pilot plant proved the concept and opened the way to the construction of a commercial demonstrator in the 0.5–1.5 MW range. Power stations of this type would be ideal for many small island nations, and for local needs in larger countries (such as China) where the electrical grid systems are not yet fully developed. Two large systems have been installed on the coast of Norway.

Floating devices function on the principle that wave motion can be converted into reciprocal motion[6]. Some of the systems currently being refined, such as Salter's nodding duck (*Figure 21.22*), seem to hold considerable promise as energy sources for the future[3,42].

Currents

Although ships' captains have used ocean currents for decades to make speedy passage, the rivers in the sea have only once been harnessed to generate electricity, off northwest Iceland. The potential is there – the Gulf Stream[4], for example, carries 30 million m³/s of water past Miami with a velocity of 2.5 m/s. The problem is that all this power, five times larger than the flow of all the world's rivers, is too diffuse to harness easily by conventional means. Scientists of the Woods Hole Oceanographic Institution have, nevertheless, calculated that an array of turbines stretched across the current in the Florida Straits would produce 1000 million W/day, as much as two conventional nuclear power stations[4].

Thermal energy

Closer to reality is the conversion of the ocean's thermal energy into electricity[3,6,27]. The principle of ocean thermal-energy conversion (or OTEC) is simple, and has been around for over 100 years. OTEC uses the difference in temperature between warm surface water and cold deep water to power a turbine and generate electricity (*Figure 21.23*). The potential for applying this principle is greatest in the tropical regions, between 10°N and 10°S, where warm surface waters (25–30°C) overlie cold waters (4–7°C) situated at depths of 500–1000 m.

The first pilot OTEC plant was built by the French OTEC pioneer, Georges Claude, in Cuba in 1930. Pilot plants have since been built or proposed in the Ivory Coast, in Tahiti, in Nauru, in Japan, and elsewhere. The first closed-cycle Mini-OTEC plant to produce power was built in Hawaii in 1979. Although this successful experimental plant generated an electrical output of 50 kW, about 80% of this energy was needed to pump up the cold water for the system[23], leaving a net output of 10–15 kW. Plants like this would be doing well to produce an annual return on capital of 1%, which is unlikely to be commercially attractive to investors.

21.23

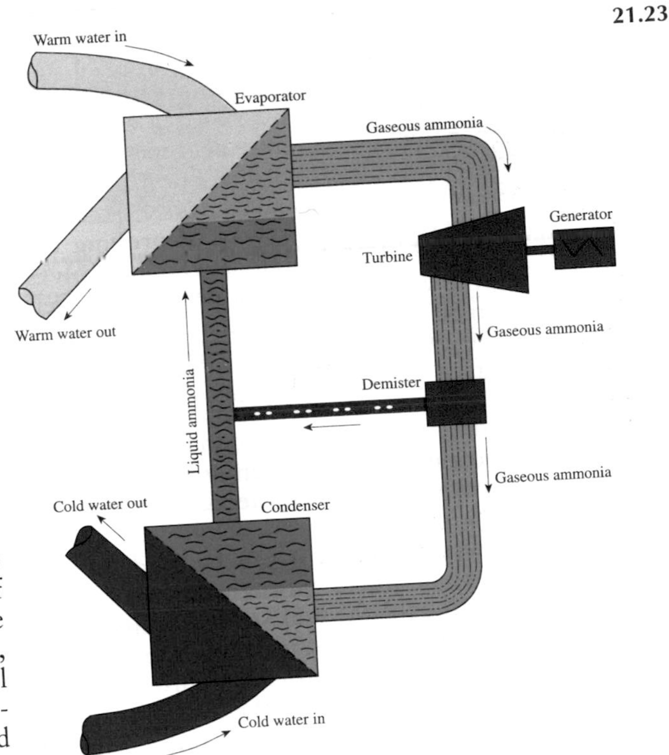

Figure 21.23 Sketch of the workings of an OTEC plant. Most modern OTEC plants are closed-cycle systems, like that shown here, in which warm surface water vaporises an intermediate fuel, such as ammonia; the vapour powers a turbine and is then condensed back into liquid by the cool waters pumped up from the depths. These are called closed-cycle systems because the working fluid is recirculated. (© Living Tapes Ltd.)

aggregate production is from Japar
deposits are close to shore, care m
ensure that the natural offshore sup
beaches are not destroyed by dredg
around the UK, for instance, dredgi
being confined to specified concess
why little offshore dredging is allov
US, even though there are substanti
sand and gravel off the US coasts[10].
is from depths of less than 45 m,
extended to 50–60 m over the next fe

Mineral ores

Most sand and gravel is made of mixt
and feldspar, the main rock-formin
the continents. But the precise mir
beach depends on the mineral make u
rocks, the local climate (which dict
rocks weather), how the eroded m
transported to the sea, and the proces
subjected to by sea waves and currents

On beaches and offshore, the contin
waves and tides helps to concentra
heavy minerals in deposits known as
Placers generally form toward the base
sand deposits, at the interface with th
rock, and in depressions and channels.
ety of types of dredgers can be used to
unconsolidated mineral ores, which can
different types depending on the loc
Beaches around the world have been
many minerals, including diamonds (Nar
(Alaska and Nova Scotia), and chromite
Offshore, diamond placers are mined of
cassiterite (for tin) off Malaysia, Indo
Thailand, and (in the past) off (
England[8,10,16]. Other minerals, such as chr
chromium), rutile (for titanium), ilmenite
and titanium), magnetite (for iron), zirco
conium), monazite (for rare earths), and
(for tungsten), have been or are curren
dredged in various places around the worl
Lanka and Australia[10,16]. Most dredging
depths of less than 50 m, but the Japar
dredged cassiterite from up to 4000
Beaches have even been mined for their m
mineral, quartz; pure quartz is the basic co
of glass sand, mined from the beach in
North Island, New Zealand, for instance[8].

On tropical islands fringed with coral r
white sands consist not of quartz but of co
ments, made of calcium carbonate, the ba
stituent of cement. Calcium carbonate in t
of shell remains is also common in places
continental shelves, and has been dredged
(e.g., off Iceland) for the production of ce
Extraction of carbonate by the demolition
reefs on some tropical islands helps build

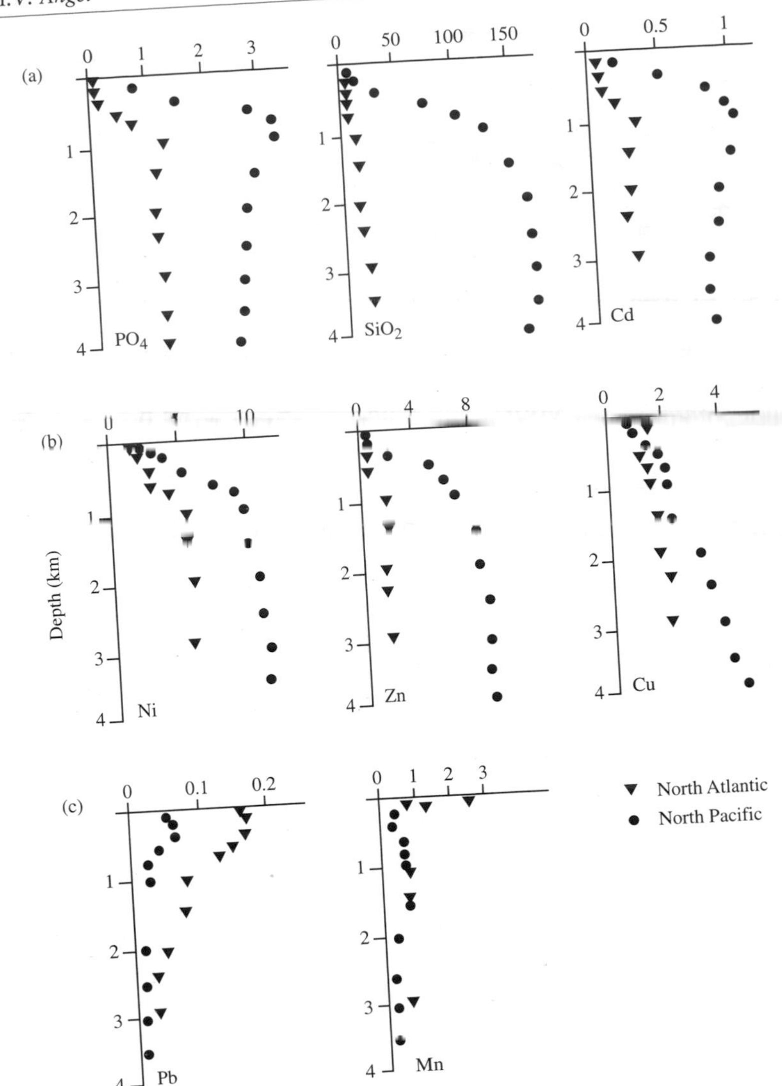

Figure 22.4 Concentration profiles (nmol/kg) of dissolved substance at two stations (t) at 34°N 66°W in the North Atlantic and (l) at 32°N 145°W in the North Pacific. (a) Marked reductions in concentration in the near-surface waters as a result of biological utilisation; these are categorised as bio-limited substances. (b) Some reduction in concentration in the near-surface waters, but the influence of biological processes on the profiles is relatively small; these are categorised as bio-intermediate substances. (c) No reduction in concentration in the surface waters, and biological processes play little or no role in determining the shape of the profiles; these are categorised as bio-unlimited substances (redrawn from Kester *et al.*[1]).

In ocean waters, chemical constituents can be classified by their concentration profiles into being bio-limited, bio-intermediate, and bio-unlimited (*Figure 22.4*). If wastes contaminated with metals, such as copper, cobalt, nickel, lead, and even mercury, were to be discharged directly onto the deep-ocean floor, even with the remarkable sensitivity of modern chemical analytical techniques it is doubtful if increases in dissolved concentrations of these metals would be detectable, and so there would be no biological impact. This hypothesis can easily tested by a serendipitous 'experiment' that has already been conducted. Prior to 1983, when all ocean dumping of radioactive waste was banned by international agreement, drums containing low-level radioactive waste were routinely dumped over abyssal plains at one or two licensed sites, one being in the Bay of Biscay. A few of the drums have been recovered and found to be leaking, but no contamination of the surrounding sediments was detected. Biological studies revealed little change in the communities around the drums, nor was it evident whether these minimal changes had been induced by the physical presence of the drums or by the impact of any leakage of radioisotopes. While not advocating the resumption of the disposal of radioactive wastes in the ocean, a fuller investigation of this old low-level dump site could provide much clearer evidence of the environmental acceptability of disposing other types of waste materials into the deep ocean at a few licensed sites.

Similarly, any pathogens in the waste would need to survive for two or three centuries if they were to return back into the surface waters by the slow stirring of thermohaline circulation. High hydrostatic pressures certainly prevent the metabolism of micro-organisms in sewage, but studies are needed to establish how long the pathogens remain viable. Even if some pathogens do remain viable,

22.5

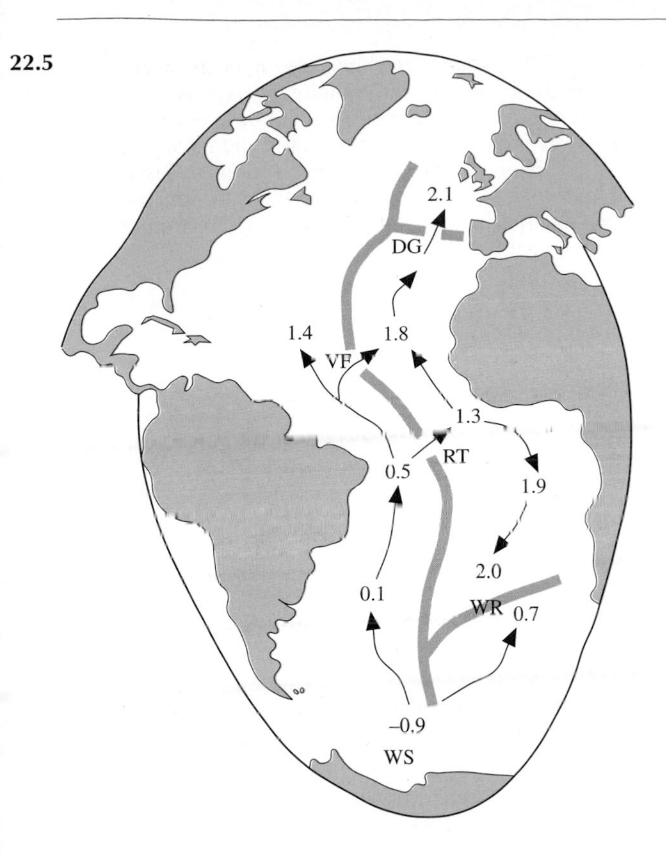

Figure 22.5 Temperatures of bottom water in the Atlantic, showing how from its source in the Weddell Sea (WS) it flows northward up the western side, its northward flow in the east being blocked by the Walvis Ridge (WR) off southwest Africa. Water flows along the Mid-Atlantic Ridge through the Romanche Trench (RT) on the equator and the Vema Fracture Zone (VF) at 10°N. Eventually, the water flows into the Western European Basin through Discovery Gap (DG), opposite the Straits of Gibraltar, at 1–2 Sverdrups (i.e., 1–2 x 10^6 m^3/s), by which time the geothermal flux from the Earth and mixing has warmed it to 2.1°C. The hatching indicates the approximate position of the mid-ocean ridges.

they would have to be concentrated by many orders of magnitude to be infectious and to be transferred by some as yet unknown process back to humans.

Oxygen Demand

Would oxygen concentrations be reduced significantly if the deep ocean were to be used for the disposal of organically enriched waste, such as sewage sludge? Most deep waters in the oceans are rich in dissolved oxygen; the bottom waters are formed at the surface of polar seas, when sea ice is forming. Sea water freezes at –1.9°C and, since the colder the water the more gas will dissolve in it, where the bottom water is being formed in the Weddell Sea and to the west of Greenland, the dissolved oxygen concentrations exceed 8 ml/l. The flow of bottom water northward in the Atlantic can be traced by temperature (*Figure 22.5*), where the bottom topography plays a key role in determining the patterns of flow. The concentrations of dissolved oxygen reflect the age of the deep water (*Figure 22.6*). These concentrations are enhanced in the northwestern Atlantic by the bottom water formation off Greenland, but in the northeastern Atlantic concentrations over the Porcupine Abyssal Plain to the west of Ireland remain at about 5.5 ml/l. In the North Atlantic, the bottom water mixes with the overlying waters to form North Atlantic deep water; this water mass supplies bottom waters to the other major oceans. The greater age of these bottom waters is reflected in their relatively low dissolved oxygen content. So, the deep waters of the North Pacific and Indian Oceans are more vulnerable to oxygen stress. However, microbial degradation rates are slowed by the cool *in situ* temperatures and high hydrostatic pressures, so if the waste piles up, much of the organic matter will remain buried and will not be oxidised.

If we assume the worst-case scenario, that the full chemical oxygen demand is realised, will the supply of oxygen in deep water be sufficient to cope with the demand? First, we need to know how much reduction in oxygen availability can be tolerated by the abyssal ecosystems. If, for example, the biological systems at abyssal depths in the northeastern Atlantic can tolerate a reduction of 10% (i.e., the *in situ* oxygen concentrations being reduced from 5.5 to 5.0 ml/l), then the supply of bottom water will allow the disposal of 150 million tonnes of the type of sludge dumped in Barrow

22.6

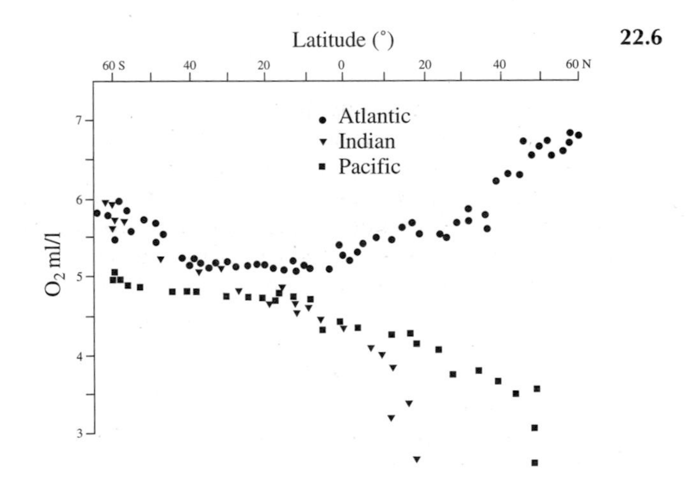

Figure 22.6 Latitudinal distribution of dissolved oxygen (ml/l) in bottom waters of the western regions of the three main oceans (modified from Mantyla and Reid[3]). Note the rise in oxygen concentration in the bottom waters of the northwestern Atlantic resulting from bottom water formation off Greenland. There is no such poleward rise in the oxygen concentrations in the northeastern Atlantic.

22.7

Figure 22.7 *Eurythenes gryllus,* a large amphipod species, which can grow to lengths of over 12 cm and is one of the most abundant and voracious species attracted to baits on the sea bed. These sorts of scavengers will probably become more abundant around an organically enriched dump site, and may prove to be an important vehicle for the dispersal of some of the contaminants. (© Heather Angel.)

Deep in the Thames Estuary. During the Quaternary (i.e., the last 2 million years), the concentrations of oxygen dissolved in ocean bottom-water fluctuated beyond this range as a result of the climatic oscillations between glacial and interglacial conditions.

Biological Impact

How much damage might be caused to the deep-living communities and would it be a serious loss of biodiversity? If dumped in one pile, the material would totally destroy the community beneath it, just as everything natural is destroyed at a land-fill site. Such an impact would be trivial compared with the effects of major geological events (Chapter 9), and would probably be preferable to dispersing the waste widely. The waste would probably be invaded by large numbers of opportunistic species, as happened in the shallow waters off Garroch Head (*Figure 22.3*), and in turn these dense concentrations of 'detritivores' would attract large numbers of predators and scavengers, in much the same way that scavengers are attracted to baited cameras and traps (*Figures 22.7* and *22.8*; see also Chapter 13). These mobile animals may play a greater role in dispersing the contaminants than would the water currents, which at abyssal depths are generally weak, except in regions where meso-scale eddies generate benthic storms (Chapter 4).

Oceanographers have still not sampled even a hundred millionth of the total area of the ocean bed, so we still know very little about the broad-scale distribution patterns of deep-living species. However, preliminary data for the larger animals imply that most benthic species have widespread distributions, both within and between the ocean basins. There seems to be no reason to doubt the assumption that the smaller species are equally widespread, but this needs to be verified. However, there are some obvious exceptions, such as the specialised inhabitants of hydrothermal vents and seeps (Chapters 10 and 13), and in hadal environments (>6000 m) at the bottom of ocean trenches. The proximity of these and any other special environments would have to be accounted for if any selection is made of sites for such waste disposal.

The Future

The question that urgently needs to be addressed is whether continuing with the extensive use of land-fill sites is likely to be environmentally more damaging and ultimately more expensive than introducing other waste-management options, including the use of a limited number of sites on the ocean floor. What should not be disposed of in the ocean?

- Anything that can be considered to be a resource.
- Any persistent synthetic organic substance which is totally unknown in the natural marine or terrestrial environment (for example, organohalo-

22.8

Figure 22.8 The remains of a mackerel used as bait in a trap set on the sea bed for 2 days at a depth of 4500 m off the Cape Verde Islands. Three species of amphipods, including *Eurythenes gryllus,* were in the trap and had taken part in the rapid consumption of the bait. (Courtesy of the Southampton Oceanography Centre, UK.)

gens and PCBs) and is truly essential to manufacture should be chemically destroyed.

- Radioactive isotopes? Maybe there is a case for reconsidering how the isotopes produced for weapons and peaceful purposes are to be contained and disposed of. Some of the present repositories are vulnerable to natural catastrophes and terrorist acts.
- How about carbon dioxide? Can humans slow down the rate of increase in the atmosphere (which promises to change our climate) by discharging carbon dioxide into the deep ocean, where it will dissolve? The Japanese are actively developing technologies to do so, but there are potential dangers which require research. Adding more carbon dioxide to deep waters will lower their pH. If these more acidic waters erode and destabilise calcareous sediments at the base of the continental slopes, the result might be massive failures of the continental margins (Chapter 9). There may be substantial biological impact – an effective way of anaesthetising aquatic animals is to squirt carbonated water into their container. Would the disposal of so much carbon dioxide into the deep ocean put benthic communities to sleep permanently, and what would be the scale of this impact – over 10, 100, or even a 1000 km? And what scale of impact might we consider to be acceptable in the context of even greater problems associated with global warming? The scale of such impacts is of serious concern, because terrestrial ecologists repeatedly find it impossible to extrapolate the results of experiments conducted in plots of a few metres to forecast what will happen at scales of tens and hundreds of kilometres.

These are complex issues which need further research and large-scale experiments to assess. But will these be conducted before the urgency to find solutions to waste problems becomes so overwhelming that decisions have to be made (because of socio-economic pressures on the land) in the absence of proper scientific evaluations based on sound ecological and biogeochemical principles?

General References

Kullenberg, G. (ed.) (1986), *The Role of the Oceans as a Waste Disposal Option*, NATO ASI Series, D. Reidel Publishing Co, Dordrecht, 725 pp.

Spencer, D.W. (1991), *Report of a Workshop to Determine the Scientific Research Required to Assess the Potential of the Abyssal Ocean as an Option for Future Waste Management*, Woods Hole Oceanographic Institution, 111 pp.

References

1. Kester, D.R., Burt, W.V., Capuzzo, J.M., Park, P.K., Ketchum, B.H., and Duedall, I.W. (1985), *Wastes in the Ocean*, Vol. 5, *Deep-Sea Waste Disposal*, Wiley Interscience, 346 pp.
2. Mantyla, A.W. and Reid, J.W. (1983), Abyssal characteristics of the world ocean waters, *Deep Sea Res.*, **30A**, 805–833.
3. McAllister, D.E. (1993), How much land is there on Earth? For people? For nature?, *Global Biodivers.*, 3, 6–7.
4. Pearson, T.H. (1986), Disposal of sewage in dispersive and non-dispersive areas: contrasting case histories in British coastal waters, in *The Role of the Oceans as a Waste Disposal Option*, Kullenberg, G. (ed.), NATO ASI Series, D. Reidel Publishing Co, Dordrecht, pp 577–595.

Some Commonly Used Words and Terms

Depth Zones

Abyssal. A subdivision of the benthic zone encompassing the ocean floor between a depth of 2–6 km.

Abyssopelagic. Open-ocean (oceanic) environment below 4 km depth.

Aphotic. The dark region of the ocean that lies below sunlit surface waters.

Benthic. That part of the ocean adjoining the sea bed.

Epipelagic. The upper region of the ocean extending to a depth of about 200 m.

Euphotic. The surface layer of the ocean that receives enough light to support photosynthesis.

Eutrophic. That with an abundance of nutrients.

Littoral. The benthic zone between the highest and lowest normal water marks; the intertidal zone.

Neritic. The water that overlies the continental shelf, generally of water depth less than 200 m.

Oceanic. The waters beyond the shelf break, generally of water depth greater than 200 m.

Pelagic. All water in the oceans, including the neritic zone and oceanic zones.

Sublittoral. That portion of the benthic environment extending from low tide to a depth of 200 m, often taken as the surface of the continental shelf.

Subneritic. The benthic environment extending from the shoreline across the continental shelf to the shelf break.

Plankton, Bacteria, and Marine Animals

Aerobic bacteria. Bacteria that undergo respiration in the presence of free oxygen (O_2).

Anaerobic bacteria. Bacteria that undergo respiration in the absence of free oxygen (O_2).

Autotroph. Plants and bacteria that synthesise food from inorganic nutrients.

Benthos. Organisms that live on or within the sea bed.

Coccolithophores. Microscopic, single-celled plant plankton having exo-skeletons composed of tiny, calcareous plates or discs called coccoliths.

Demersal organisms. Organisms which rest on the sea bed, but swim and feed in the water column.

Diatoms. Microscopic, unicellular phytoplankton possessing silica valves.

Dinoflagellates. Microscopic unicellular phytoplankton that propel themselves using tiny, whip-like flagella.

Epifauna. Animals that live on the sea bed, either attached or moving freely over it.

Epiflora. Plants that live in contact with the sea bed.

Eutrophication. The process whereby water becomes anoxic through decomposing organic matter.

Foraminifera. Planktonic and benthic protozoans that have a skeleton or shell composed of calcium carbonate ($CaCO_3$).

Holoplankton. Plants and animals that are plankton for their entire life.

Infauna. Animals that live within or burrow through the substrate (sand or mud).

Macroplankton. Large plankton (such as jellyfish and Sargassum weed).

Meiofauna. Small species of animals that live in the spaces among particles in a marine sediment.

Meroplankton. Planktonic larval forms of organisms that are members of the benthos or nekton as adults.

Microplankton. Plankton of length 0.06–1 mm.

Nanoplankton. Plankton of length <50 µm.

Nektobenthos. Those members of the benthos that are active swimmers and spend much time off the bottom.

Nekton. Pelagic animals, such as adult squids, fish, and mammals, that are active swimmers to the extent they can determine their position in the ocean by swimming.

Phytoplankton. Plant plankton, the primary producers of the oceans.

Picoplankton. The smallest plankton, with a body length of <2 µm.

Plankton. Organisms that float or have weak swimming abilities.

Sulfur-oxidizing bacterium. Any bacteria which use energy released by oxidation to synthesise organic matter chemosynthetically.

Ultraplankton. Plankton with body length <5 µm.

Zooplankton. Animal plankton.

Further definitions may be found in the Glossaries of:

- Pinet, P.R. (1992), *Oceanography. An Introduction to the Planet Oceanus*, West Pub. Co., New York, 570 pp.
- Thurman, H.V. (1994), *Introductory Oceanography*, 7th edn, Macmillan, New York, 550 pp.